Fixed Point Theory

Mathematics and Its Applications

Managing Editor:

M. HAZEWINKEL
Department of Mathematics, Erasmus University, Rotterdam, The Netherlands

Editorial Board:

R. W. BROCKETT, *Harvard University, Cambridge, Mass., U.S.A.*
Yu. I. MANIN, *Steklov Institute of Mathematics, Moscow, U.S.S.R.*
G.-C. ROTA, *M.I.T., Cambridge, Mass, U.S.A.*

Volume 7

Vasile I. Istrăţescu

Fixed Point Theory
An Introduction

D. REIDEL PUBLISHING COMPANY
Dordrecht : Holland / Boston : U.S.A. / London : England

Library of Congress Cataloging in Publication Data

Istrăţescu, Vasile I.
   Fixed point theory.

   (Mathematics and its applications ; v. 7)
   Bibliography: p.
   Includes index.
   1. Fixed point theory. I. Title. II. Series: Mathematics and its applications (D. Reidel Publishing Company) ; v. 7.
QA329.9.I85   515.7'24   81–5224
ISBN 90–277–1224–7   AACR2

Published by D. Reidel Publishing Company
P.O. Box 17, 3300 AA Dordrecht, Holland

Sold and distributed in the U.S.A. and Canada
by Kluwer Boston Inc.,
190 Old Derby Street, Hingham, MA 02043, U.S.A.

In all other countries, sold and distributed
by Kluwer Academic Publishers Group,
P.O. Box 322, 3300 AH Dordrecht, Holland

D. Reidel Publishing Company is a member of the Kluwer Group

Printed in The Netherlands

To my parents
Paraschiva and Ion (Ilie) Istrățescu

Editor's Preface

Growing specialization and diversification have brought a host of monographs and textbooks on increasingly specialized topics. However, the 'tree' of knowledge of mathematics and related fields does not grow only by putting forth new branches. It also happens, quite often in fact, that branches which were thought to be completely disparate are suddenly seen to be related.

Further, the kind and level of sophistication of mathematics applied in various sciences has changed drastically in recent years: measure theory is used (non-trivially) in regional and theoretical economics; algebraic geometry interacts with physics; the Minkowsky lemma, coding theory and the structure of water meet one another in packing and covering theory; quantum fields, crystal defects and mathematical programming profit from homotopy theory; Lie algebras are relevant to filtering; and prediction and electrical engineering can use Stein spaces.

This series of books, *Mathematics and Its Applications*, is devoted to such (new) interrelations as *exempla gratia*:

- a central concept which plays an important role in several different mathematical and/or scientific specialized areas;
- new applications of the results and ideas from one area of scientific endeavor into another;
- influences which the results, problems and concepts of one field of enquiry have and have had on the development of another.

With books on topics such as these, of moderate length and price, which are stimulating rather than definitive, intriguing rather than encyclopaedic, we hope to contribute something towards better communication among the practitioners in diversified fields.

The present book furnishes good example of a central concept (technique) with multitudes of different uses. Fixed points and fixed point theorems have always been a major theoretical tool in fields as widely apart as differential equations, topology, economics, game theory, dynamics, optimal control, and functional analysis. Moreover, more or less recently, the usefulness of the concept for applications increased enormously by the development of accurate and efficient techniques for computing fixed points, making fixed point methods a major weapon in the arsenal of the applied mathematician.

The unreasonable effectiveness of mathematics in science . . .

> Eugene Wigner

Well, if you knows of a better 'ole, go to it.

> Bruce Bairnsfather

What is now proved was once only imagined.

> William Blake

As long as algebra and geometry proceeded along separate paths, their advance was slow and their applications limited.

But when these sciences joined company, they drew from each other fresh vitality and thenceforward marched on at a rapid pace towards perfection.

> Joseph Louis Lagrange

Krimpen a/d IJssel
March, 1979

Michiel Hazewinkel

Table of Contents

Foreword

This book is intended as an introduction to fixed point theory and its applications. The topics treated range from fairly standard results (such as the Principle of Contraction Mapping, Brouwer's and Schauder's fixed point theorems) to the frontier of what is known, but we have not tried to achieve maximal generality in all possible directions. We hope that the references quoted may be useful for this purpose.

The point of view adopted in this book is that of functional analysis; for the readers more interested in the algebraic topological point of view we have added some references at the end of the book. A knowledge of functional analysis is not a prerequisite, although a knowledge of an introductory course in functional analysis would be profitable. However, the book contains two introductory chapters, one on general topology and another on Banach and Hilbert spaces. As a special feature of these chapters we note the study of measures of noncompactness; first in the case of metric spaces, and second in the case of Banach spaces.

Chapter 3 contains a detailed account of the Contraction Principle, perhaps the best known fixed point theorem. Many generalizations of the Contraction Principle are also included. We note here the connection between ideas from projective geometry and contractive mappings. After presenting some ways to compute the fixed points for contractive mappings, we discuss several applications in various areas.

Chapter 4 presents Brouwer's fixed point theorem, perhaps the most important fixed point theorem. After some historical notes concerning opinions about Brouwer's proof – which have been influential for the future of the fixed point theory (Alexander and Birkhoff and Kellogg) – we present many proofs of this theorem of Brouwer, of interest to different categories of readers. Thus we present an elementary one, which requires only elementary properties of polynomials and continuous functions; another uses differential forms; still another uses differential topology; and one relies on combinatorial topology. These different proofs may be used in different ways to compute the fixed points for mappings. In this connection, some algorithms for the computation of fixed points are given. The chapter ends with some applications, among which we mention here those concerning economic equilibrium prices.

Chapter 5 is a natural continuation of Chapter 4 and presents the generalizations of Brouwer's theorem obtained by Schauder as well as the new results connected with Schauder's generalization of Brouwer's theorem. Various important and interesting contributions due to Krasnoselskii, Rothe, Altman, Ky Fan, F. Browder and also results due to Darbo and Sadovskii are presented. The contributions of Darbo and Sadovskii concern a new direction of extending the classical fixed point theorems, namely, using the so-called measures of noncompactness. We also present in this chapter some applications, among them the proof by Lomonosov of the existence of nontrivial hyperinvariant subspaces for bounded linear operators on Banach spaces commuting with nonzero bounded linear compact operators.

In Chapter 6 we present some results about mappings which do not increase distance – the so called nonexpansive mappings. First we present some results on the extension of nonexpansive mappings, with a simple example to show that, without certain restrictions on the mappings or on the space, fixed points do not exist for nonexpansive mappings. Further we present some results about the existence of fixed points for nonexpansive mappings or related classes of mappings on certain classes of Banach spaces, as well as results about the convergence of the iterates. An example and a method for computing fixed points for such mappings close this chapter.

Chapter 7 discusses results about fixed points for sequences of mappings. First we give some results for the case of contraction mappings, and second, for condensing mappings.

In Chapter 8 we present the elements of a theory of duality mappings and their connections with monotonic and nonexpansive operators as well as some surjectivity theorems which have many and useful applications in the theory of partial differential equations.

Chapter 9 contains results centering around the Markov–Kakutani results, including a beautiful result of Ryll–Nardzewski, as well as some results about the connection between invariant means on semigroups and fixed point theorems.

As a natural extension of the theory of fixed points for single-valued mappings, in Chapter 10, the case of set-valued mappings is considered. First we give some results about the Pompeiu–Hausdorff metric and results about set-valued contraction mappings. Some results concerning the extension of Brouwer and Schauder as well as Tichonov results for set-valued mappings are included.

Probabilistic metric spaces were introduced by Karl Menger in 1942, and since then the interest about these spaces has been growing constantly. We present in Chapter 11 some results about fixed points for contractive mappings on probabilistic metric spaces (abbreviated as PM-spaces) as well as some results concerning measures of noncompactness on these spaces and fixed points for certain classes of mappings.

Finally Chapter 12 contains results concerning topological degree. First the case of finite dimensional spaces is treated, in which we have the so-called Brouwer's degree; next this degree is extended to certain perturbations of the identity operator on Banach spaces. We include also the famous example of Leray which shows that it is not possible to define the degree for arbitrary perturbations.

Next, we present briefly the results obtained by extending the degree concept to certain perturbation of the identity by k-set contraction mappings. Using some approximation theorems (which are also of independent interest) we prove uniqueness of the topological degree. Next we give some algorithms for the computation of Brouwer's topological degree based on Stenger's formula. Some applications of the degree are noted at the end of this chapter.

We have tried to indicate the original location of the various results we have learned, and the references (which contain in turn references to many earlier results) may be used to obtain further information; when a result is not ascribed to anyone we do not make any claim to originality.

We wish to acknowledge with thanks conversations and correspondence on Functional Analysis, Operator Theory and Fixed Point Theory from which we have benefitted. The author wishes to acknowledge especially his debt to Professor Michiel Hazewinkel for his interest in the work, as well as for the suggestions concerning the material contained in this book.

My appreciation goes also to the editors of the D. Reidel Publishing Company for their attention to this volume.

Vasile I. Istrățescu

Chapter 1

Topological Spaces and Topological Linear Spaces

1.1. METRIC SPACES

The notion of convergence is of fundamental importance in analysis as well as in many chapters of mathematics. Thus we have the convergence of sequences of real numbers, the convergence of sequences of complex numbers, the convergence of sequences of functions, etc. It is worth remarking that in the case of functions we have many types of convergence, for example: pointwise convergence; uniform convergence; the convergence in measure, etc.

Also, when we define the convergence of a sequence of numbers or functions we use points which are 'near' to our points. This vague notion of 'nearness' can be made precise using some functions which are called generally 'metrics' or 'distances'.

DEFINITION 1.1.1. Let S be a nonempty set and $d : S \times S \to R$.

The function d is called a metric on S (or distance) iff the following properties hold:

1. $d(x, y) = 0$ iff $x = y$,
2. $d(x, y) = d(y, x)$ for all $x, y \in S$,
3. $d(x, z) \leq d(x, y) + d(y, z)$ for all $x, y, z \in S$.

The number $d(x, y)$ is called the distance between x and y and the pair (X, d) is called a metric space. For simplicity we write X and we say that X is a metric space.

PROPOSITION 1.1.2. *For any $x, y \in S$, $d(x, y) \geq 0$.*

Proof. Let x, y, z be arbitrary points in S. In this case we have

$$d(x, z) \leq d(x, y) + d(y, z)$$

and thus for $x = z$ we obtain

$$0 \leq d(x, y) + d(y, x) = 2\,d(x, y)$$

and the assertion follows.

If S_1 is a subset of a metric space S then we can define a metric on S_1, simply by the relation

$$d_1(x, y) = d(x, y)$$

and d_1 is called the induced metric on S_1.

We give now some examples of metric spaces.

Example 1.1.3.   Let E_n(or R^n) $= \{x = (x_1, x_2, \ldots, x_n), x_i \in \mathbf{R}, \mathbf{R}$ the set of real numbers$\}$ and let d be defined as follows: if $y = (y_1, y_2, \ldots y_n)$ then

$$d(x, y) = \left(\sum_1^n |x_i - y_i|^p\right)^{1/p} = d_p(x, y).$$

where p is a fixed number in $[1, \infty)$. The fact that d is a metric follows from the well-known Minkowski inequality. Also another metric on S considered above can be defined as follows:

$$d(x, y) = \sup_i \{|x_i - y_i|\} = d_\infty(x, y).$$

Example 1.1.4.   Let S be the set of all sequences of real numbers $x = (x_i)_1^\infty$ such that for some fixed $p \in [0, \infty)$

$$\sum_1^\infty |x_i|^p < \infty.$$

In this case, if $y = (y_i)$ is another point in S, we define

$$d(x, y) = \left(\sum |x_i - y_i|^p\right)^{1/p} = d_p(x, y).$$

and from Minkowski's inequality it follows that this is a metric on S.

Example 1.1.5.   Let $S = L^2_{[0, 1]} = \{f, \int_0^1 |f|^2 dt < \infty\}$ and for any two functions (classes of functions) f, g we define

$$d_2(f, g) = \left(\int_0^1 |f - g|^2 \, dt\right)^{1/2}$$

and from Minkowski's inequality for integrals, it follows that this is a metric on S.

Example 1.1.6.   Let $S = C_{[0, 1]}$ be the set of all continuous complex-valued functions on $[0, 1]$. We define, for any f, g in S

$$d(f, g) = \sup \{|f(t) - g(t)| : t \in [0, 1]\}$$

and it is easy to see that this is a metric on S.

Example 1.1.7.   For the set S of all real numbers, which in what follows will be denoted by **R**, we define

$$d(x, y) = |x - y|$$

for any two real numbers x, y. Then $(\mathbf{R}, d)$ is a metric space.

Example 1.1.8.   Let S be the set **Q** of all rational numbers in **R** and the metric induced by d. Then $(\mathbf{Q}, d_1)$ is a metric space. The notion of convergence in metric spaces is defined as follows:

DEFINITION 1.1.9.   A sequence $\{x_n\}$ in a metric space (X, d) converges to an element x of X if, for any $\varepsilon > 0$, there exists N_ε such that for all $n \geq N_\varepsilon$,

$$d(x_n, x) \leq \varepsilon.$$

An important class of sequences in metric spaces are the so-called Cauchy sequences. These sequences are defined as follows:

DEFINITION 1.1.10.   A sequence $\{x_n\}$ in a metric space (X, d) is called a Cauchy sequence if, for any $\varepsilon > 0$, there exists N_ε such that for all $n, m \geq N_\varepsilon$,

$$d(x_n, x_m) \leq \varepsilon.$$

It is easy to see that any sequence which converges is a Cauchy sequence.

An important class of metric spaces in which the converse is also true is the class of so-called 'complete metric spaces' and formally this class is introduced in the following

DEFINITION 1.1.11.   A metric space (X, d) in which any Cauchy sequence $\{x_n\}$ has the property that it converges to a point of X, is called complete.

Example 1.1.12.   All the spaces in Examples 1.1.5–1.1.6 are complete metric spaces; the metric space in Example 1.1.7 is not complete.

For any $r > 0$ and $x \in X$, X a metric space, we define

1. $S_r(x) = \{y, d(x, y) \leq r\}$ the disc with centre x and radius r,

2. $\overset{\circ}{S}_r(x) = \{y, d(x, y) < r\}$ the open disc with centre x and radius r,

3. $\partial S_r(x) = \{y, d(x, y) = r\}$ the boundary of the disc with centre x and radius r or the circumference of centre x and radius r.

DEFINITION 1.1.13.   Let (X, d) be a metric space and G a subset of X. The point $x \in G$ is said to be interior to G if there exists an open disc $S_x(X) \subset G$. The set G is called open if all its points are interior points or is the empty set.

DEFINITION 1.1.14.   A set F in a metric space is called closed if the set

$$C_F = \{x, x \in X, x \notin F\}$$

is an open set.

For any set in a metric space X, the diameter is the number

$$d(A) = \sup \{d(x, y), x, y \in A\}$$

and the distance from a point $x \in X$ to the set A is the number

$$d(x, A) = \inf \{d(x, y); y \in A\}.$$

We define now the fundamental notion of neighbourhood of a point in a metric space.

DEFINITION 1.1.15.   If X is a metric space and $x \in X$ is an arbitrary point then a neighbourhood of x is any set which contains an open set containing x.

The important notion of continuity is introduced as follows:

DEFINITION 1.1.16.   If X and Y are two metric spaces and $f: X \to Y$ is any function, then we say that f is continuous at $x \in X$ if, for any neighbourhood V of $f(x)$, there exists a neighbourhood U of x such that for all $z \in U$, $f(z) \in V$. The function f is continuous on X if it is continuous at each point of X.

We have the following characterization of continuous functions at x:

THEOREM 1.1.17.   *If X and Y are two metric spaces and $f : X \to Y$ is any function, then f is continuous at $x \in X$ if and only if (contracted to 'iff' following a suggestion of Halmos) for any sequence $(x_n) \subset X$ converging to x, the sequence $(f(x_n))$ converges to $f(x)$.*

Proof. Since the proof is the same as for functions defined on $[0, 1]$ we omit it.

DEFINITION 1.1.18.   If X is metric space and S is any set in S then the closure of S is the intersection of all closed sets containing S and is denoted by $\bar{S}$.

Example 1.1.19.   There exist metric spaces for which

$$\bar{\bar{S}}_r(x) \neq S_r(x).$$

For, if we take X as

$$X = \{x, 0 \leq x \leq 1\} \cup \{e^{i\theta}, 0 \leq \theta \leq \pi/2\},$$

then clearly $S_1(0) = X$ and

$$\mathring{S}_1(0) = \{x, 0 \leqq x < 1\}.$$

This gives that

$$\bar{\mathring{S}}_1(0) = \{x, 0 \leqq x \leqq 1\}.$$

For the connection between complete metric spaces and incomplete metric spaces (i.e. metric spaces which are not complete) we mention the following result, which can be proved exactly as for the case $X = \mathbf{Q}$ the set of all rational numbers:

First we give the following

DEFINITION 1.1.20. If (X, d) is a metric space then the subset X_1 is dense in X if for any x in X there exists a sequence (x_n), $x_n \in X_1$ such that

$$\lim d(x_n, x) = 0.$$

Then we have

THEOREM 1.1.21. *If (X, d) is an incomplete metric space then there exists a complete metric space $(X^\sim, d^\sim)$ such that for some function f we have*

1. $\qquad f : X \to X^\sim$,

2. $\qquad d(x, y) = d^\sim(f(x), (f(y))$,

3. $\qquad f(X)$ *is dense in $X^\sim$.*

We recall that any function f with Property 2 is called an isometry.

1.2. COMPACTNESS IN METRIC SPACES. MEASURES OF NONCOMPACTNESS

As is well known, for any bounded set E of real numbers there exists, for any sequence $(x_n) \subset E$, a convergent subsequence (this is the so called Bolzano-Weierstrass theorem), and any closed and bounded set can be characterized as having the following equivalent properties:

1. E has the property that for every sequence $(x_n) \subset E$ there exists a convergent subsequence to an element of E,
2. For any family $(V_i)_{i \in I}$ of open sets such that $E \subset \bigcup_{i \in I} V_i$ there exist $i_1, \ldots, i_m$ such that $E \subset \bigcup_1^m V_{i_j}$ (this is the so called Lebesgue–Borel Lemma).

These two properties suggest the definition of compact sets in metric spaces.

DEFINITION 1.2.1.   Let (X, d) be a metric space and E a subset of X. E is called compact if, whenever E is included in the union of a family of open sets $(V_i)_{i \in I}$ then there exist $V_{i_1}, \ldots, V_{i_m}$ from the family, such that

$$E \subset \bigcup_1^m V_{i_j}.$$

We say for short that E is compact if every open cover of the set has a finite subcover of the set.

There exist other important formulations of compactness and we give here some of them. First we give the following

DEFINITION 1.2.2.   If X is a set and X_1 is a subset of X then a family $\mathscr{F}$ of subsets of X is said to have the finite intersection property with respect to X_1 if X_1 meets the intersection of every finite subcollection of $\mathscr{F}$ and we say that $\mathscr{F}$ has the full intersection property with respect to X_1 if X_1 meets the intersection of all members of $\mathscr{F}$.

The following theorem gives a characterization of compactness using the above notion.

THEOREM 1.2.3.   *A set E in a metric space (X, d) is compact iff for any family $\mathscr{F}$ of closed sets with the finite intersection property with respect to E then $\mathscr{F}$ has the full intersection property with respect to E.*

Proof. Suppose first that E is compact and let $\mathscr{F}$ be a family with the finite intersection property with respect to E. If $\mathscr{F}$ does not have the full intersection property with respect to E then E is in $\bigcup C_F, F \in \mathscr{F}$ and, since E is compact, there exist $F_1, \ldots, F_m$ such that $E \subset C_{F_1} \cup \ldots \cup C_{F_m}$, i.e., $E \cap \{\bigcap_1^m F_i\} = \emptyset$ which is a contradiction. In a similar way we prove the converse assertion.

For the following characterization we need the notion of an accumulation (or cluster) point of a set in a metric space.

DEFINITION 1.2.4.   Let (X, d) be a metric space and E be a set in X. A point x is called an accumulation point of E (or limit point, or cluster point) if, in every neighbourhood $V \ni x$, there exists at least one point of E.

The set of all accumulation points of a set E is called the derived set and is denoted by E'. For any set $E \subset X$ we denote by $\mathrm{Cl}\, E$ or $\bar{E}$, the smallest closed set in X which contains E. Then we have the following

THEOREM 1.2.5. *For any subset E of a metric space (X, d), E is closed iff*

$$\bar{E} = E.$$

Proof. Since the proof is simple we omit it.

The characterization of compact sets in metric spaces using the above notions is as follows:

THEOREM 1.2.6. *A set E in a complete metric space X is compact iff the following conditions are fullfilled:*

1.   *E is closed, and*
2.   *each sequence $\{x_n\} \subset E$ has a cluster point in E.*

Proof. Since we prove in the next paragraph a more general theorem the proof of this theorem is omited.

From the above theorem we have

COROLLARY 1.2.7. *If s and Y are metric spaces and $f : X \to Y$ is continuous then, for any compact subset E of X, $f(E) = \{y, y = f(x), x \subset E\}$ is a compact subset of Y.*

COROLLARY 1.2.8. *Every compact subset of a metric space is closed.*

DEFINITION 1.2.9. *A set E in a metric space is bounded if its diameter is bounded.*

In this case we have obviously,

COROLLARY 1.2.10. *In a metric space every compact set is bounded.*

As is well known the properties stated in the Corollaries 1.2.8 and 1.2.10 are characteristic for compact sets in $\mathbf{R}^m$. We now give an example to show that in arbitrary metric spaces this is no longer true.

Example 1.2.11. We consider the space l^2 of all infinite sequences $x = (x_i)_1^\infty$ with the property that

$$\sum_1^\infty |x_i|^2 = \|x\|^2 < \infty$$

and with the metric defined as follows: if $y = (y_i)$ is another point of l^2 then

$$d(x, y) = \left(\sum_1^\infty |x_i - y_i|^2 \right)^{1/2}.$$

Let

$$S_1(0) = \{x, \|x\| \leq 1\}$$

which is obviously closed and bounded. It is not compact. Indeed, let $\{e_i\}$ be the sequence in l^2 defined by

$$e_i = (0, 0, \ldots, 0, 1_i, 0, \ldots)$$

and it is clear that

$$d(e_i, e_j) = 2^{1/2}, \qquad i \neq j.$$

The following theorem gives an important characterization of compact sets in complete metric spaces:

THEOREM 1.2.12.   *Let (X, d) be a complete metric space and X_1 be a subset of X. Then X_1 is a compact set iff, for any $\varepsilon > 0$, there exists a finite number of sets, $A_1, \ldots, A_m$ such that*

1. $d(A_i) < \varepsilon, \qquad i = 1, \ldots, m,$

2. $X_1 \subset \bigcup_1^m A_i.$

Proof. First we show that the condition is necessary. Indeed, if X_1 is a compact set and $\varepsilon > 0$ is given, then for each point x in X_1 we consider the set $\mathring{S}_\varepsilon(x)$ and clearly $\{\mathring{S}_\varepsilon(x) : x \in X_1\}$ is an open cover of X_1. Then there exists $x_1, \ldots, x_m$ such that

$$X_1 \subset \bigcup_1^m \mathring{S}_\varepsilon(x_i)$$

and since the diameter of $\mathring{S}_\varepsilon(x_i)$ is not greater than 2ε the assertion is proved if we begin with $\varepsilon/2$.

To prove that the condition is sufffcient we need a new characterization of complete metric spaces given by G. Cantor.

THEOREM 1.2.13.   *Let (X, d) be a metric space. Then (X, d) is complete iff for any sequence $\{F_n\}$ of closed sets in X having the following properties:*

1. $F_n \supset F_{n+1}, \qquad n = 1, 2, 3, \ldots,$

2. $d(F_n) \to 0$

then $\bigcap F_n$ is nonempty.

Proof. Suppose that (X, d) is complete and let $\{F_n\}$ be a sequence of closed sets satisfying the above conditions. Let $x_n \in F_n$ for each n. Since $\lim d(F_n) = 0$ for $\varepsilon > 0$ there exists N_ε such that for $n \geq N_\varepsilon$ $d(F_n) < \varepsilon$. From Property 1, it follows that, for $n, m \geq N_\varepsilon$, we have

$$d(x_n, x_m) < \varepsilon$$

and thus $\{x_n\}$ is a Cauchy sequence. Let $x_0 = \lim x_n$. Since

$$\lim d(x_n, x_m) = 0$$

we have that

$$d(x_n, x_0) = 0$$

i.e. $x_0 \in F_n$ for all n.

Suppose now that (X, d) possesses the stated property. Suppose that $\{x_n\}$ is a Cauchy sequence in X and we define the sets,

$$F_n = \overline{\{x_n, x_{n+1}, \ldots\}}$$

which obviously have the Properties 1 and those of the theorem. Thus $\bigcap F_n$ is nonempty and $x_0 \in \bigcap F_n$ satisfies the condition $\lim d(x_n, x_0) = 0$, i.e. (X, d) is a complete metric space.

Now the proof of Thorem 1.2.12 is as follows: suppose that $\{x_n\}$ is a sequence in X_1. Let $\varepsilon_n \to 0$ be a sequence of positive numbers. For each n let $X_1^1, \ldots, X_n^m$ be the sets such that

1. $X_1 \subset \bigcup X_n^i$,

2. $d(X_n^i) < \varepsilon_n \qquad i = 1, 2, \ldots, m$.

Since $\{X_1^1, \ldots, X_n^m\}$ is a finite collection of sets there exists a set, say X_1^1, such that an infinite number of terms of the sequence $\{x_n\}$ are in this set. Let $\{x_{n_1}\}$ be the sequence with this property. Similarly we find a set, say X_2^2, in the family $\{X_2^i\}$ with the property that an infinite number of terms of the above sequence $\{x_{n_1}\}$ are in this set. Let $\{x_{n_2}\}$ be the set of those terms. We can continue in this way and we obtain a subsequence $\{x_{n_i}^i\}$ of the sequence $\{x_n\}$ such that the sequence of the sets $\{X_n^i\}$ contains an infinity of terms of the sequence $\{x_n\}$. We define the following sequence of closed sets $\{F_n\}$,

$$F_1 = \overline{X_1^1}, \ldots, F_m = \overline{X_{1_1}^1} \cap \ldots \cap \overline{X_{n_m}^m}, \ldots$$

which satisfies the conditions of Theorem 1.2.13 and we let x_0 be the common point of these sets. Let $x_{n_m} \in F_m$, $x_{n_m} \in \{x_n\}$ and clearly

$$d(x_{n_m}, x_0) < \varepsilon_m$$

and this gives us that $\lim x_{n_m} = x_0$ and the assertion is proved.

In a similar way we can prove the following useful characterization of compact sets in complete metric spaces.

THEOREM 1.2.14.   *A set X_1 in a complete metric space (X, d) is compact iff, for any $\varepsilon > 0$, there exists a finite set $F_\varepsilon = \{x_1, \ldots, x_m\}$ such that, for any $x \in X_1$, there exists $x_i \in F_\varepsilon$ such that $d(x, x_i) < \varepsilon$.*

DEFINITION 1.2.15.   A set X_1 in a metric space is called pecompact iff the closure X_1 is a compact set.

The above characterizations of compact sets in a metric space suggest the definition of a function on the family of bounded sets of a metric space which measure the noncompactness of a set. Such a function is called a 'measure of noncompactness'. It is worth noting that the first measure of noncompactness was introduced by K. Kuratowski in 1930. Other measures of noncompactness were defined later and for an interesting discussion of these we refer the reader to a paper of Daneš (1976).

DEFINITION 1.2.16.   (Kuratowski's measure of noncompactness). Let $\mathscr{F}_b(X)$ be the family of all bounded sets in a complete metric space (X, d). For each $A \in \mathscr{F}_b(X)$ we define the number $\alpha(A)$ by the relation

$$\alpha(A) = \inf \{\varepsilon > 0 \text{ there exists a finite covering of } A \text{ with sets}$$
$$\text{having a diameter less than } \varepsilon\}.$$

DEFINITION 1.2.17.   (Hausdorff's measure of noncompactness). For any $A \in \mathscr{F}_b(X)$ we define the number $\chi(A)$ by the relation

$$\chi(A) = \inf \{\varepsilon > 0 \text{ there exists a finite set } F_\varepsilon = \{x_1, \ldots x_n\} \text{ such}$$
$$\text{that for any } x \in A \text{ there exists } x_i \in F_\varepsilon \text{ with the property}$$
$$d(s, x_i) < \varepsilon\}.$$

From these definitions it is easily seen that,

1. $0 \leq \alpha(A) \leq d(A) = $ diameter of A,

2. $0 = \alpha(A)$ iff A is a precompact set,

3.   for any $A, B \in \mathscr{F}_b(X)$, $\alpha(A \cup B) = \max \{\alpha(A), \alpha(B)\}$,

4. $\alpha(A) = \alpha(\bar{A})$ ($\bar{A}$ denotes the closure of A)
             and similarly for the measure of noncompactness χ.

Using the measures of noncompactness, the result of Theorem 1.2.13 can be generalized as follows:

THEOREM 1.2.18. (K. Kuratowski). *Let (X, d) be a metric space. Then (X, d) is complete iff, for any sequence of closed sets $\{F_n\}$ having the following properties*:

1. $F_n \supset F_{n+1}$, $\qquad n = 1, 2, 3, \ldots,$

2. $\alpha(F_n) \to 0$

the set $\bigcap F_n$ is nonempty.

Proof. Obviously we have $\alpha(\bigcap F_n) = 0$ and thus $\bigcap F_n$ is a compact set. It remains to prove that it is nonempty. From the definition of the measure of noncompactness it follows that for any $\varepsilon > 0$ there exists $S_{\alpha(F_1) + \varepsilon/2}(x_1)$ which contains an infinite number of elements of F_1 and also an infinite number of elements of $F_2, F_3, \ldots$. We continue and we choose $S_{\alpha(F_2) + \varepsilon/2}(x_2)$ which contains an infinite number of elements of $F_3, F_4, \ldots$ and this set is a subset of $S_{\alpha(F_1) + \varepsilon/2}$.

In this way we obtain a sequence of sets $\{S_{\alpha(F_i) + \varepsilon/2^i(x_i)}\}$ with diameter less than $2\alpha(F_n) + \varepsilon/2^{i-1}$. It is clear that

$$d(x_i, x_j) \leq \alpha(F_i) + \varepsilon/2^i$$

and thus $\{x_n\}$ is a Cauchy sequence. Let $x_0 = \lim x_n$. Clearly this is an element of $\bigcap F_n$. The rest of the proof is exactly as for Theorem 1.2.13.

Remark 1.2.20. A similar result holds for the Hausdorff measure of noncompactness.

Other properties of measures of noncompactness are connected with the algebraic structure of the sets and we give the corresponding results for the case of Banach spaces.

1.3. BAIRE CATEGORY THEOREM

R. Baire has proved one of the most interesting theorems of analysis – and one which has a great number of applications. Some of them are given in the chapter on Hilbert and Banach spaces.

DEFINITION 1.3.1. A set X_1 in a metric space X is said to be nowhere dense if its closure contains a nonempty open set of X.

DEFINITION 1.3.2. A set X_1 in a metric space is said to be of the first category if it is the union of any countable family of nowhere dense sets: any set which is not of the first category is said to be of the second category.

Baire's famous theorem is given as:

THEOREM 1.3.3.   *Any complete metric space is not a set of the first category.*

We give the proof of the following assertion, which is equivalent to Baire's theorem:

THEOREM 1.3.4.   *For any complete metric space X, the intersection of every countable collection of open dense subsets is dense in X.*

Proof. Let $\{G_i\}_1^\infty$ be the collection of open sets and for each n, $\bar{G}_n = X$. Let G be any open set in X and, to prove that $\bigcap G_n$ is dense in X, it is sufficient to show that $\bigcap G_n$ has a point in G.

If d denotes the metric of X and since G_1 is dense in X, $G \cap G_1$ is a nonempty set. Thus we find x_1 and r_1 such that

$$S_{r_1}(x_1) \subset G \cap G_1, \qquad 0 < r_1 < 1.$$

Suppose now that $x_1, x_2, \ldots, x_{n-1}; r_1, r_2, \ldots, r_{n-1}$ are given and we choose x_n and r_n as follows: since G_n is dense in X, $G_n \cap S_{r_{n-1}(x_{n-1})}$ is nonempty and thus we find x_n and r_n such that

$$S_{r_n}(x_n) \subset G_n \cap S_{r_{n-1}(x_{n-1})} \qquad 0 < r_n < 1/n.$$

We can continue in this way and we obtain a sequence $\{x_n\}$ in X.

Now, for $k, r > n$, x_k and x_r are in $S_{r_n}(x_n)$ and thus $d(x_k, x_r) < 2/n$, which gives that $\{x_n\}$ is a Cauchy sequence. Since X is complete we can write $x = \lim x_n$. But x_r is in $S_{r_n}(x_n)$ for $r > n$ and thus x is in each $S_{r_n}(x_n)$, i.e. in each G_n. Thus $x \in G$ and the assertion of the theorem is proved.

We give now another useful formulation of Baire's category theorem.

THEOREM 1.3.5.   *For any complete metric space X, if $X = \bigcup_1^\infty F_n$ where each F_n is a closed set then at least one of these closed subsets contains a nonempty open set.*

1.4. TOPOLOGICAL SPACES

Let X be a nonempty set. One of the most important notions in general topology, and one which is of fundamental importance for the study of spaces and maps, is the notion of 'topology'. This is introduced in

DEFINITION 1.4.1.   A topology on a nonempty set X is a family D of subsets of X and nominally open sets, satisfying the axioms

    1.   the empty set $\emptyset$ is open,

2.   the set X is open,

3.   the union of arbitrary family of open sets is open,

4.   if $G_1, \ldots, G_m$ are open sets $(m < \infty)$ then $\bigcap_1^m G_i$ is open.

We say that (X, D) is a topological space or simply that X is a topological space.

Some examples are called for.

Example 1.4.2. For any space X we define

$$D = \{\emptyset, X\}$$

and it is clear that this is a topology; X with this topology is called an indiscrete space and D an indiscrete topology.

Example 1.4.3. For any space X we define

$$D = \{E, E \text{ a subset of } X\}$$

and it is obvious that this is a topology; X with this topology is called a discrete space and the topology D the discrete topology.

Example 1.4.4. Let (X, d) be a metric space and D be the family of all open sets in X including $\emptyset$ and X. Then clearly we have a topology and X is a topological space. The topology D defined in this way is named 'the topology induced by the metric d'.

DEFINITION 1.4.5. Let (X, D) be a topological space and $x \in X$. A neighbourhood of x is any set which contains an open set containing x.

Clearly this definition is in accordance with Definition 1.1.14.

Suppose now that we have a set X and D_1, D_2 two topologies on X.

The following definition introduces a partial order in the set of all topologies on X.

DEFINITION. 1.4.6. If D_1 and D_2 are two topologies on the set X then we say that D_1 is smaller than D_2 and we write this as $D_1 \leq D_2$ if all open sets of D_1 are open sets for D_2.

Sometimes we say that D_1 is coarser than D_2 or that D is finer than D_1.

Example 1.4.7. The discrete topology is the finest topology on X and the indiscrete topology is the smallest topology on X.

The following definition gives a method for the construction of new topological spaces.

DEFINITION 1.4.8.   Let (X, D) be a topological space and X_1 be any subset of X. A topology on X_1 can be defined as follows: an open subset is any subset of X_1 which can be represented as the intersection of X_1 with an open set of X.

DEFINITION 1.4.9.   Let (X, D) be a topological space; a basis for D is any family G of subsets of D such that every open set is a union of open subsets in G. A sub-basis of D is any family G_1 of open subsets of D such that the finite intersections of open sets in G_1 are a basis for D.

Example 1.4.10.   Let (X, d) be a metric space and we consider for each point $x \in X$ the sets

$$\overset{\circ}{S}_r(x) = \{x, d(x, y) < r\}$$

and the family $\{\overset{\circ}{S}_r(x); x \in X, r > 0\}$ is a sub-basis of the topology of X (i.e. the topology induced by d).

In connection with the structure of bases of a topology on a space, two important classes of topological spaces are considered.

DEFINITION 1.4.11.   A topological space (X, D) is said to satisfy the second axiom of countability iff its topology has a countable basis.

If (X, D) is a topological space a local basis of X at a point x is any family of neighbourhoods G of x such that every neighbourhood of x contains a member of G.

Now the second class of topological spaces is introduced in

DEFINITION 1.4.12.   A topological space (X, D) is said to satisfy the first axiom of countability iff it has a countable local basis at each and every point.

For various classes of topologies which are considered in applications it is of interest to give methods for the construction of such topologies.

To this end, first we give

THEOREM 1.4.13.   *Let X be an arbitrary set and F be a family of sets whose union is X. Then there exists a unique topology D such that F is a sub-base for D and D is the weakest topology on X containing F.*

Proof. Let D contain X, the empty set $\varnothing$ and all unions of finite intersections of elements of F. Then D is clearly a topology and obviously F is a sub-base for D. Thus the existence is proved.

We prove now the uniqueness. Suppose that we have another topology, say D', such that F is a sub-base for D'. Let x be a point of X and G' be a neighbourhood of x in the D'-topology. Since F is a sub-base for D' we find a finite collection of sets whose intersection $F^\sim$ satisfies the relation

$$x \in F^\sim \subset G'.$$

Clearly $F^\sim$ is an open set in the D-topology and thus G' is a neighbourhood of x in the D-topology. This gives the fact that any set which is open in the D'-topology is open in the D-topology. Similarly we have that any open set of D is an open set of D'. Thus $D = D'$.

Remark 1.4.14. The topology D constructed in the proof of Theorem 1.4.13. is called the 'topology generated by F'.

Now we are ready to introduce a method for the construction of topologies.

DEFINITION 1.4.15. Let X be an arbitrary nonempty set and $F = \{D_i\}_{i \in I}$ $i \in I$ be a family of topologies on X. The sup F is the topology generated by

$$\{\bigcup D_i, i \in I\}.$$

An important class of topological spaces can be considered by introducing the so-called separation axioms. The reader interested in this topic is referred to the books by Dugundji (1954), Bourbaki (1962, 1966), and Kelley (1955).

DEFINITION 1.4.16. A topological space X is called a Hausdorff space if, for each pair of distinct points x_1, x_2 in X, there exist two open sets G_1 and G_2 such that

1. $x_i \in G_i,$ $i = 1, 2$

2. $G_1 \cap G_2 = \varnothing.$

Obviously we have

THEOREM 1.4.17. *Any metric space is a Hausdorff space.*

We define now the important notion in topological spaces: the notion of convergence.

For this we need the notion of a 'directed set'.

DEFINITION 1.4.18. A partially ordered set is a set with the following property : there exists a binary relation'$\leq$'such that

1. $a \leq b, b \leq c \to a \leq c$ (transitivity),

2. $a \leq a$ for all a (reflexivity)

which holds for a, b, c in the set.

Also $a \leq b$ means also $b \geq a$. Two elements p and q of a partially ordered set are called comparable iff $a \leq b$ or $b \leq a$; in the contrary case they are incomparable.

DEFINITION 1.4.19.   A partially ordered set X is called totally ordered if the following axioms are satisfied:

1.   all members of X are comparable,

2.   if $a \leq b$ and $b \leq a$ then $a = b$.

DEFINITION 1.4.20.   A directed set is a partially ordered set X satisfying the following condition: for any x, y in X there exists $z \in X$ such that $x \leq z$ and $y \leq z$. We say for short that X is directed by the relation '$\leq$'.

DEFINITION 1.4.21.   Let X be a partially ordered set and X_1 be a subset of X. We say that X_1 is majorized if there exists a point x_0 in X such that $x \leq x_0$ for all $x \in X_1$; x_0 is called a majorant for X_1.

We define the fact that X_1 is minorized similarly. If X_1 is majorized and there exists a majorant x_0 such that $x_0 < x^\sim$ where $x^\sim$ is any majorant of X_1 then x_0 is obviously unique and is called the supremum of X_1 (shortly, lub X_1) and similarly we define the infimum of X_1 (and we write for this glub X_1).

A set is inductively ordered if each totally ordered subset has a majorant (sometimes we say upper bound).

THEOREM 1.4.22.   (Zorn's Lemma). *For each inductively ordered set there exist maximal elements (i.e. elements x with the property that whenever $y > x$ then necessarily $y = x$).*

For the proof we refer to any book on set theory.

Now we are ready to consider the fundamental notion: the convergence.

As is well known, a sequence of elements in a set X is a function defined on the set of integers. It is easy to see that the set of integers is directed by the relation '$<$' defined in the usual way. This suggests the following

DEFINITION 1.4.23.   A net of elements in a set X is a function defined on a directed set, say D, with values in X.

Shortly we write this as $(x_\delta : \delta \in D)$, D being the directed set, or more simply (x_δ).

To define the convergence for nets we need the following

DEFINITION 1.4.24.   If $(x_\delta : \delta \in D)$ is in X, and X_1 is a subset of X then we say that the net is eventually in X_1 if there exists $\delta_0 \in D$ such that for all $\delta \in D, \delta > \delta_0, x_\delta \in x_1$. The net is frequently in X_1 if, for any $\delta_0 \in D$, there exists $\delta > \delta_0$ such that $x_\delta \in x_1$.

Now the notion of convergence for nets is introduced as follows:

DEFINITION 1.4.25. The net $(x_\delta : \delta \in D)$ converges to a point $x \in X$ if for any neighbourhood V of x, (x_δ) is eventually in V.

We write this in abbreviated form as

$$\lim x_\delta = x.$$

Of course, when $D = \{$the set of integers$\}$, we have the usual convergence for sequences.

DEFINITION 1.4.26. Let X, Y be two topological spaces and $f : X \to Y$. We say that f is continuous at x_0 if for any neighbourhood V of $f(x_0)$ there exists a neighbourhood U of x_0 such that for any $z \in U, f(z) \in V$. The function f is said to be continuous on X if it is continuous at each point $x \in X$.

Using the notion of convergence for nets we can characterize the continuity as follows

THEOREM 1.4.27. *Let X, Y be two topological spaces and $f : X \to Y$ be any function. The function f is continuous at x iff for any net (x_δ) converging to x, the net $(f(x_\delta))$ converges to $f(x)$*

Proof. Since the proof is immediate we omit it.

Now, using the concept of continuous functions we can define some useful and interesting topologies.

DEFINITION 1.4.28. Let X be a topological space and Y be an arbitrary nonempty set. If $f : Y \to X$ is a function then the topology generated by

$$\{f^{-1}(G), \quad G \text{ open in } X\}$$

is named 'the weak topology defined by f'

This topology is denoted by $w(Y, f)$.

Using a family of functions $F, f \in F, f : Y \to X$ we define the weak topology $w(Y, F)$ as the topology generated by

$$\{\bigcup w(Y, f), \ f \in F\}.$$

This method of constructing topologies is very useful for the case of product spaces where the 'product topology' or 'Tichonov topology' was first defined by A.N. Tichonov.

Let I be an arbitrary index set and for each $i \in I$ we have a topological space X_i. We can define the Cartesian product ΠX_i and, for each $i \in I$, we have the projection

$$p_j(\{x_i\}) = x_j.$$

DEFINITION 1.4.29.   The 'product topology' (= Tichonov topology) on ΠX_i is the topology defined by the family of $F = \{p_j : j \in I\}$.

Now following Wilansky we define the notion of subnet.

DEFINITION 1.4.30.   Let D and D' be two directed sets and $f : D' \to D$ a function. We say that f is finalizing if, for any $\delta \in D, f(\delta') > \delta$, i.e. there exists δ'_0 such that $\delta' > \delta'_0$ implies $f(\delta') > \delta$.

DEFINITION 1.4.31.   Let $(x_\delta : \delta \in D)$ be a net. A subnet of the net $(x_\delta : \delta \in D)$ is a net $(x_{f(\delta')} : D')$ where D' is a directed set and $f : D' \to D$ is finalizing.

Remark 1.4.32.   The notion of a subnet represents an analogy of the notion of subsequence of a sequence.

The notion of accumulation point (or cluster point, or limit point) is introduced as follows:

DEFINITION 1.4.33.   Let (x_δ) be a net. A point x is said to be an accumulation point of the net if for any neighbourhood X of the point x, (x_δ) is frequently in V.

The characterization of compact sets using nets is as follows:

THEOREM 1.4.34.   *Let X be a topological space and X_1 be a subset of X. The following assertions are equivalent*:

1.   *X_1 is compact,*

2.   *Every net in X_1 has an accumulation point in X_1,*

3.   *Every net in X_1 has a subnet which converges to a point in X_1.*

Proof. Suppose that 1 is true. In this case if (x_δ) is a net in X_1 we consider the following sets

$$\overline{\{x_\delta : \delta > \delta_0\}}$$

where the bar denotes the closure, i.e. the smallest closed set containing the set $\{x_\delta : \delta > \delta_0\}$. It is clear that this family has the finite intersection property and thus X_1 has a common point with all the above sets. This clearly gives the existence of the accumulaion point of the net. The assertions $2 \to 3 \to 1$ can be proved by imitating, more-or-less, the proofs in the case of subsets on **R** and we omit them.

1.5. LINEAR TOPOLOGICAL SPACES. LOCALLY CONVEX SPACES

We give now the notion of linear space.

DEFINITION 1.5.1. Let S be a nonempty set. S is a linear space over K ($\mathbf{R}$ or $\mathbf{C}$) if there exist two functions: ' $+$ ' and ' $\cdot$ ' defined on $S \times S$ and on $K \times S$ respectively and satisfying the following axioms

1. $x + (y + z) = (x + y) + z,$

2. $x + y = y + x,$

3.   There exists $0 \in S$ such that $0 + x = x$ for all $x \in S,$

4. $0 \cdot x = 0,$

5. $(a + b)x = a \cdot x + b \cdot x,$

6. $a(x + y) = a \cdot x + a \cdot y,$

7. $a(b \cdot x) = (ab) \cdot x,$

8. $1 \cdot x = x$

for all $x, y, z \in S$ and all $a, b \in K.$

We say that S is a complex linear space if $K = \mathbf{C}$ and we say that S is a real linear space if $K = \mathbf{R}$. The elements of S are called 'vectors' or points. The function ' $+$ ' is called the addition and the function ' $\cdot$ ' is called the multiplication with scalars.

DEFINITION 1.5.2.   A linear subspace S_1 of a linear space S (a subspace of S for short) is any subset of S which is a linear space with the addition and multiplication with scalars of L

Example 1.5.3.   (1) The space S in Example 1.1.3 is a real linear space under the addition and multiplication with scalars defined as

1. $x + y = (x_1 + y_1, x_2 + y_2, \ldots, x_n + y_n)$

2. $a \cdot x = (ax_1, ax_2, \ldots, ax_n)$

This space is denoted by $\mathbf{R}^n$.

(2) the space in Example 1.1.4 is also a linear real space for

1. $x + y = ((x_i + y_i))_1^\infty,$

2. $a \cdot x = ((ax_i))_1^\infty.$

This space is denoted by l^p. Also we use the same notation when we have sequences of complex numbers. In this case we have complex linear spaces.

(3) The space in Example 1.3.5 is real or complex depending upon whether we consider real-valued or complex-valued functions.

(4) Obviously if S is the space in Example 1.1.4, the set S_1 defined as

$$S_1 = \{x, x = (0, 0, \ldots, 0, x_{m+1}, x_{m+2}, \ldots)\}$$

is a linear subspace of S.

Suppose now that A and B are two subsets in a linear space S. We define their sum as the set

$$A + B = \{a + b, a \in A, b \in B\}$$

and the multiplication with $a \in K$ as the set

$$aA = \{a \cdot x, x \in A\}.$$

Let $x_1, x_2, \ldots, x_m$ be m elements of S (we suppose that $m < \infty$).

DEFINITION 1.5.4.   A linear combination of the elements x_i is any element of S of the form

$$z = a_1 x_1 + a_2 x_2 + \cdots + a_m x_m$$

where a_i are in K.

DEFINITION 1.5.5.   Let S be a linear space over K and $x_1, \ldots, x_m$ be m elements of S ($m < \infty$). These elements are said to be linearly independent if, for any linear combination $z = a_1 x_1 + \cdots + a_m x_m$ for which $z = 0$, it follows that $a_1 = a_2 = \cdots = a_m = 0$.

DEFINITION 1.5.6.   Let S be a linear space over K. The subset X_1 of S is said to be a linearly independent set if any finite subset of X_1 is linearly independent.

Example 1.5.7.   Let $S = C_{[0,1]}$ be the space of all continuous functions on $[0,1]$. In this case the set

$$X = \{1, t^2, t^3, \ldots, t^n, \ldots\}$$

is linearly independent.

One of the most important and interesting class of subsets of a linear space is the class of convex sets.

DEFINITION 1.5.8.   Let S be a linear space and C be a subset of S. The set C

is said to be convex if, for any $x, y \in C$, the element $z_t = tx + (1 - t)y \in C$ for all $t \in (0,1)$.

DEFINITION 1.5.9.   Let E be a set in a linear space S and define

$$\text{conv } E = \{z, z = \sum_1^n a_i x_i, a_i \geq 0, \sum_1^n a_i = 1\}.$$

The set conv E is called the 'convex hull' of E.

THEOREM 1.5.10.   *For any set E, conv E is a convex set.*
   Proof. Let z_1 and z_2 be two elements of conv E. In this case

$$z_1 = \sum_1^n a_i x_i, \quad a_i \geq 0 \quad \sum_1^n a_i = 1,$$

and

$$z_2 = \sum_1^m b_j y_j, \quad b_j \geq 0 \quad \sum_1^m b_j = 1.$$

It is clear that for any $t \in (0, 1)$, $tz_1 + (1 - t)z_2 \in \text{conv } E$ since

$$t \sum_1^n a_i + (1 - t) \sum_1^m b_j = 1.$$

THEOREM 1.5.11.   *For any set E, conv E is the smallest convex set of S containing E.*
   Proof. Obviously conv E contains E and, by the above remark, it is a convex set. Let C be another convex set which contains E.
   In this case C contains all elements of the form

$$\sum_1^n a_i x_i, \quad a_i \geq 0, \quad \sum_1^n a_i = 1, \quad x_i \in E.$$

We prove the assertion by induction on n. Clearly the assertion is true for $n = 2$. Suppose that it is true for all $k \leq n$.
   Let $z = \sum_1^{n+1} a_i x_i, a_i \geq 0, \sum_1^{n+1} a_i = 1$. Since we can write

$$z = a_{n+1} x_{n+1} + \sum_1^n a_i x_i$$

$$= a_{n+1} x_{n+1} + (1 - a_{n+1}) \cdot \sum_1^n a_i/(1 - a_{n+1}) x_i,$$

this gives us that $z \in C$. Thus every convex set C contains conv E and hence the assertion is proved.

In the following definition we consider some important classes of sets.

DEFINITION 1.5.12.   A set E in a linear space is called:

1.   symmetric if

$$E = \{-x, x \in E\} = -E,$$

2.   balanced if

$$tE = \{tx, x \in E\} \subset E \quad \text{for all } |t| \leq 1,$$

3.   absorbing if for each $x \in S$ there exists $\varepsilon_x > 0$ such that $tx \in E$ for all $|t| \leq \varepsilon_x$,

4.   affine if

$$tE + (1 - t)E \subset E \quad \text{for all } t,$$

5.   the line through x and y if

$$E = \{tx + (1 - t)y, t \in R\},$$

6.   the segment joining x and y if

$$E = \{tx + (1 - t)y, t \in [0, 1]\}.$$

Example 1.5.13.   Let $S = \mathbf{R}^2$ and

$$C = \{(x, y), x^2 + y^2 \leq 1\}.$$

Then C is symmetric, convex, balanced and absorbing.

Simple examples show, however, that the above classes of sets are all distinct.

We consider now an important class of functions on linear spaces.

DEFINITION 1.5.14.   Let S be a linear space over K. A seminorm on S is a function $p : S \to \mathbf{R}$ satisfying the following axioms:

1. $p(x + y) \leq p(x) + p(y),$

2. $p(ax) = |a| p(x),$

3. $p(x) \geq 0$

        for all $x, y \in S$ and $a \in K$.

Remark 1.5.15.   Axiom 3 is a consequence of Axioms 1 and 2. Indeed, we can take $x = -y$ and we obtain

$$0 = p(0) \leq p(x) + p(-x) = 2p(x).$$

Let S be a linear space and p be an arbitrary seminorm on S. If we define

$$B_p(1) = \{x, \, p(x) \leq 1\}$$

then it is not difficult to see that $B_p(1)$ has the following properties: it is convex, balanced and absorbing (i.e. a balloon). We show now that there is a strong connection between seminorms and balloons. For this we need the notion of the gauge or the Minkowski functional of a balloon.

DEFINITION 1.5.16.   Let A be a balloon in a linear space S. The gauge of the balloon A ($=$ Minkowski functional) is defined as

$$p_A(x) = \inf\{t > 0, \, x \in tA\}.$$

The properties of the gauges are given in

THEOREM 1.5.17.   *Let p_A be the gauge of the balloon A. Then the following assertions hold:*

1. P_A *is a seminorm on* S,
2. $\{x, \, p_A(x) < 1\} \subset A \subset \{x, \, p_A(x) \leq 1\}.$

Proof. The properties $p_A(x) \geq 0$ and $p_A(ax) = |a|p_A(x)$ are obvious. To prove the inequality

$$p_A(x + y) \leq p_A(x) + p_A(y)$$

let $\varepsilon > 0$ and

$$x \in [p_A(x) + \varepsilon]A$$
$$y \in [p_A(y) + \varepsilon]A.$$

Since A is a convex set we have

$$x + y \in [p_A(x) + p_A(y) + 2\varepsilon]A.$$

and since $\varepsilon > 0$ is arbitrary we obtain the desired inequality. The second assertion follows directly from the definition of p_A.

From this theorem we obtain the following interesting property of balloons:

COROLLARY 1.5.18.   *Let S be a linear space and let A be an arbitrary balloon. Then there exists the balloon A_1 such that*

$$A_1 + A_1 \subset A.$$

Proof.   We can take as A_1 the following set,

$$A_1 = \{x, p_A(x) < \tfrac{1}{2}\}$$

We consider now the notion of 'topological linear space'.

DEFINITION 1.5.19.   Let S be a linear space over K and D be a topology on S. We say that (S, D) is a topological linear space if the functions ' $+$ ' and ' $\cdot$ ' are continuous. In this case D is called a linear topology.

Example 1.5.20.   Given that S is a linear space and letting $f : S \to \mathbf{R}$ be a linear functional, then $D = w(\mathbf{R}, f)$ is a linear topology on S.

DEFINITION 1.5.21.   Two linear topological spaces over K are said to be isomorphic if there exists a biunivocal linear map $i : S_1 \to S_2$ which is a homeomorphism (i.e. i and i^{-1} are continuous).

THEOREM 1.5.22.   *Let S be a linear space. A topology τ on S is a linear topology on S if and only if τ is translation-invariant and possesses an o-neighbourhood base B with the following properties:*

1.   *for each $G \in B$ there exists $U \in B$ such that $U + U \subset G$,*

2.   *$G \in B$ is balanced,*

3.   *there exists $a \in K$, $|a| < 1$ such that $G \in B$ then $aG \in B$.*

Proof.   Let G be any neighbourhood of o in S and since ' $+$ ' is continuous there exists U, a neighbourhood of o and $\varepsilon > o$, such that $aU \in G$ for all $|a| \leqq \varepsilon$. In this case $V = \bigcup \{aU, |a| \leqq \varepsilon\}$ is a neighbourhood of o and clearly it is balanced. Since $V \in G$, the family of all balanced o-neighbourhoods is a base at o. The continuity of ' $\cdot$ ' implies that V is absorbing. From the continuity of ' $+$ ' it follows that conditions 1 and 3 are also satisfied.

Conversely, if τ is any translation-invariant topology on S with the properties 1–3 above, then τ is a linear topology on S. Since the proof is not difficult we omit it.

DEFINITION 1.5.23.   A linear topology D on a linear space S is called 'locally convex' if every neighbourhood of o includes a convex neighbourhood of o.

A locally convex topology D on a linear space S is determined by a family of seminorms $\{p_\alpha : \alpha \in I\}$.

Indeed, if $\{p_\alpha : \alpha \in I\}$ is a family of seminorms then, for each $\alpha \in I$, we consider the set

$$G = \{x, p_\alpha(x) \leqq 1\}$$

and let $\{1/nG\}$ be a family of sets where n runs through all positive integers and G ranges over all finite intersections of sets G_α. Then this family, B, is a base at o for a locally convex topology τ which is called the topology generated by the family of seminorms $\{p_\alpha : \alpha \in I)$. The family of seminorms is called the generating family for τ.

Now, if S is a locally convex space, we can consider the gauges of all balloons which are o-neighbourhoods. Let $\{p_\alpha\}$ be this family, which is a family of seminorms on S and the topology generated by this family is exactly the original one.

DEFINITION 1.5.24. A complete metrizable locally convex space is called a Fréchet spacè (abbreviated as (F)-space).

For an excellent treatment of topological linear spaces as well as locally convex spaces we refer to Bourbaki, (1966), and Grothendieck (lectures at São Paulo, 1954).

Chapter 2

Hilbert Space and Banach Spaces

2.1. NORMED SPACES. BANACH SPACES

Let S be a linear topological space over K.

DEFINITION 2.1.1. A norm on S is any function $f : S \to \mathbf{R}$ with the following properties:

1. $f(x + y) \leq f(x) + f(y)$,

2. $f(ax) = |a| f(x)$, $\qquad a \in K$,

3. $f(x) = 0$ $\qquad$ if and only if $x = 0$.

In what follows we use the notation $f(x) = \|x\|$.

Example 2.1.2. Let $S = C_{[0,\,1]}$ be the space of all continuous functions on $[0, 1]$. Let $f \in C_{[0,\,1]}$ and we set

$$f \to \|f\| = \sup \{|f(t)|, \, t \in [0, 1]\}.$$

Then clearly

$$f \to \|f\|$$

is a norm on $C_{[0,\,1]}$.

Example 2.1.3. Let S be as in Example 2.1.2. and define

$$f \to \|f\|_1 = \int_0^1 |f(t)| \, dt$$

then it is easy to see that

$$f \to \|f\|_1$$

is a norm on $C_{[0,\,1]}$ and, for each $f \in C_{[0,\,1]}$, we have

$$\|f\|_1 \leq \|f\|.$$

Example 2.1.4. Let S be the space l^p with $p \in [1, \infty)$. If we define for any

$x \in l^p$

$$x \to \|x\|_p = \left(\sum_1^\infty |x_i|^p \right)^{1/p}$$

then from the well-known Minkowski inequality it is easily seen that this is a norm on l^p.

Example 2.1.5. Let S be the space of all bounded sequences denoted in what follows by the letter m or denoted as l^∞.

If we define for each $x \in m$,

$$x \to \|x\| = \sup \{|x_i|, x = (x_i)\}$$

then this is a norm on m.

Example 2.1.6. Let $(\Omega, \mathcal{B}, \mu)$ be a measure space, and suppose for simplicity that $\mu(\Omega) < \infty$. Let $p \in [1, \infty)$ and S be the set of all functions (classes of functions) on Ω such that

$$\|f\|_p = \left(\int_\Omega |f(s)|^p \, d\mu \right)^{1/p} < \infty$$

In this case the function

$$f \to \|f\|_p$$

is a norm on S. In what follows we use for this space the notation $L^p(\Omega, \mathcal{B}, \mu)$ or simply L^p.

Example 2.2.7. We consider the space m. In this space we have two important subspaces, namely the subspaces c and c_0:

$$c = \{\text{the set of all convergent sequences}\},$$

$$c_0 = \{\text{the set of all sequences converging to zero}\}.$$

It is obvious that with the norm induced by the norm of m, we obtain a norm on these spaces, i.e. the function

$$x \to \sup \{|x_i|, x = (x_i)\}$$

is a norm on the spaces c and c_0.

Example 2.1.8. Let K^m be the set of all elements of the form $x = (x_1, x_2, \ldots, x_m)$ with $x_i \in K$ and for each fixed $p \in [1, \infty)$ we define

$$\|x\|_p = \left(\sum_1^m |x_i|^p \right)^{1/p}.$$

Then

$$x \to \|x\|_p$$

is a norm on K^m.

An important class of normed spaces are the so called 'Banach spaces'. Their definition is given in

DEFINITION 2.1.9.   A normed space S with the norm $\| , \|$ is called a Banach space if S is a complete metric space for the metric d defined on S by the formula,

$$d(x, y) = \|x - y\|.$$

Remark 2.1.10.   If X is a normed space with the norm $\| , \|$ then we can define a distance on X by the formula

$$d(x, y) = \|x - y\|.$$

All properties of the distances are obvious except possibly the following

$$d(x, z) \leq d(x, y) + d(y, z).$$

This property follows immediately from property 1 of the norms.

The spaces in examples 2.1.2, 2.1.4–2.1.8 are all Banach spaces. It is also true that any normed space S can be embedded in a Banach space $S^\sim$ such that S is isometric with a dense subspace of $S^\sim$. The method of proving this assertion is similar to the method of completion of the rationals.

Let X, Y be two Banach spaces over the same field K.

DEFINITION 2.1.11.   A function $T : X \to Y$ is called a linear operator if

1. $T(x + y) = T(x) + T(y),$

2. $T(a \cdot x) = aT(x)$

for all $x, y \in X$ and all $a \in K$.

DEFINITION 2.1.12.   The linear operator $T : X \to Y$ is termed bounded if there exists a number $M < \infty$ such that

$$\|T(x)\| \leq M \|x\|$$

for all $x \in X$.

Remark 2.1.13.   We write $a \cdot x$ for $a \in K$ and $x \in X$ briefly as ax and for any linear operator we sometimes write $T(x)$ simply as Tx.

The following theorem gives a useful characterization of bounded operators.

THEOREM 2.1.14.   *Let X, Y be two Banach spaces and $T : X \to Y$ be a linear operator. Then T is bounded if and only if T is continuous.*

For this we need the following result:

THEOREM 2.1.15.   *Let X, Y be two Banach spaces and $T : X \to Y$ be a linear operator. If T is continuous at some point $x_0 \in X$ then T is continuous on X.*
Proof. Since the proof is easy we omit it.

Now the proof of Theorem 2.1.14. is as follows:

First we remark that it is obvious that, if T is bounded, then T is continuous. Now, suppose that T is continuous and not bounded. In this case, for each n there exists x_n, $\|x_n\| = 1$ such that $\|Tx_n\| \geqq n$. We consider the sequence $\{y_n\}$ with $y_n = x_n/n$.

It is clear that $y_n \to 0$ and thus, since T is continuous, we have $Ty_n \to 0$. But $\|Ty_n\| = 1/n \ \|Tx_n\| \geqq 1$ for all n. This contradiction proves the assertion.

DEFINITION 2.1.16.   Let X be a Banach space and $Y = K$. Then every linear operator $T : X \to Y$ is called a linear functional.

The set of all continuous linear functionals on X is called the dual of X and is denoted by X^*. Similarly we can define the higher duals of the space X by

$$X^{*(n)} = (X^{*(n-1)})^*, \qquad n = 1, 2, \ldots.$$

For any two Banach spaces X, Y we set $L(X, Y) = \{T : X \to Y \text{ linear and continuous}\}$. If $X = Y$ we set $L(X, X) = L(X)$.

DEFINITION 2.1.17.   Let X, Y be two Banach spaces and $T \in L(X, Y)$. The function defined on $L(X, Y)$ by the relation

$$T \to \|T\| = \sup \{\|Tx\|, \|x\| \leqq 1\}$$

is a norm on $L(X, Y)$.

The fact that the above function is indeed a norm is not difficult to verify. The following result gives more information about this function.

THEOREM 2.1.18.   *The space $L(X, Y)$ (which is a linear with the '$+$' and '$\cdot$' defined as for functions) is a Banach space with the norm defined in 2.1.17.*
Proof. In fact we have only to prove that $L(X, Y)$ is a complete metric space.

Let $\{T_n\}$ be a Cauchy sequence in $L(X, Y)$. In this case for each $x \in X$, the sequence $\{T_n x\}$ is also a Cauchy sequence and since Y is a Banach space we can define for each $x \in X$ the linear operator $T : X \to Y$ by the relation

$$Tx = \lim T_n x.$$

To prove the completeness of $L\{X, Y\}$ it remains to show that $T \in L(X, Y)$. Let $\varepsilon > 0$ and N_ε such that for $n, m \geq N_\varepsilon$,

$$\| T_n - T_m \| \leq \varepsilon/2.$$

In this case for any $x \in X$, $\|x\| \leq 1$ we have that

$$\| T_n x - T_m x \| \leq \varepsilon/2$$

and thus for such x we have,

$$\|(T - T_n)x\| = \|(T - T_m)x + (T_m - T_n)x\| \leq \varepsilon/2 + \varepsilon/2 = \varepsilon.$$

This gives that $\| T_n - T \| \to 0$, and this implies that $T \in L(X, Y)$ and $T = \lim T_n$.

One of the most powerful tools of functional analysis is the extension theorem of Hahn–Banach for real Banach spaces and the extension theorem of Bohnenblust–Sobczyc for complex Banach spaces. In what follows we give these important results as well as some applications. For this we need some definitions.

DEFINITION 2.1.19.   Let X be a real Banach space and X_1 be a subspace of X. Let $f : X_1 \to \mathbf{R}$ be a linear functional on X_1 and we suppose that there exists a linear functional $F : X \to \mathbf{R}$ such that $f(x) = F(x)$ for all $x \in X_1$. In this case we say that F is an extension of f (sometimes we say that f is the restriction of F to X_1).

THEOREM 2.1.20.   (The Hahn–Banach Extension Theorem). *Let X be a Banach space over reals and X_1 be a linear subspace and also $f : X_1 \to \mathbf{R}$ with the property that, for some seminorm $p : X \to \mathbf{R}$,*

$$f(x) \leq p(x) \quad \text{for all } x \in X_1.$$

In this case there exists an extension $F : X \to \mathbf{R}$ of f such that

$$F(x) \leq p(x)$$

for all $x \in X$.

Proof.   We suppose, without loss of generality, that f is not identically zero on X_1. Let x_0 be an arbitrary point which is not in X_1 and we consider the space generated by X_1 and x_0 (i.e. the smallest linear subspace of X which contains X_1 and x_0).

It is obvious that this is

$$X^{\sim} = \{x + ax_0\}$$

with $x \in X_1$ and $a \in \mathbf{R}$. It is not difficult to see that every element z of $X^{\sim}$ has the (unique) form

$$z = x + ax_0.$$

In this case it follows that the number a is uniquely determined and, for any element z of the above form, we define

$$f^{\sim}(z) = f(x) + ac$$

where c is a number which will be determined so that

$$f^{\sim}(z) \leq p(z).$$

for all $z \in X^{\sim}$. We show now that we can choose c such that the above inequality holds. Suppose first that such a number exists. In this case,

$$f(x/a) + c \leq p(x/a + x_0); \qquad x \in X_1 \quad \text{and} \quad a > 0,$$

and

$$f(x/a) + c \geq -p(-x/a - x_0); \qquad x \in X_1 \quad \text{and} \quad a < 0$$

and further

$$f(x) - p(x - x_0) \leq c \leq -f(y) + p(y + x_0)$$

for all $x, y \in X_1$. But

$$f(x) + f(y) = f(x + y) \leq p(x + y) \leq p(x - x_0 + x_0 + y)$$
$$\leq p(x - x_0) + p(x_0 + y)$$

for all $x, y \in X_1$. If we set

$$m = \sup \{f(x) - p(x - x_0)\}$$

and

$$M = \inf \{-f(y) + p(x_0 + y)\}$$

we have obviously $m \leq M$ and we can take as c any number between m and M.

To prove the theorem we denote by F the family of all linear functionals which are extensions of f. Thus for each element $f^\sim$ in F there exists a linear subspace $X_{f^\sim}$ such that:

1. $\qquad X_1 \subset X_{f^\sim}$,

2. $\qquad f(x) = f^\sim(x) \quad \text{for } x \in X_1.$

We define now a partial order in F as follows: if $g^\sim$ and $h^\sim$ are elements of F we say that

$$g^\sim \leqq h^\sim$$

iff $h^\sim$ is an extension of $g^\sim$. We show now that F with this partial order is inductively ordered. Let $F^\sim$ be a totally ordered part of F. In this case on the space which contains all the elements of the spaces associated with the functionals in $F^\sim$, say V, there exists a functional which is an extension of f. Indeed, let x be arbitrary in this space and thus, from the definition of V, there exists a subspace $X_k \ni x$ and a linear functional $k^\sim$ which is an extension of f. Obviously $k^\sim(x) \leqq p(x)$ and we define a linear functional at the point x by the relation $l^\sim(x) = k^\sim(x)$. Since $F^\sim$ is totally ordered this is a good definition of $l^\sim$. Obviously $l^\sim$ is an extension of f. Thus we can apply Zorn's Lemma and we obtain a maximal element, say F^*.

Let X_{F^*} be the corresponding subspace. This space clearly contains X_1. We show that X_{F^*} is exactly X. Indeed, in the contrary case there exists an element, say y_0, which is in X and not in X_{F^*}. By the first step of the proof we can find an extension of F^* to this space and this is obviously an extension of f. This contradicts the fact that F^* is maximal and the theorem is proved.

We now give some useful applications of the theorem.

COROLLARY 2.1.21. *Let X be a Banach space over the reals and X_1 be a linear subspace of X. Let $f : X_1 \to \mathbf{R}$ be a continuous linear functional; then there exists an extension F of f such that $\|f\| = \|F\|$.*

Proof. First we remark that the norm of f is computed on X_1.
We consider the seminorm

$$p(x) = \|f\| \, \|x\|$$

(which is actually a norm) and from the Hahn–Banach theorem we obtain that there exists an extension of f, say F, with the property that

$$F(x) \leqq \|f\| \, \|x\|.$$

This gives us that $\|F\| \leqq \|f\|$. We show now that in fact we have an equality. Indeed,

$$\|F\| = \sup_{\|x\| \leq 1} |F(x)| \geq \sup_{\substack{x \in X_1 \\ \|x\| \leq 1}} |F(x)| = \sup_{\substack{x \in X_1 \\ \|x\| \leq 1}} |f(x)| = \|f\|$$

and the assertion is proved.

COROLLARY 2.1.22. *Let X be a real Banach space and $x \in X$ be an arbitrary element in X. Then there exists an element $f \in X^*$ with the following properties:*

1. $\|f\| = 1$,

2. $f(x) = \|x\|$.

Proof. Let X_1 be the space generated by x and define on this space the linear and continuous functional g by the relation

$$g(ax) = a\|x\|.$$

It is clear that the norm of g is 1 and by the Hahn–Banach theorem we find an extension, say f, to the whole of X. Clearly f satisfies the Corollary.

COROLLARY 2.1.23. *Let X be a real Banach space and X_1 be a linear subspace. Then X_1 is dense in X if and only if, for any $f \in X^*$ which is zero on X_1, necessarily $f = 0$.*

Proof. Let $\bar{X}_1$ be the closure of X_1. Clearly we can suppose that this is not X. Let x_0 be a element of X not in $\bar{X}_1$. For the space generated by $\bar{X}_1$ and x_0 we can define the linear functional $f\tilde{}$ by

$$f(x + ax_0) = a$$

and it is not difficult to see that this is a continuous linear functional. We can extend this to the whole space X and thus, if X_1 is not dense, then there exists a nonzero linear functional which is zero on X_1 and the Corollary is proved.

COROLLARY 2.1.24. *Let X be a real Banach space and x be an arbitrary element of X. In this case*

$$\|x\| = \sup_{\|f\| = 1} |f(x)|.$$

Proof. The assertion follows from Corollary 2.1.22.

We now give the Bohnenblust–Sobczyc extension theorem.

THEOREM 2.1.25. (Bohnenblust–Sobczyc Theorem). *Let X be a complex*

Banach space and X_1 be a subspace of X. Let $f : X \to \mathbf{C}$ be a linear and continuous functional satisfying the inequality

$$|f(x)| \leq \|f\| \|x\|$$

for all $x \in X_1$. Then there exists a linear functional $F : X \to \mathbf{C}$ with the following properties:

1. $F(x) = f(x)$ for all $x \in X_1$,

2. $\|F\| = \|f\|$.

Proof. Let f be a linear functional on the complex space X. We consider X as a real Banach space and the problem concerns the connection between the dual of X as a real space and as a complex space. First we remark that every f in the dual of X as a complex space is of the form

$$f(x) = \operatorname{Re} f(x) - i \operatorname{Re} f(ix)$$

and this is a consequence of the fact that every complex number z has the form

$$z = \operatorname{Re}(z) - i\operatorname{Re}(iz).$$

Further if $u : X \to \mathbf{R}$ then

$$f(x) = u(x) - iu(ix)$$

is an element of the dual of X (X is a complex space). Indeed, it is sufficient to prove that

$$f(ix) = if(x)$$

for all $x \in X$. We have,

$$f(ix) = u(ix) - i(-x) = i(u(x) - iu(ix)) = if(x)$$

and the assertion is proved.

Suppose now that X is a Banach space over K and consider the duals X^*, X^{**}, etc.

Let $x \in X$ and, for each $f \in X^*$, we can define the linear functional

$$F_x(f) = f(x)$$

and from the extension theorem we see that $\|F_x\| = \|x\|$. It is natural to ask if there exist spaces for which the elements of X^{**} are exactly only of this form. Thus we are led naturally to consider a class of Banach spaces for which this is true.

DEFINITION 2.1.26. Let X be a Banach space. Then X is said to be reflexive if the elements of X^{**} are of the form F_x with x in X.

As an important example of Banach space which is reflexive we have the l^p and L^p space for $p \in (1, \infty)$. This will be proved in 2.5.11.

Also we remark that the mapping

$$x \to F_x$$

is a linear isomorphism of X into a part of X^{**}. The above mapping is called the 'natural embedding' of X into its second conjugate. Since $\|x\| = \|F_x\|$ this isomorphism is isometric.

DEFINITION 2.1.27. A metric space is separable if there exists a countable set which is dense in the space.

THEOREM 2.1.28. *Let X be a Banach space such that X^* is separable. Then X is separable.*

Proof. Let $\{x_n^*\}$ be a countably dense set in X^*. We consider the set $\{x_n\}$ of elements in X such that $\|x_n\| \leq 1$ and $|x_n^*(x_n)| \geq \|x_n^*\|/2$. It is clear that the set of all linear combinations of elements $\{x_n\}$ with rational coefficients is a countable set.

Suppose that this is not dense in X. We find an element x^* in X^* such that $x^*(x_n) = 0$ for all $n = 1, 2, \ldots$. Since $\{x_n^*\}$ is dense we obtain that $x^* = \lim x_{n_i}^*$. Since,

$$\|x^* - x_{n_i}^*\| \geqq \|(x^* - x_{n_i}^*)(x_{n_i}) = |x_{n_i}^*(x_{n_i})| > \tfrac{1}{2} \|x_{n_i}^*\|$$

which is a contradiction.

We mention without proof the following

THEOREM 2.1.29. *Every closed linear subspace of a reflexive space is again reflexive.*

2.2. HILBERT SPACES

An important class of Banach spaces is the class of Hilbert spaces and this class is characterized by the fact that the norm has some very useful additional properties connected with the algebraic structure of the space.

First we define the important notion of inner product or scalar product.

DEFINITION 2.2.1. Let X be a linear space over K. A function defined on $X \times X$ with values in K is said to be a scalar product (or an inner product) if the following axioms are satisfied:

1.   for all $a \in K$, $b \in K$ and $x, y, z \in X$,

$$\langle ax + by, z \rangle = a \langle x, z \rangle + b \langle y, z \rangle,$$

2.   for all $x, y \in X$,

$$\langle x, y \rangle = \langle y, x \rangle^*$$

(for any complex number z, z^* is its complex conjugate)

3.   for any $x \in X$ and $x \neq 0$,

$$\langle x, x \rangle > 0.$$

PROPOSITION 2.2.2.   *For any $x \in X$,*

$$\langle x, 0 \rangle = 0.$$

Proof. Obvious.

From Axioms 1 and 2 we obtain that, for all $a, b \in K$ and $x, y, z \in X$,

$$\langle x, ay + bz \rangle = a^* \langle x, y \rangle + b^* \langle x, z \rangle.$$

Let X be a linear space and suppose that there exists a scalar product $\langle , \rangle$ on X. Then we can define the function,

$$x \to \|x\| = \langle x, x \rangle^{1/2}$$

and the basic properties of this function (we make here an abuse of notation) are given in

THEOREM 2.2.3.   *The function*

$$x \to \|x\|$$

is a norm on X.

For this we need the following result which has also an independent interest.

THEOREM 2.2.4.   (Cauchy inequality). *Let X be a linear space and suppose that on X there exists a scalar product $\langle , \rangle$. In this case for any $x, y \in X$,*

$$|\langle x, y \rangle|^2 \leq \|x\| \cdot \|y\|.$$

Proof. Indeed, we prove that for any $x, y \in X$,

$$|\langle x, y \rangle| \leq \tfrac{1}{2} \{ \|x\|^2 + \|y\|^2 \}.$$

We consider the real number s such that

$$\langle e^{is}x, y \rangle = |\langle x, y \rangle|$$

and thus we have,

$$
\begin{aligned}
0 \leq \|e^{is}x - y\|^2 \\
= \langle e^{is}x, e^{is}x \rangle + \langle y, y \rangle - \langle e^{is}x, y \rangle - \langle y, e^{is}x \rangle \\
= \|x\|^2 + \|y\|^2 - 2|\langle x, y \rangle|
\end{aligned}
$$

and this is equivalent to the required inequality.

Now to prove the Cauchy inequality: first, we suppose that $\|x\| = \|y\| = 1$ and in this case we have,

$$|\langle x, y \rangle|^2 \leq \tfrac{1}{2}\{1 + 1\} = 1 = \|x\| \cdot \|y\|.$$

For the general case we can suppose without loss of generality, that $x \neq 0$ and $y \neq 0$. In this case we have for

$$x^\sim = x/\|x\|, \qquad y^\sim = y/\|y\|$$

that

$$|\langle x^\sim, y^\sim \rangle| \leq 1$$

and this obviously gives the Cauchy inequality.

Now the proof of Theorem 2.2.3 is as follows: First we remark that the properties:

1. $\qquad \|x\| \geq 0 \quad$ and $\quad \|x\| = 0 \quad$ if and only if $x = 0$,

2. $\qquad \|ax\| = |a| \, \|x\|$

are obvious.

We prove now the triangle inequality, i.e. the inequality

$$\|x + y\| \leq \|x\| + \|y\|.$$

Indeed, we have for all $x, y \in X$,

$$
\begin{aligned}
\|x + y\|^2 &= \langle x + y, x + y \rangle \\
&= \langle x, x \rangle + \langle y, y \rangle + \langle x, y \rangle + \langle y, x \rangle \\
&= \|x\|^2 + \|y\|^2 + 2\,\mathrm{Re}\langle x, y \rangle \langle \|x\|^2 + \|y\|^2 + \\
&\quad + 2\|x\| \cdot \|y\| \rangle \\
&= (\|x\| + \|y\|)^2
\end{aligned}
$$

and this clearly gives the assertion.

We give now some identities for elements in linear spaces with a scalar product.

THEOREM 2.2.5.   *Let X be a linear space with scalar product $\langle\,,\,\rangle$ and for any $x \in X$, let*

$$x \to \|x\| = \langle x, x \rangle^{1/2}.$$

Then for all x, y in X the following identities hold:

1. $\|x + y\|^2 + \|x - y\|^2 = 2\{\|x\|^2 + \|y\|^2\}$,

2. $\|x + y\|^2 - \|x - y\|^2 = 2\{\langle x, y \rangle + \langle y, x \rangle\}$,

3. $\|x + y\|^2 - \|x - y\|^2 + i\|x + iy\|^2 - i\|x - iy\|^2 = 4\langle x, y \rangle$.

Proof. These are direct consequences of some computations which we omit.

Now we are in a position to introduce the notion of Hilbert space.

DEFINITION 2.2.6.   A Banach space X is said to be a Hilbert space if on X there exists a scalar product $\langle\,,\,\rangle$ such that the norm of X is exactly the norm defined by the relation

$$x \to \|x\| = \langle x, x \rangle^{1/2}.$$

We give now some useful examples of Hilbert spaces.

Example 2.2.7.   Let l^2 be the set of all elements of the form

$$x = (x_1, x_2, \ldots)$$

such that

$$\|x\| = \left(\sum_1^\infty |x_i|^2 \right)^{1/2} < \infty.$$

We can define a scalar product on l^2 by

$$\langle x, y \rangle = \sum_1^\infty x_i y_i^*$$

and, using the Cauchy inequality for series, it is easy to see that this is, indeed, a Hilbert space.

Example 2.2.8.   For $(\Omega, \mathscr{B}, \mu)$ the space L^2 is a Hilbert space with the scalar product defined as follows,

$$\langle f, g \rangle = \int_{\Omega} f(s)g(s)^* d\mu.$$

Example 2.2.9.   Let I be an arbitrary family. Let $F(I)$ be the set of all finite parts of I. We define a partial order on $F(I)$ in the following manner: for any J_1 and J_2 in $F(I)$ we say that

$$J_1 \leq J_2 \quad \text{if} \quad J_1 \subset J_2.$$

We remark that $F(I)$ with this order is a directed set. We now define the summability with respect to the set I. Let $\{x_i, i \in I\}$ be a family of complex numbers. We say that this family is square summable with the sum x, if $\{\sum_{i \in J \in F(I)} |x_i|^2, F(I)\}$ is a net converging to x.

Now we can construct a Hilbert space l_1^2 as the set of all elements of the form $\{x_i\}_{i \in I}$ which are square summable. Then we can define as above a scalar product by the net $\{\sum_{i \in J} x_i y_i^*, F(I)\}$ and we remark that when I is the set of integers this is exactly the space in Example 2.2.7.

The usefulness of this space will now be shown using the notion of base of a Hilbert space.

First we need the following

DEFINITION 2.2.10.   Two elements of a Hilbert space H are said to be orthogonal if

$$\langle x, y \rangle = 0.$$

DEFINITION 2.2.11.   Consider $\{x_i\}_{i \in I}$, I an index family, of elements of a Hilbert space H. The family is said to be orthonormal if

$$\langle x_i, x_j \rangle = \delta_{i,j}$$

where $\delta_{i,j}$ is the Kronecker symbol.

In the set of all orthonormal families we can introduce an order by inclusion: if F and G are two orthonormal families we say that $F \leq G$ if G includes F.

DEFINITION 2.2.12.   A basis of a Hilbert space H is any maximal orthonormal family of elements of H.

First we remark that the definition has sense since, as is easy to prove, such families exist. This is an easy consequence of Zorn's Lemma.

LEMMA 2.2.13.   *A family* $\mathscr{F} = \{l_\alpha\}_{\alpha \in I}$ *of orthonormal elements is a basis for* $\varkappa$ *iff* $\langle x, l_\alpha \rangle = 0 \ \forall \alpha \in I$ *then* $x = 0$.

Proof. If $\mathcal{F} = \{l_\alpha\}_{\alpha \in I}$ is a basis and $x \neq 0$ is such that

$$\langle x, l_\alpha \rangle = 0 \qquad \forall \; \alpha \in I$$

then $\mathcal{F}_1 = \{l_\alpha\}_{\alpha \in I} \cup \{x/\|x\|\}$ is an orthonormal family containing $\mathcal{F}$. This is a contradiction and thus $x = 0$. The converse assertion is obvious. Let $\{x_1, \ldots, x_n\}$ be a finite set of elements in $\varkappa$ such that $\langle x_i, x_j \rangle = \delta_{ij}$,

LEMMA 2.2.14.   *For any* $x \in \varkappa$,

$$\inf_{c_1, \ldots, c_n \in K} \left\| x - \sum_1^n c_i \varkappa_i \right\| = \left\| x - \sum_{i=1}^n \langle x, x_i \rangle x_i \right\|.$$

Proof. We have

$$0 \leq \left\| x - \sum_{i=1}^n c_i \varkappa_i \right\|^2 = \|x\|^2 - \left\langle x, \sum_{i=1}^n c_i \varkappa_i \right\rangle -$$

$$- \left\langle \sum_{j=1}^n c_j \varkappa_j, x \right\rangle + \sum_{j=1}^n |c_j|^2 = \|x\|^2 - \sum_{i=1}^n \bar{c}_i \langle x, \varkappa_i \rangle -$$

$$- \sum_{j=1}^n c_j \langle x_j, \varkappa \rangle + \sum_{j=1}^n |c_j|^2$$

$$= \|x\|^2 - 2 \operatorname{Re} \left(\sum_i c_i \overline{\langle \varkappa, x_i \rangle} \right) + \sum_{j=1}^n c_j^2$$

$$= \sum_{i=1}^n |c_i - \langle x, x_i \rangle|^2 + \|x\|^2 - \sum_{i=1}^n |\langle x, \varkappa_i \rangle|^2$$

and thus the norm of $\|x - \sum_{i=1}^M c_i x_i\|$ is minimal if $\langle x, x_i \rangle = c_i$ and this gives the assertion.

COROLLARY 2.2.15.   *For any* $x \in \varkappa$ *and any orthonormal family* $\mathcal{F} = \{l_\alpha\}_{\alpha \in I}$

$$\|x\|^2 \geq \sum_{\alpha \in I} |\langle x, l_\alpha \rangle|^2.$$

(this is sometimes called the 'Bessel inequality').

DEFINITION 2.2.16.   Two Hilbert spaces H and $H^\sim$ are called isomorphic-isometric if there exists a map $T : H \to H^\sim$ such that

1. $T(H) = H^\sim$,

2. $\langle Tx, Ty \rangle = \langle x, y \rangle$ for all $x, y \in H$.

THEOREM 2.2.17. *Let H be a Hilbert space and $\{e_i\}_{i \in I}$ be an arbitrary basis of H. In this case H is isomorphic-isometric with l_I^2.*

The proof of the theorem follows from the following result

THEOREM 2.2.18. *Let H be a Hilbert space and $\{e_i\}_{i \in I}$ be an arbitrary basis of H. Then for any $x \in H$,*

$$\|x\|^2 = \sum_{i \in I} |\langle x, e_i \rangle|^2.$$

Proof. From the Bessel inequality we see that

$$\|x\|^2 \geq \sum |\langle x, e_i \rangle|^2.$$

Since $\{e_i\}$ is a basis we can consider the element

$$x^{\sim} = \sum_{i \in I} \langle x, e_i \rangle e_i$$

which is in H and clearly all the Fourier coefficients of the element $x - x^{\sim}$ are zero. This gives that $x = x^{\sim}$ because $\{e_i\}$ is a basis for H. Thus $x = \sum_{i \in I} \langle x, e_i \rangle e_i$ and this gives the above equality.

Now the proof of Theorem 2.2.17 is as follows: for $x \in H$ we define

$$Tx = (\langle x, e_i \rangle)_{i \in I}$$

and clearly this has all the required properties.

Using the existence and the decomposition of an element with respect to a basis we can prove the following very important decomposition of Hilbert spaces.

First we give a re-formulation of the result obtained in the proof of Theorem 2.2.18. as

THEOREM 2.2.19. *Let H be a Hilbert space and $\{e_i\}_{i \in I}$ be an arbitrary basis of H. Then any element $x \in H$ can be written as*

$$x = \sum_{i \in I} \langle x, e_i \rangle e_i.$$

Now we can give the decomposition theorem.

THEOREM 2.2.20. *Let H be a Hilbert space and M be a closed linear subspace of H. Then there exists a closed linear subspace $N = M^{\perp}$ such that every element $x \in H$ has the decomposition*

$$x = x_M + x_N$$

with

$$\langle x_M, x_N \rangle = 0$$

and the above decomposition is unique.

Proof. Let $\{e_i^M\}_{i \in I_M}$ be a basis of M. Using Zorn's Lemma we can extend this to a basis $\{e_i\}_{i \in I}$ of H. Let $x \in H$, then by the above result we have that

$$x = \sum \langle x, e_i \rangle e_i = \sum_{i \in I_M} \langle x, e_i \rangle e_i + \sum_{i \in II_M} \langle x, e_i \rangle e_i$$

and it is clear that

$$x = x_M + x_N$$

with

$$x_M = \sum_{i \in I_M} \langle x, e_i \rangle e_i, \; x = \sum_{i \in II_M} \langle x, e_i \rangle e_i$$

and it is easy to see that this decomposition is unique.

Remark 2.2.21. We write the above decomposition as

$$H = M \oplus M^\perp$$

and the subspace $M^\perp$ is called the orthogonal complement of the subspace M. Another interesting proof of the above decomposition can be given using the fact that every Hilbert space is uniformly convex. (For a proof along these lines see V. Istrăţescu (1978).)

Using the above decomposition we can obtain the following result concerning the dual space of a Hilbert space known as the theorem of Fréchet–Riesz.

THEOREM 2.2.22. (Fréchet–Riesz (1909)). *Let H be a Hilbert space and let $x^* \in H^*$ be a linear and continuous functional on H. Then there exists a unique element $a \in H$ such that:*

1. $x^*(x) = \langle x, a \rangle,$

2. $\|x^*\| = \|a\|.$

Proof. We can suppose, without loss of generality, that $x^* \neq 0$ since otherwise we can take $a = 0$.

Let $M = \{x, f(x) = 0\}$ and clearly this is a closed subspace of H and by the above theorem we find that $M^\perp$ possesses the described properties.

We show first that $M^\perp$ is one-dimensional. Suppose, on the contrary, that in $M^\perp$ there exist two linearly independent vectors, say z_1 and z_2. In this case the vector

$$z = f(z_2)z_1 - f(z_1)z_2$$

is obviously in $M^\perp$ and since

$$f(z) = f(z_1)f(z_2) - f(z_1)f(z_2) = 0$$

it is also in M. Thus $z = 0$ and this contradiction proves the assertion.

Let x be arbitrary in H. We can thus write

$$x = x_1 + x_2$$

with $x \in M$ and $x \in M^\perp$. Since $M^\perp$ is one-dimensional we can choose an element of norm 1, say x_0. In this case $x_2 = bx_0$ where $\|x_2\| = |b|$. Now we have,

$$f(x) = f(x_1) + f(x_2) = f(x_2) = bf(x_0).$$

From the properties of the spaces M and $M^\perp$ we obtain that

$$\langle x, x_0 \rangle = \langle x_1 + x_2, x_0 \rangle = \langle x_2, x_0 \rangle = b$$

and, since $f(x_0)$ is a fixed number, we can consider the element

$$a = f(x_0)x_0$$

and we have that

$$f(x) = \langle x, a \rangle$$

which proves the first assertion. The second follows from the first in a simple way.

COROLLARY 2.2.23. *Every Hilbert space is reflexive.*

Proof. Follows from the Fréchet–Riesz representation theorem that the canonical embedding is onto and thus it is the reflexivity of the space H.

2.3. CONVERGENCE IN X, X^* AND $L(X)$

Let X be a Banach space and (x_n) be a sequence of elements of X. Concerning the convergence we have many possibilities, all depending on the topology considered.

In what follows we consider only the most commonly used type of convergence for elements in X and for elements in X^*.

DEFINITION 2.3.1. The sequence (x_n) converges strongly to x if $\lim \|x_n - x\| = 0$.

DEFINITION 2.3.2. The sequence (x_n) converges weakly to x if

$$\lim x^*(x_n) = x^*(x)$$

for all $x^* \in X^*$.

It is obvious that every sequence which converges strongly to x converges weakly to x. The following example shows that the converse assertion is not true.

Example 2.3.3. Let H be a separable Hilbert space and $\{e_n\}$ be the basis for H. In this case, since for each e_j,

$$\lim \langle x, e_j \rangle = 0$$

it follows that $\{e_n\}$ is weakly convergent to zero. But from

$$\|e_i - e_j\|^2 = 2$$

we get that this sequence has no convergent subsequences.

On the space $L(X)$ of all bounded linear operators on X we can introduce several notions of convergence. Some of these are listed below.

DEFINITION 2.3.4. Let $\{T_n\}$ be a sequence of elements in $L(X)$. The sequence $\{T_n\}$ converges in norm to T if $\lim \|T_n - T\| = 0$. We say sometimes that we have uniform convergence.

DEFINITION 2.3.5. The sequence $\{T_n\}$ converges strongly to T if for any $x \in X$, $\lim \|T_n x - Tx\| = 0$.

In this case we say that we have strong convergence.

DEFINITION 2.3.6. The sequence $\{T_n\}$ converges weakly to T if for each $x \in X$ and each $x^* \in X^*$, $\lim x^*(T_n x) = x^*(Tx)$.

In this case we say that we have weak convergence.

It is not difficult to see the relations between the types of convergence defined above.

Another interesting convergence refers to the elements in X^*.

First we recall that the weak*-topology is the topology on X^* whose neighbourhood base at $x^* = 0$ are the sets of the form

$$V(0, \varepsilon, x_1, \ldots, x_n) = \{x^*, |x^*(x_i)| < \varepsilon, i = 1, \ldots, n\}$$

where $\varepsilon > 0$ and $x_1, \ldots, x_n$ is a finite set of elements of X.

Concerning the weak and weak* convergence of elements we have the following theorem which gives necessary and sufficient conditions. Since the proof is easy we omit it.

THEOREM 2.3.7.   *Let X be a Banach space and let $(x_n), (x_n^*)$ be the sequences of elements in X and X^* respectively. Then:*

1.   *(x_n) converges weakly to x if and only if*

     a.   *$\sup \|x_n\| < \infty$,*

     b.   *$\lim x^*(x_n) = x^*(x)$*

for $x^* \in D^*$ where D^* is a dense subset in X^*,

2.   *(x^*) converges weakly* to x if and only if*

     a.   *$\sup \|x_n^*\| < \infty$,*

     b.   *$\lim x_n^*(x) = x^*(x)$*

for $x \in D$ where D is a dense subset in X.

2.4. THE ADJOINT OF AN OPERATOR

Let X and Y be two Banach spaces over K. Let $g \in Y^*$ and for each $x \in X$ we define the functional on X by the relation

$$f(x) = g(Tx)$$

where $T \in L(X, Y)$, the set of all bounded linear operators on X with values in Y with the norm

$$T \rightarrow \|T\| = \sup_{\|x\| \leq 1} \|Tx\|.$$

In this case it is clear that

$$|f(x)| \leq \|g\| \, \|T\| \, \|x\|$$

and thus $f \in X^*$. Thus for each functional $g \in Y^*$ we can define a linear functional $f \in X^*$ using the operator T. It is easy to see that we have in this way a linear operator:

$$T^* : Y^* \rightarrow X^*$$

and we write

$$f = T^*g.$$

Since from the above estimate we have

$$\|T^*\| \leq \|T\|$$

it follows that this is a bounded linear operator.

We show now that we have, in fact, an equality. Let x be an arbitrary point in X and consider a linear functional $g \in Y^*$ with the following properties:

1. $\qquad \|g\| = 1,$

2. $\qquad g(Tx) = \|Tx\|$

which exists by the extension theorems.

In this case we have,

$$\|Tx\| = g(Tx) = (T^*g)(x) \leq (\|T^*g\| \, \|x\| \leq \|T^*\| \, \|x\|$$

and thus

$$\|T^*\| \geq \|T\|$$

from which it follows that $\|T\| = \|T^*\|$.

2.5. CLASSES OF BANACH SPACES

In applications some particular features of the corresponding Banach space play an important role. In what follows we consider several classes of Banach space which have some importance in fixed point theory. In this respect it is worth mentioning that many fixed point theorems have a more general form only on such spaces.

The first class which we consider was introduced by J. A. Clarkson in 1936 under the name 'uniformly convex spaces'. The next class we consider was also defined by Clarkson – and independently by M. G. Krein – under the name 'strictly convex spaces'.

In many papers this class is called also 'rotund spaces'.

We also give a brief account of these classes as well as some generalizations. We should also mention that the class of Banach spaces with 'normal structure' was introduced by M. S. Brodski and D. P. Milman (1948).

DEFINITION 2.5.1.   A Banach space X is said to be uniformly convex if and only if, for each ε in $(0, 2]$, there exists $\delta(\varepsilon) > 0$ such that for all x, y in X, $\|x\| \leq 1$, $\|y\| \leq 1$ and $\|x - y\| \geq \varepsilon$ then $\|x + y\| \leq 2(1 - \delta(\varepsilon))$.

The function on $(0, 2]$

$$\varepsilon \to \delta(\varepsilon)$$

is said to be the modulus of convexity of X. More generally, following Lindenstrauss, we define the modulus of convexity of a Banach space as the function, on $[0, 2]$ defined by the relation

$$\delta_X(\varepsilon) = \tfrac{1}{2}\inf\{2 - \|x + y\| : \|x\| \leq 1, \|y\| \leq 1, \|x - y\| \geq \varepsilon\}.$$

and it is clear that a Banach space X is uniformly convex if $\delta_X(\varepsilon) > 0$ for every $\varepsilon > 0$.

Also we define the modulus of smoothness of a Banach space as follows:

DEFINITION 2.5.2.   The modulus of smoothness of a Banach space X is the function

$$\rho_X(\tau) = \sup_{\|x\| \leq 1, \|y\| = \tau} \{\|x + y\| + \|x - y\| - 2\}$$

and we say that X is uniformly smooth if $\rho_X(\tau) = 0(\tau)$ for $\tau \to 0$. As suggested by the above definitions, we define the local modulus of convexity of a Banach space and also the local modulus of smoothness of a Banach space thus:

DEFINITION 2.5.3.   Let X be a Banach space and $x \in X$, $\|x\| = 1$. The local modulus of convexity of X at x is the function

$$\delta_{X,x}(\varepsilon) = \inf_{\|y\| \leq 1, \|x - y\| \geq \varepsilon} \{2 - \|x + y\|\}$$

and we say that X is locally uniformly convex at x if $\delta_{X,x}(\varepsilon)$ is > 0 for all $\varepsilon > 0$.

DEFINITION 2.5.4.   Let X be a Banach space and $x \in X$, $\|x\| = 1$. Then the local modulus of smoothness of X at x is the function

$$\rho_{X,x}(\tau) = \tfrac{1}{2} \sup_{\|y\| = \tau} \{\|x + y\| + \|x - y\| - 2\}$$

and we call X locally uniformly smooth at x if $\rho_{X,x}(\tau) = 0(\tau)$ for $\tau \to 0$.

The following theorem gives a characterization of uniformly convex spaces:

THEOREM 2.5.5.   *Let X be a Banach space. Then X is uniformly convex if and only if*

1. $\|x_n\| \leq 1$, $\|y_n\| \leq 1$,

2. $\lim \|x_n + y_n\| = 1$

then $\lim \|x_n - y_n\| = 0$.

Proof. First we remark that the condition is necessary, for if not, for some $\varepsilon > 0$, we find a sequence of integers $\{m_k\}$ such that

$$\|x_{m_k} - y_{m_k}\| \geq \varepsilon$$

and since

$$\lim \|x_{m_k} + y_{m_k}\| = 2$$

we have a contradiction for, since X is uniformly convex,

$$\|x_{m_k} + y_{m_k}\| \leq 2(1 - \delta(\varepsilon)).$$

This contradiction proves the assertion. Conversely, if the condition is satisfied, and for $\varepsilon > 0$ there is no $\delta(\varepsilon)$ such that if $\|x - y\| \geq \varepsilon$ then $\|x + y\| \leq 2(1 - \delta(\varepsilon))$, we find the sequences $\{x_n\}$ and $\{y_n\}$ with the following properties:

1. $\|x\| \leq 1, \qquad \|y\| \leq 1,$
2. $\lim \|x_n + y_n\| = 2.$

Now since $\|x_n - y_n\| \geq \varepsilon$ we have a contradiction.

THEOREM 2.5.6.   *Let X be a uniformly convex space. Then each closed and quotient subspace is uniformly convex.*

Proof. The first assertion is obvious. We prove now the second assertion. Let $\varepsilon > 0$ and $x^\sim$, $y^\sim$ be two elements of the quotient space Q of X. In this case if

1. $\|x^\sim\| \leq 1, \qquad \|y^\sim\| \leq 1$
2. $\|x^\sim - y^\sim\| > \varepsilon$

we find $x \in x^\sim$ and $y \in y^\sim$ such that

$$\|x\| \leq 1 + s, \qquad \|y\| \leq 1 + s$$

where s is arbitrarily small. Since $\|x - y\| \geq \|x^\sim - y^\sim\| \geq \varepsilon$ we have that

$$\|x^\sim + y^\sim\| \leq \|x + y\| \leq 2(1 - \delta(\varepsilon/(1 + s))(1 + s)$$

and if we define

$$\delta^\sim(\varepsilon) = \lim_{s \to 0^+} \delta(\varepsilon/(1 + s)) \geq \delta(\varepsilon/2)$$

and this is the required modulus of convexity.

The following result has many applications.

THEOREM 2.5.7.   *Let X be a uniformly convex space and $\{x_n\}$ be a sequence with the following properties:*

1. $\lim \|x_n\| = 1$,

2. $\lim \|x_n + x_m\| = 2$
             (i.e. for any $\varepsilon > 0$ there exists N_ε such that for all $n, m > N_\varepsilon, |\|x_n + x_m\| - 2| \leq \varepsilon$).

then the sequence $\{x_n\}$ is convergent.

Proof. Obviously it is sufficient to prove that $\{x_n\}$ is a Cauchy sequence. We have two cases:

1. $\|x_n\| \leq 1$ *for all n.* Let $\varepsilon > 0$ and $\delta(\varepsilon)$ from the definition of uniformly convex spaces. From condition 2 we found N_ε such that for all $n, m \geq N_\varepsilon$,

$$\|x_n - x_m\| \leq \varepsilon$$

since X is uniformly convex.

Thus in this case $\{x_n\}$ is a Cauchy sequence.

2. *If $\|x_n\| \leq 1$ is not true for all n* we can suppose, without loss of generality, that $\|x_n\| \neq 0$ for all n. We can define the sequence $\{y_n\}$ by $y_n = x_n/\|x_n\|$. This sequence satisfies case 1 and thus it is a Cauchy sequence. This may easily be seen to imply that $\{y_n\}$ is a Cauchy sequence.

The following property can be used to give a short proof of orthogonal decomposition of Hilbert spaces.

THEOREM 2.5.8.   *Let X be a uniformly convex space and let C be any closed and convex subset of X. Then there exists a unique element of C with minimal norm.*

Proof. Let $d = \inf\{\|x\|, x \in C\}$. If $d = 0$ the assertion is obvious. Suppose that $d > 0$. We find a sequence $\{x_n\} \subset C$ such that

$$\lim \|x_n\| = d.$$

Since C is convex, for all n, m we have that

$$d \leq \|\tfrac{1}{2}(x_n + x_m)\|.$$

and we define the sequence $\{y_n\}$ where $y_n = x_n/d$. It is easy to see that this sequence satisfies the conditions of Theorem 2.5.7. Then $\lim x_n = x^\sim$ has the property that $\|x^\sim\| = d$. Thus the existence is proved. Now we prove the uniqueness. Suppose that there exists another element, say $y^\sim$ in C such that $\|y^\sim\| = d$. In this case we have, if $\|x^\sim - y^\sim\| > \varepsilon$, that

$$1 \leq \tfrac{1}{2}\|x^\sim/d + y^\sim/d\| \leq 1 - \delta(\varepsilon/d)$$

and this contradiction proves the uniqueness.

Example 2.5.9.   (1) Every Hilbert space is uniformly convex.
Proof. Since for any x, y in the Hilbert space H,

$$\|x + y\|^2 + \|x - y\|^2 = 2\{\|x\|^2 + \|y\|^2\}$$

and thus, if $\|x\| \leq 1$, $\|y\| \leq 1$ and $\|x - y\| \geq \varepsilon$,

$$\|x + y\|^2 \leq 2 - \varepsilon^2.$$

Let

$$\varepsilon \to \delta(\varepsilon) = 1 - (1 - (\varepsilon/2)^2)^{1/2}$$

and thus

$$\|x + y\| \leq 2(1 - \delta(\varepsilon))$$

and the assertion is proved.

(2) The L^p and l^p spaces are uniformly convex for $p \in (1, \infty)$.

This result was proved by Clarkson, and in proving it we follow the presentation of Hewitt and Ross (1970).

First we prove the following basic inequalities of Clarkson:

(α) For any f, $g \in L^p$, $p \in [2, \infty)$

$$\|(f + g)/2\|_p^p + \|(f - g)/2\|_p^p \leq \tfrac{1}{2}(\|f\|_p^p + \|g\|_p^p)$$

(β) For any f, $g \in L^p$, $p \in (1, 2]$, $1/p + 1/q = 1$,

$$\|(f + g)/2\|_p^q + \|(f - g)/2\|_p^q \leq \tfrac{1}{2}(\|f\|_p^p + \|g\|_p^p)^{1/(p-1)}.$$

For the proof we consider the function

$$f : [0, 1] \to \mathbf{R}, \qquad f(x) = ((1 + x)/2)^p + ((1 - x)/2)^p$$

and we remark that

$$f(x) \leq \tfrac{1}{2}(1 + x^p).$$

Indeed, if

$$g(x) = f(x) - \tfrac{1}{2}(1 + x^p) \cdot 2^p/x^p$$

on $(0, 1]$ it is clear that this has the property that $g'(x) > 0$ for all $x \in (0, 1)$. Since $g(1) = 0$ we have that $g(x) < 0$ which is equivalent to the stated property of f.

We show now that the above inequality holds also for the case of complex numbers, i.e., for any complex numbers z_1 and z_2,

$$|(z_1 + z_2)/2|^p + |(z_1 - z_2)/2|^p \leq \tfrac{1}{2}(|z_1|^p + |z_2|^p)$$

for $p \in [2, \infty)$.

We remark that if $z_2 = 0$ then we have to prove

$$|z|^p/2^{(p-1)} \leq |z|^p/2$$

which is true since $p \in [2, \infty)$.

Thus we can suppose, without loss of generality, that $z_2 \neq 0$ and, moreover, that $|z_1| \geq |z_2|$.

The required inequality is equivalent to the following inequality,

$$|\tfrac{1}{2}(1 + z_2/z_1)^p + |\tfrac{1}{2}(1 - z_2/z_1)|^p < \tfrac{1}{2}(1 + |z_2/z_1|^p).$$

Let $z_2/z_1 = re^{i\vartheta}$ and thus the above inequality becomes

$$|(1 + re^{i\vartheta})/2|^p + |(1 - re^{i\vartheta})/2|^p \leq \tfrac{1}{2}(1 + r^p).$$

It is clear that we can consider the case when $\vartheta \in [0, \pi/2]$ and we then consider the function

$$f(\vartheta) = |1 + re^{i\vartheta}|^p + |1 - re^{i\vartheta}|^p.$$

We prove that this function has a maximum at the point $\vartheta = 0$.
Since we can write,

$$f(\vartheta) = |1 + r^2 + 2r \cos \vartheta|^{p/2} + |1 + r^2 - 2r \cos \vartheta|^{p/2}$$

and thus

$$\begin{aligned} f'(\vartheta) &= (p/2)(1 + r^2 + 2r \cos \vartheta)^{p/2}(-2r \sin \vartheta) + \\ &\quad + (p/2)(1 + r^2 - 2r \cos \vartheta)(2r \sin \vartheta) \\ &= -pr \sin \vartheta[1 + r^2 + 2r \cos \vartheta]^{(p/2-1)} - \\ &\quad - [1 + r^2 - 2r \cos \vartheta]^{(p/2-1)}. \end{aligned}$$

Since $p \geq 2$ we have that $f'(\vartheta) < 0$ for all $\vartheta \in [0, \pi/2]$ and this gives us

$$1/2^p f(\vartheta) \leq 1/2^p f(0) = \tfrac{1}{2}(1 + r^p)$$

which is equivalent to the inequality we required.

Now we shall discuss the case when $p \in (1, 2]$. First we prove that for all $x \in [0, 1]$ and $p \in (1, 2]$,

$$(1 + x)^q + (1 - x)^q \leq 2(1 + x^q)^{1/(p-2)}$$

where $1/p + 1/q = 1$.

We remark initially that this inequality is obvious in the case $x = 0$. Suppose now that we have the following inequality,

$$(1 + y^q)^{p-1} \leq \tfrac{1}{2}\{(1 + y)^p + (1 - y)^p\} \qquad 0 \leq y \leq 1$$

and this is equivalent to the following inequality

$$2^p(1 + y^q) \leqq 2[(1 + y)^p + (1 - y)^p]^{1/(p-1)}.$$

From this we obtain, since $(1 + y)^q \neq 0$,

$$(1 + (1 - y)/(1 + y))^q + (1 - (1 - y)/(1 + y))^q$$
$$< 2\{1 + (1 - (1 - y)/(1 + y))^p\}^1$$

and for $x = (1 - y)/(1 + y)$ we obtain the required inequality.
Thus it remains to prove the inequality

$$(1 + y^q)^{(p-1)} \leqq \tfrac{1}{2}\{(1 + y)^p + (1 - y)^p\}$$

Since for $y \in (0, 1)$ we have,

$$\frac{1}{2}\left[(1 + y)^p + (1 - y)^p - \left(\frac{m}{1 + y^q}\right)^{(p-1)}\right]$$

$$= \frac{1}{2}\left[\sum_0^\infty \binom{p}{i}y^i + \sum_0^\infty \binom{p}{i}(-1)^i y^i\right] -$$

$$- \sum_0^\infty \binom{p-1}{i}y^{qi}$$

$$= \sum_0^\infty \left[\binom{p}{2i}y^{2i} - \binom{p-1}{i}y^{qi}\right]$$

$$= \sum_0^\infty \left[\binom{p}{2i}y^{2i} - \binom{p-1}{2i-1}y^{q(2i-1)} - \binom{p-1}{2i}y^{2qi}\right].$$

Since $p > 1$ these series are absolutely convergent and they converge
uniformly for $x \in [0, 1]$. It remains to show that the coefficients in these
series are positive. Let us consider the jth term of this series,

$$p(p - 1) \cdots (p - (2j - 1))/(2j!)y^{2j} -$$
$$- (p - 1) \cdots (p - (2j - 1))/(2p - 1)! \times y^{q(2j-1)} -$$
$$- (p - 1) \cdots (p - (p - 2j))/(2j!)y^{2qj}$$
$$= p(p - 1) \cdots (2j - 1 - p)/(2j!)y^{2j} -$$
$$- (p - 1) \cdots (2j - 1 - p)/(2j - 1) \times$$
$$\times y^{q(2j-1)} + (p - 1) \cdots (2j - p)/(2j)!y^{2jq}$$
$$= y^{2j}(2 - p) \cdots (2j - p)/(2j - 1)! \times$$

$$\times [p(p-1)/2j(2j-p)-(p-1)/(2j-p)y^{q(2j-1)-2j}+$$
$$+(p-1)/2jy^{2jq-2j}]$$
$$= y^{2j}(2-p)\cdots(2j-p)/(2j-1)! \times$$
$$\times [(1-y^{(2j-p)}/(p-1))/[(2j-p)/(p-1)]-$$
$$-[[1-y^{(2j)/(p-1)}]/[2j/(p-1)]]]$$

which is clearly positive for $1 < p \leqq 2$.

Exactly as above, this inequality can be extended to complex numbers, and thus we obtain the following inequality: for any complex numbers z_1 and z_2 and any fixed $p \in (1, 2]$

$$|z_1+z_2|^q+|z_1-z_2|^q \leqq 2\{|z_1|^p+|z_2|^p\}^{1/(p-1)}.$$

Now the proof of the basic inequalities of Clarkson are as follows:

Let f and g be two functions defined a.e. (i.e. almost everywhere) with respect to the measure μ and take x for which these are defined.

In this case we have,

$$|(f(x)+g(x))/2|^p+|(f(x)-g(x))/2|^p$$
$$\leqq \tfrac{1}{2}\{|f(x)|^p+|g(x)|^p\}$$

and this gives the inequality (α). For the second inequality we first apply Minkowski's inequality and we obtain

$$\||(f+g)/2|^q\|_{p-1}+\||(f-g)/2|^q\|_{p-1}$$
$$\leqq \||(f+g)/2|^q+|(f-g)/2|^q\|_{p-1}.$$

Since the right-hand member of this inequality is in fact

$$\{\int(|(f+g)/2|^q+|(f-g)/2|^q)^{p-1}d\mu\}^{1/(p-1)},$$

by the application of the inequality obtained for z_1 and z_2 and fixed $p \in (1, 2]$ we see that it is majorized by

$$\{2^{p-1}\int(|f|^p+|g|^p)d\mu\}^{1/(p-1)}$$

which is

$$\tfrac{1}{2}\{\|f\|_p^p+\|g\|_p^p\}$$

and thus Clarkson's inequalities are proved.

From the above inequalities it follows that L^p, for $p \in (1, \infty)$ are uniformly convex spaces; the case of l^p spaces is a particular example of this for the case in which the measure is defined on the integers and for each i, $\mu\{i\} = 1$.

We now prove a very interesting property of uniformly convex spaces:

THEOREM 2.5.10   *Every uniformly convex space is reflexive.*

Proof. Let X be a uniformly convex space and X^*, X^{**} be the first and the second dual of X. Consider now an element F of $X^{**}, \| F \| = 1$. From the definition of the norm, there follows the existence, for each n, of an element $f_n \varepsilon X^*$ such that $\| f_n \| = 1$ and $|F(f_n)| \geq 1 - 1/n$. In this case, by Helly's theorem, for each n we find an element $x_n \in X$ such that

1. $\qquad \| x_n \| \leq 1 + 1/n,$

2. $\qquad F(f_k) = f_k(x_n), \qquad k = 1, 2, \ldots, n.$

We remark that $\lim \| x_n \| = 1$ since

$$\| x_n \| \geq |f_k(x_n)| = |F(f_k)| \geq 1 - 1/n.$$

Also, we have that

$$\lim \| x_n + x_m \| = 2.$$

Indeed, for $n > m$ we have that

$$\| x_n + x_m \| \geq |f_m(x_n) + f_m(x_m)| = 2|F(f_m)| \geq 2(1 - 1/m)$$

and the assertion follows.

Since the sequence $\{x_n\}$ satisfies the conditions of Theorem 2.5.7., it is a convergent sequence. Let $x = \lim x_m$. Since we have

$$F(f_k) = f_k(x_n)$$

it follows that

$$F(f_k) = f_k(x)$$

for $k = 1, 2, 3, \ldots .$

Now the set

$$\{x, F(f_n) = f_n(x)\}$$

is obviously closed and convex. Then by the theorem, which will be proved for more general spaces (Theorem 2.5.25), it follows that there exists at most one point in this set with the property $\| x \| = 1 = \| F \|$.

Let now $f_0 \in X^*$ and by Helly's theorem for F and $f_0, f_1, \ldots, f_n$ we find that x satisfies the relations $F(f_k) = f_k(x)$. For $k = 0$ we have that $F(f_0) = f_0(x)$ and the reflexivity of X is proved.

The following theorem is easy to prove.

THEOREM 2.5.11. *Let X and Y be two uniformly convex spaces. In this case the Banach space X × Y with the norm*

$$(x, y) \to \|(x, y)\| = (\|x\|^p + \|y\|^p)^{1/p}$$

is uniformly convex. (Here p is in $(1, \infty)$).

We give now some classes of spaces which contain the class of uniformly convex spaces.

DEFINITION 2.5.12   (A. R. Lovaglia (1955)). Let X be a Banach space. We say that X is locally uniformly convex if, for any $x \in X$, $\|x\| = 1$ and $\varepsilon > 0$, there exists $\delta_x(\varepsilon) > 0$ such that, for all y, $\|y\| \leq 1$ and $\|x - y\| \geq \varepsilon$, $\|x + y\| \leq 1(1 - \delta_x(\varepsilon))$.

Obviously every uniformly convex space is locally uniformly convex. There exists an example of a Banach space which is locally uniformly convex and is not isomorphic with a uniformly convex space.

Another interesting extension of the notion of uniformly convex spaces was given by Ky Fan and I. Glicksberg (1958) and is as follows:

DEFINITION 2.5.13.   A Banach space X is said to be k-rotund if

$$\lim_{n_1 \cdots n_k \to \infty} 1/k(\|x_{n_1} + x_{n_2} + \cdots + x_{n_k}\|) = 1$$

implies that

$$\lim_{n, m \to \infty} \|x_n - x_m\| = 0.$$

Obviously, for $k = 2$, we have the notion of uniformly convex space given by Clarkson. Another extension is as follows:

DEFINITION 2.5.14.   (V. Istrățescu). A Banach space is said to be k-uniformly convex if, for any $x_1, x_2, \ldots, x_k$ with the properties

1. $\|x_i\| \leq 1$,

2. $\min_{i \neq j} \|x_i - x_j\| \geq \varepsilon$,

there exists a $\delta(\varepsilon)$ such that

$$\|x_1 + x_2 + \cdots + x_k\| \leq k(1 - \delta(\varepsilon)).$$

We can introduce the following extension of the notion of locally uniformly convex spaces as follows:

DEFINITION 2.5.15 (V. Istrăţescu).   A Banach space X is said to be k-locally uniformly convex if, for any $x \in X$, $\|x\| = 1$ and $\varepsilon > 0$, there exists $\delta_x(\varepsilon) > 0$ such that, whenever $x_1, x_2, \ldots, x_{k-1}$ has the following properties:

1. $\|x_i\| \leq 1$,

2. $\|x - x_i\| \geq \varepsilon$

then

$$\|x + x_1 + \cdots + x_{k-1}\| \leq k(1 - \delta_x(\varepsilon)).$$

Remark 2.5.16.   Other extensions of the notion of uniformly convex spaces can be given using the weak convergence. For further details and information, see V. Istrăţescu (1980), Lovaglia (1955), Šmulian (1938).

As is well known and easy to prove, in every Hilbert space H the following property holds: let $\{x_n\}$ be a sequence such that

1. $x_n \rightharpoonup x$,   ($\rightharpoonup$ weak convergence).

2. $\|x_n\| \to \|x\|$

then $\lim \|x_n - x\| = 0$.

This result has been extended to the case of L^p spaces by F. Riesz and J. Radon. We show now that there exists a more general result which is valid for the k-locally uniformly convex spaces and, clearly, this contains the Riesz–Radon result.

THEOREM 2.5.17.   *Let X be a k-locally uniformly convex space and $\{x_n\}$ be a sequence of elements satisfying the following properties:*

1. $x_n \rightharpoonup x$,

2. $\|x_n\| \to \|x\|$

then $\lim \|x_n - x\| = 0$.

Proof. Suppose that the assertion of the theorem is not true. In this case there exists a sequence of integers $\{n_i\}$ such that $\|x_{n_i} - x\| > \varepsilon$. Let $f \in X^*$ such that

1. $\|f\| = 1$,

2. $f(x) = 1$

which exists as a consequence of the Extension Theorems. Now, since X is

k-locally uniformly convex we have

$$1 = \lim f((x + x_{n_1} + \cdots + x_{n_{k-1}})/k) \leqq 1 - \delta_x(\varepsilon),$$

which is a contradiction, and the assertion is proved.

Remark 2.5.18.   The above result shows that the class of k-locally uniformly convex spaces is contained in the class of spaces with the property (H) considered by Ky Fan and I. Glicksberg: a Banach space is said to have the property (H) if, whenever $\{x_n\}$ is a sequence such that

1. $x_n \rightharpoonup x$,

2. $\lim \|x_n\| = \|x\|$,

then $\lim \|x_n - x\| = 0$.

Now we consider the class of 'strictly convex spaces' introduced by Clarkson and by Krein.

DEFINITION 2.5.19.   A Banach space X is said to be strictly convex if and only if every point x, $\|x\| = 1$ is an extreme point of the set

$$\{x, \|x\| \leqq 1\}.$$

The following theorem gives some characterizations of this class of spaces.

THEOREM 2.5.20.   *For a Banach space X the following assertions are equivalent*:

1.   *X is strictly convex,*

2.   *for $x, y \in X$,*

         $\|x\| = \|y\| \leqq 1, \quad x \neq y \qquad$ then $\|x + y\| < 2$,

3.   *if $x, y \in X$ and $\|x + y\| = \|x\| + \|y\|$*

         then $x = ay \quad$ for some a.

Proof. We prove the sequence of implications,

$$1 \to 2 \to 3 \to 1.$$

Let us suppose 1 and that x, $y \in X$ with the property $\|x\| = \|y\| \leqq 1$. In this case, if $\|x + y\| = 2$, then $z = \frac{1}{2}(x + y)$ has the property $\|z\| = 1$ and clearly is not an extreme point. This contradiction proves that $1 \to 2$. Let us suppose that 2 is true and let x, $y \in X$ with $\|x + y\| = \|x\| + \|y\|$. We remark that we can suppose, without loss of generality, that $\|x\| \le \|y\|$. In this case,

$$2 \ge \| \, x/\|x\| + y/\|y\| \, \|$$
$$\ge \| \, x/\|x\| + y\|x\| \, \| - \| \, y/\|x\| - y/\|y\| \, \|$$
$$= \|x + y\|/\|x\| - \|(\|y\| - \|x\|)\|y\|/(\|x\| \, \|y\|) = 2,$$

and thus

$$\| \, x/\|x\| + y/\|y\| \, \| = 2.$$

Since 2 is true, this gives us that,

$$x/\|x\| = y/\|y\|$$

and thus $2 \to 3$.

Suppose now that 3 is true. Suppose further that 1 is not true and thus that there exists an element z, $\|z\| = 1$ such that $z = \frac{1}{2}(x + y)$. Clearly $\|x\| = \|y\| = 1$ and in this case

$$\|x + y\| = 2 = \|x\| + \|y\|$$

which gives us that $x = y$. This contradiction proves that $3 \to 1$.

The following theorem gives examples of strictly convex spaces.

THEOREM 2.5.21.   *Every locally uniformly convex space is strictly convex.*
    Proof. This follows directly from the definition of locally uniformly convex spaces and from the above theorem.

COROLLARY 2.5.22.   *Every uniformly convex space is strictly convex.*

The following example of Lovaglia's shows that there exist Banach spaces which are strictly convex and not locally uniformly convex.
    Indeed, define $C_{[0,1]}$ with the norm

$$f \to \|f\| = \sup_t |f(t)|$$

and choose a sequence $\{t_n\}$ in $[0, 1]$ and $\{t_n\}$ dense in $[0, 1]$.
    We define a new norm on $C_{[0,1]}$ by

$$\|f\|_1 = \left\{ \|f\|^2 + \sum_1^\infty |f(t_n)|^2/2^{2n} \right\}^{1/2}$$

and it is easy to see that this norm is equivalent to $\|,\|$.
    From Minkowski's inequality it follows that $\{C_{[0,1]}, \|,\|_1\}$ is a strictly

convex space. We prove that $\{C_{[0,1]}, \|\,,\,\|_1\}$ is not locally uniformly convex. Consider for this $g(t) = 3^{1/2}/2$ for $t \in [0, 1]$ and the sequence

$$g_n(t) = \begin{cases} 3^{1/2}/2 & 1/n \le t \le 1 \\ 3^{1/2}/2 \cdot nt & 0 \le t \le 1/n \end{cases}$$

and we remark that $\|g_n\|_1 \to 1$, $\|g\|_1 = 1$, $\lim \|g_n - g\|_1 = 1$.
But

$$\|g_n - g\| = \max |g_n(t) - g(t)| = 3^{1/2}/2$$

for all n. This is in contradiction with the result of Theorem 2.5.19.

THEOREM 2.5.23 *In every strictly convex space, any closed and convex set has at most one point with the minimum norm.*
Proof. This follows directly from the definition of strictly convex spaces.

The following theorem gives a method for the construction of strictly convex spaces.

THEOREM 2.5.24 (Klee). *Let X be a Banach space. Then there exists an equivalent norm on X, say $\|\,,\,\|_1$ such that $(X, \|\,,\,\|_1)$ is a strictly convex space if and only if there exists another Banach space Y which is strictly convex and also there exists a one-to-one continuous linear operator $T : X \to Y$.*
Proof. Let X be such that there exists an equivalent norm $\|\,,\,\|_1$ such that $(X, \|\,,\,\|_1)$ is a strictly convex space. In this case we can take $T = I$ (the identity operator $I(x) = x$).
Conversely, if there exists a strictly convex space Y and a continuous linear one-to-one $T: X \to Y$ then we define

$$\|x\|_1 = \|x\| + \|Tx\|$$

which, as is easy to see, satisfies our theorem.

COROLLARY 2.5.25 (Clarkson). *On every separable Banach space there exists an equivalent norm $\|\,,\,\|_1$ such that the space possessing this norm is strictly convex.*
Proof. Let $\{x_n^*\}$ be a sequence of linear functionals in X^*, then this sequence has the property that it is dense in the set

$$\{x^*, x^* \in X^*, \|x^*\| \le 1\}.$$

We define the operator $Tx \to l^2$ by the relation

$$Tx = \{x_n^*(x)/2^{2n}\}$$

and clearly this is continuous and one-to-one. The assertion follows now from the above theorem.

Remark 2.5.26.   Interesting methods for the construction of equivalent norms on Banach spaces with properties connected with strict convexity, locally uniform convexity, etc., were proposed by Clarkson, Day, Asplund, Kadets, and others.

We consider now an important class of Banach spaces which has many applications in fixed point theory.

First we give the following:

DEFINITION 2.5.27.   Let X be a Banach space and C be a closed, bounded and convex set in X. A point $x \in C$ is called diametral if

$$\text{diam } C = \sup \{ \|x - y\|, \, y \in C \}.$$

DEFINITION 2.5.28.   A closed convex and bounded set C in a Banach space X is said to have 'normal structure' whenever, given any closed and convex subset C_1 of C containing more than one point, there exists a non-diametral point $x \in C_1$.

The following theorem gives examples of sets with normal structure.

THEOREM 2.5.29.   *Every compact set C in a Banach space has normal structure.*

Proof. Let C be a compact set which is also convex. Suppose that C does not have normal structure (of course we can suppose, without loss of generality, that diam $C > 0$). Then for any $x_1 \in C$ there exists $x_2 \in C$ such that diam $C = \|x_1 - x_2\|$. Since C is a convex set we have that $\frac{1}{2}(x_1 + x_2) \in C$. Further we find x_3 in C such that diam $C = \|x_3 - (x_1 + x_2)/2\|$ and we can continue in this way to obtain a sequence $\{x_n\}$ with the following property:

$$\text{diam } C = \|x_{n+1} - (x_1 + x_2 + \cdots + x_n)/n\| \qquad n \geq 2.$$

But in this case we have,

$$\text{diam } C = \|x_{n+1} - (x_1 + x_2 + \cdots + x_n)/n\|$$

$$\leq (\sum_{1}^{n} \|x_{n+1} - x_i\|)$$

$$\leq \text{diam } C$$

and we get

$$\text{diam } C = \|x_{n+1} - x_i\|, \qquad 1 \leq i \leq n.$$

In this case the sequence $\{x_n\}$ has no convergent subsequences and this contradiction proves the theorem.

THEOREM 2.5.30 *In every uniformly convex space, every closed convex and bounded set has normal structure.*

Proof. We can suppose, without loss of generality, that C is contained in $\{x, \|x\| \leq 1\}$. Let $\varepsilon > 0$ and x_1 be an arbitrary point C_1. We find $x_2 \in C_1$ (C_1 is a closed and convex subset of C) such that $\|x_2 - x_1\| \geq \operatorname{diam} C_1/2$. In this case for any $z \in C_1$ we have

$$\|z - (x_1 + x_2)/2\| = \|(z - x_1)/2 + (z - x_2)/2\|$$
$$= \operatorname{diam} C_1 \|(z - x)/\operatorname{diam} C_1 +$$
$$+ (z - x)/\operatorname{diam} C_1\|$$
$$< \operatorname{diam} C_1(1 - \delta(\varepsilon)),$$

and this proves the assertion.

DEFINITION 2.5.31.   A Banach space X is said to have 'normal structure' if each closed, convex and bounded set in X has normal structure.

From the above result we have

COROLLARY 2.5.32.   *Every uniformly convex space has normal structure.*

The following result was obtained by Belluce and Kirk (1967):

THEOREM 2.5.33.   *Let X, Y be two Banach spaces with normal structure and consider $X \times Y$ with the norm*

$$\|(x, y)\| = \sup \{\|y\|, \|y\|\}.$$

In this case $X \times Y$ has normal structure.

Proof. Let K be a closed, bounded and convex set in $X \times Y$. In this case if p_X and p_Y denote the natural projections of $X \times Y$ on X (respectively Y), we define the sets

$$K_X = p_X(K), \qquad K_Y = p_Y(K).$$

In this case K_X and K_Y are closed, bounded and convex sets, and they thus have normal structure. Let x and y be a non-diametral points of K_X (respectively of K_Y). We construct a non-diametral point of K as follows:
Let U_x and U_y be points in $p_X^{-1}(x) \cap K$ and in $p_Y^{-1}(y) \cap K$, respectively.

We set

$$U_1 = (x, v)$$
$$U_2 = (w, y)$$

where $v \in p_X^{-1}(U_x)$ and $w \in p_Y^{-1}(U_y)$. We prove that

$$m = \tfrac{1}{2}(U_x + U_y) = (m_1, m_2)$$

where

$$m_1 = \tfrac{1}{2}(x + w), \qquad m_2 = \tfrac{1}{2}(v + y)$$

is the required non-diametral point of K. Indeed, for z arbitrary in K, $z = (z_1, z_2)$ we have,

$$\|m_1 - z_1\| = \|\tfrac{1}{2}(x + w) - \tfrac{1}{2}(z_1 + z_1)\|$$
$$\leq \|(x - z_1)/2\| + \tfrac{1}{2}\|(w - z_1)\|$$

and since x is non-diametral,

$$\|m_1 - z_1\| \leq \operatorname{diam} K_X - \varepsilon_1$$

for some $\varepsilon_1 > 0$. Similarly we get,

$$\|m_2 - z_2\| \leq \operatorname{diam} K_Y - \varepsilon_2$$

for some $\varepsilon_2 > 0$. Thus we have,

$$\|m - z\| \leq \sup \{\operatorname{diam} K_X - \varepsilon, \operatorname{diam} K_Y - \varepsilon_2\}$$

and for $\varepsilon = \min \{\varepsilon_1, \varepsilon_2\}$ we get

$$\|m - z\| \leq \sup \{\operatorname{diam} K_X - \varepsilon, \operatorname{diam} K_Y - \varepsilon\} = \operatorname{diam} K - \varepsilon$$

and the assertion is proved.

L. P. Belluce and W. A. Kirk (1967) have introduced a more general concept than normal structure; namely, the notion of complete normal structure.

For this we need some notations.

For K and H, H bounded, subsets in a Banach space X, let

$$r_x(H) = \sup \{\|x - y\|, y \in H\},$$
$$r(H, K) = \inf \{r_x, x \in K\},$$
$$C(H, K) = \{x, x \in K, r_x(H) = r(H, K)\}.$$

The set $C(H, K)$ is sometimes called the Chebyshev centre of H in X.

DEFINITION 2.5.34. Let K be a bounded, closed and convex set in a Banach space X. Then K is said to have a complete normal structure (abbreviated as c.n.s.) if every closed convex set W of K which contains more than one point satisfies the following condition:
'For every net $\{W_\alpha : \alpha \in D\}$ of subsets of W which have the property that $r(W_\alpha, W) = r(W, W)$ for $\alpha \in D$, then the closure of $\bigcup_{\alpha \in D} C(W_\alpha, W)$ is a nonempty subset of W'.

Belluce and Kirk have also proved that in every uniformly convex space every bounded and closed convex set has a c.n.s. and also that every compact set in arbitrary Banach spaces has a c.n.s. when it is convex.

Remark 2.5.35. There exists an example, given by R. C. James, of a weakly compact convex set which does not possess normal structure.

2.6. MEASURES OF NONCOMPACTNESS IN BANACH SPACES

In what follows we give some properties of measures of noncompactness for sets in Banach spaces. The most important one refers to the convex hull of the set.

THEOREM 2.6.1. *For any bounded set A in a Banach space X having the Kuratowski measure of noncompactness $\alpha(,)$ we have*

$$\alpha(A) = \alpha(\operatorname{conv} A).$$

Proof. We can suppose, without loss of generality, that A is a closed set. First we prove that for any two convex set C and $C^\sim$ we have the relation

$$\alpha(\operatorname{conv}(C \cup C^\sim)) \leqq \max \{\alpha(C), \alpha(C^\sim)\}$$

First we remark that C and $C^\sim$ being convex sets we have

$$\operatorname{conv}(C \cup C^\sim) \subset \bigcup_a \{aC + (1-a)C^\sim\}_{a \in [0,1]}$$

Let $\varepsilon > 0$ and $a > 0$ such that $\|x\| \leqq a$ for all $x \in C + C^\sim$ and

$$t_1 < t_2 < \cdots < t_k$$

such that $t_i - t_{i-1} < \varepsilon/2a$, $i = 2, \ldots, k$. In this case we have also

$$\operatorname{conv}(C \cup C^\sim) \bigcup \{t_i C + (1 - t_i)C^\sim + \varepsilon S_1(0)\}$$

where $S_1(0) = \{x, \|x\| \leqq 1\}$.

From this and the properties of $\alpha(\,,)$ we further obtain,

$$\alpha(\text{conv}(C \cup C^{\tilde{}})) \leq \max_i \{t_i\alpha(C) + (1 - t_i)\alpha(C^{\tilde{}}) + 2\varepsilon\}$$

$$= \max \{\alpha(C), \alpha(C^{\tilde{}})\} + 2\varepsilon$$

and, since $\varepsilon > 0$ is arbitrary, the assertion follows.

Since, for any bounded sets A_1 and A_2 such that $A_1 \subset A_2$, it follows that $\alpha(A_1) \leq \alpha(A_2)$. To prove the theorem it is sufficient to show that $\alpha(\text{conv } A) \leq \alpha(A)$. Consider a number $b > \alpha(A)$. Then there exist the subsets $A_1, A_2, \ldots, A_k$ such that:

1. $\text{diam } A_i \leq b,$

2. $A \subset \bigcup_1^k A_i.$

Since $\text{diam}(A_i) = \text{diam}(\text{conv } A_i)$ we can suppose, without loss of generality, that each A_i is convex. We now define the sets

$$E = \text{conv}(\text{conv}(\cdots \text{conv}(A_1 \bigcup A_2) \bigcup A_2) \bigcup \cdots \bigcup A_k)$$

and applying the above relation we obtain successively,

$$\alpha(E) \leq \max_i \{\alpha(A_i)\}.$$

But $\text{conv } A$ is in $\text{conv } \{\bigcup A_i\} \subset E$, and thus,

$$\alpha(\text{conv } A) \leq \max \{\alpha(A_i)\} < b$$

and since b is an arbitrary number greater than $\alpha(A)$ we have the inequality

$$\alpha(\text{conv } A) \leq \alpha(A)$$

and thus the theorem is proved.

The following result gives information about the measure of noncompactness for the sum of two sets.

First we recall that if A and B are two sets in a linear space then

$$A + B = \{a + b, a \in A, v \in B\}, \qquad aA = \{a \cdot s : s \in A\}.$$

THEOREM 2.6.2.   *For any two bounded sets in a Banach space* X.

1. $\alpha(A + B) \leq \alpha(A) + \alpha(B)$

2. $\alpha(aA) = |a|\alpha(A).$

Proof. The assertions follow from the definition of the measure of noncompactness.

The following theorem of Nussbaum (1971) gives the measure of noncompactness for a special class of sets.

THEOREM 2.6.3.   *Let X be a Banach space and consider the set*

$$S_1(0) = \{x, \|x\| \leq 1\}.$$

Then

1. $\alpha(S_1(0)) = 0$ *if X is finite-dimensional,*

2. $\alpha(S_1(0)) = 2$ *if X is infinite-dimensional.*

Proof. The first assertion is clear. The second uses a deep result, the so-called Liusternik–Schnirelman–Borsuk theorem, which asserts that if the unit sphere in an n-dimensional space is covered by n closed sets, then at least one of the sets contains a pair of antipodal points (i.e. there exists an x such that x and $-x$ are contained in that set).

See Section 12.7 for a proof of the Liusternik–Schnirelman–Borsuk theorem.

2.7. CLASSES OF SPECIAL OPERATORS ON BANACH SPACES

Let X, Y be two Banach spaces and $T \in L(X, Y)$.

DEFINITION 2.7.1.   The operator T is said to be finite-dimensional if there exists a finite set $y_1, y_2, \ldots, y_n$ of elements in Y such that, for all $x \in X$,

$$Tx = \sum_1^n x_i^*(x) y_i, \qquad x_i^* \in X^*.$$

The following assertion is an easy exercise.

THEOREM 2.7.2.   *The set of all finite-dimensional operators is a linear subspace of $L(X, Y)$.*

DEFINITION 2.7.3.   An operator $T \in L(X, Y)$ is said to be completely continuous if there exists a sequence of finite-dimensional operators $\{T_n\}$ such that $\lim \| T_n - T \| = 0$

The following theorem gives the basic properties of the set of completely continuous operators.

THEOREM 2.7.4.   *Let K be the set of all completely continuous operators on a Banach space X. Then the following assertions hold:*

1.   *K is a Banach space,*

2.   *K is a two-sided ideal in L(X).*

Proof. It is not difficult to see that K is a linear space. We prove now that K is a Banach space.

Let $\{T_n\}$ exist in K and suppose that $\lim \|T_n - T\| = 0$. Let $\varepsilon > 0$ and choose N_ε such that for all $n > N_\varepsilon$, $\|T_n - T\| \leq \varepsilon/2$. Since $T_n \in K$ we find a finite-dimensional operator, say T_n such that $\|T_n - T'_n\| \leq \varepsilon/2$ and thus,

$$\|T'_n - T\| \leq \|T'_n - T_n\| + \|T_n - T\| \leq \varepsilon/2 + \varepsilon/2 = \varepsilon$$

and this gives assertion 1.

Now, if A and B are two operators such that A is finite-dimensional, then AB and BA are finite-dimensional operators. This implies, obviously, that for any $T \in K$ and $S \in L(X)$ the operators TS and ST are in K and this is exactly the property 2.

We now introduce a new class of operators on Banach spaces related to (pre)compact sets. For some Banach spaces this class is exactly K.

DEFINITION 2.7.5.   An operator $T \in L(X)$ is said to be compact if, for any bounded set $M \subset X$, the set $\overline{TM}$ is compact (i.e. TM is a precompact set).

The following theorem gives the basic information about the class of compact operators.

THEOREM 2.7.6.   *Let X be a Banach space and let $L_c(X)$ be the set of all compact operators. Then the following assertions hold:*

1. $L_c(X)$ *is a Banach space,*

2. $L_c(X)$ *is a two-sided ideal in L(X),*

3.   *if $T \in L_c(X)$ then $T^* \in L_c(X^*)$.*

Proof. The fact that $L_c(X)$ is a linear space is easy to see.

We prove now that $L_c(X)$ is a Banach space. Let $\{T_n\}$ be in $L_c(X)$ and let $\lim \|T_n - T\| = 0$. Let $\varepsilon > 0$ and consider the set $S = \{x, \|x\| < 1\}$.

We take N_ε such that $\|T_n - T\| \leq \varepsilon/2$ and since T_n is a compact operator we find a finite set $\{y_1, y_2, \ldots, y_n\}$ such that the sets $S_{\varepsilon/2}(y_i)$ cover the set $T_n S_1(0)$. In this case the sets $S_\varepsilon(Y_i)$ cover the set $TS_1(0)$. Indeed, we have,

$$\|Tx - y_i\| \leq \|Tx - T_n x\| + \|T_n x - y_i\| \leq \varepsilon/2 + \varepsilon/2 = \varepsilon.$$

From the Hausdorff characterization of (pre)compact sets the assertion of the theorem follows.

The proof of the second assertion is very simple and we thus omit it.

Property 3 is known as the Schauder theorem. The proof is as follows. We take a cover of the set $TS_1(0)$ with sets of the form $S_{\varepsilon/2}(x)$ and since this is a (pre)compact set we can select a finite number of such sets: $S_{\varepsilon/2}(y_1), \ldots, S_{\varepsilon/2}(y_n)$.

We define now a map on X^* with values in $\mathbf{C}^n$ as follows:

$$U^*(y^*) = \{y^*(y_i)\}.$$

Obviously this is a compact map and thus the set $U^* S_1(0^*)$ can be covered with a finite number of sets of the form $S_{\varepsilon/4}(y_j^*) \, J = 1, \ldots, k$. In this case for each $y^* \in S_1(0^*)$ there exist y_j such that

$$|y^*(y_h) - y_j^*(y_h)| \leq \varepsilon/4, \qquad h = 1, 2, \ldots, n.$$

But for each $x \in S_1(0)$ there exist y_k such that $\| Tx - y_k\| \leq \varepsilon/4$ and thus we obtain,

$$|y^*(Tx) - y_h^*(Tx)| \leq |y^*(Tx) - y^*(y_k)| + |y^*(y_k) - y^*(y_k)| +$$
$$+ |y_h^*(y_k) - y_h^*(Tx)|$$
$$\leq \tfrac{3}{4}\varepsilon.$$

This implies that

$$\|T^*(y^*) - T^*(y_h^*)\| = \sup_{\|x\| \leq 1} |y^*(Tx) - y_h^*(Tx)| \leq \varepsilon$$

i.e.

$$T^* S_1(0^*) \subset \bigcup_1^h S_\varepsilon(T^* y_h^*)$$

and, since $\varepsilon > 0$ is arbitrary, from the Hausdorff characterization of (pre)compact sets it follows that T^* is a compact operator.

For the following result we need the notion of spectrum of an element in $L(X)$. For more information see N. Dunford and J. Schwartz (1958), V. Istrăţescu, 1981

DEFINITION 2.7.7. Let $T \in L(X)$. We say that T has the inverse operator

$S \in L(X)$ if

$$TS = ST = I.$$

A complex number is in the resolvent set of T, $\rho(T)$ if the operator $T_z = T - z$ [$= T - z$ has an inverse]. The complement of the set $\rho(T)$ is called the spectrum of T and is denoted by $\sigma(T)$.

THEOREM 2.7.8.   *The spectrum of a compact operator is an at most countable set having zero as its possible accumulation point.*

For a proof, as well as for other results connected with compact operators we refer to N. Dunford and J. Schwartz (1958), V. Istrǎţescu (1981).

Another interesting and important class of operators with many applications in nonlinear analysis is the class of Fredholm operators.

DEFINITION 2.7.9.   Let X and Y be two Banach spaces and $T \in L(X, Y)$. The operator T is called Fredholm operator if the following properties hold:

1.   Ker $T = \{x, Tx = 0\}$ is a finite-dimensional space,

2.   Coker $T = Y/TX$ is a finite-dimensional space.

For any operator which is a Fredholm operator we can define an integer which is called the 'index of T' (shortly ind T) as follows:

$$\text{ind } T = \dim \text{Ker } T - \dim \text{Coker } T.$$

From the properties of compact operators there follows the following theorem, which gives examples of Fredholm operators.

THEOREM 2.7.10.   *Let X be a Banach space and T be any compact operator in $L(X)$. Then the operator T-I is a Fredhom operator.*

Also using the fact that Ker and Coker are finite-dimensional subspaces we can prove the following result.

THEOREM 2.7.11.   *Let X, Y be two Banach spaces and let $T \in L(X, Y)$ be a Fredholm operator. Then the following assertions hold:*

1.   *TX is a closed subspace of Y,*

2.   *T^* is a Fredholm operator,*

3.   *ind $T^* = -$ ind T.*

The following characterization of Fredholm operators follows from the properties of Ker, Coker and of complemented subspaces and was obtained by Atkinson.

THEOREM 2.7.12.　*Let X be a Banach space and let $T \in L(X)$. Then T is a Fredholm operator if and only if $T^\sim$, the image of T in the algebra $L(X)/K(X)$ [$K(X)$ is the two-sided ideal of compact operators] is an invertible element [the definition of invertibility is essentially as given in Definition 2.7.7.].*

We are now in a position to introduce several important classes of mappings on Banach spaces or on subsets of Banach spaces.

Let X be a Banach space and let E be a bounded subset in X. We consider the measure of noncompactness $\alpha(,)$ of Kuratowski. Of course, similar definitions can be given using other measures of noncompactness, but we do not emphasize this in what follows.

DEFINITION 2.7.13.　A continuous map $f : E + E$ is called an α-set-contraction if, for any subset $A \in E$ we have,

$$\alpha(f(A)) \leq k\alpha(A)$$

where $k \in [0,1)$ and is independent of A, whenever A is noncompact.

Remark. 2.7.14. Of course the definition can be given for any complete metric space. This was done by C. Kuratowski and is as follows: suppose that (X, d_X) and (Y, d_Y) are two complete metric spaces and $F : X \to Y$. We say that F is a k-set-contraction if there exists $k \in [0, 1]$ such that the following conditions are satisfied:

1.　　　if A is bounded, then $F(A)$ is bounded,

2.　　　$\alpha(F(A)) \leq k\alpha(A)$.

Now if F is any F which is any k-set-contraction, we define

$$\gamma_F = \inf \{k, F \text{ is a } k\text{-set-contraction}\}.$$

R. Nussbaum has defined a notion which localizes the above property.

DEFINITION 2.7.15.　Let X, Y be defined as above and let $F : X + Y$. We say that F is a local strict-set-contraction if for all $x \in X$ there exists a neighbourhood $V_x \in x$ such that F/V_x is a k_x-set-contraction.

An interesting and useful class of maps was considered by V. Sadovski (1967). His class is defined as follows:

DEFINITION 2.7.16.　Let X, Y be two complete metric spaces and let

$F : X \to Y$. Then F is said to be condensing if the following conditions are satisfied:

1. F is continuous,

2.   for any set $A_1 \subset X$, $\alpha(A_1) \neq 0$ then $\alpha(F(A_1)) < \alpha(A_1)$.

Obviously we can define a notion localizing this one. This is as follows.

DEFINITION 2.7.17.   Let X, Y be as given above and let $F : X \to Y$. Then F is said to be a local condensing map if for any $x \in X$ there exists a neighbourhood $V_x \ni x$ such that F/V_x is a condensing map.

For an example of a condensing mapping which is not a k-set-contraction for any $k \in (0, 1)$ see Section 5.2.3.

Other classes of mappings can be considered using powers and the idea of localization.

DEFINITION 2.7.18.   Let X be a complete metric space. The map $F : X \to X$ is called a local power k-set-contraction if the following conditions are satisfied:

1. F is continuous,

2.   for each bounded noncompact set there exists an integer

            $n = n(\mathrm{A})$ such that

            $\alpha(F^n(A)) \leqq k\alpha(A)$

            where k is in $[0, 1)$ and independent of A.

Example 2.7.19.   Let X be a Banach space and let T be an operator which is quasi-compact (i.e. $\sup \|T^n\| < \infty$), then there exists an integer m and a compact operator Q such that $\|T^n - Q\| < 1$. Then clearly T is a local power k-set-contraction. In this case it is clear that $k = \|T^m - Q\|$.

Using quasi-compact operators we can readily construct other examples of local power k-set-contractions.

We can now define the analogous notion of condensing map.

DEFINITION 2.7.20.   Let X be a complete metric space. The map $F : X \to X$ is said to be local power condensing if the following conditions are satisfied:

1. F is continuous,

2.   for each bounded set A, $\alpha(A) > 0$, there exists an integer $n = n(A)$
            such that $\alpha(F^n(A)) < \alpha(A)$.

We may mention here that the classes of mappings considered in Definition 2.7.20 have applications in the Ergodic Theory of Markov processes.

Another class of mappings can be introduced as follows: suppose we have a complete metric space X and a mapping $F : M \to M$ where M is a bounded closed set in X. We suppose that $\alpha(M) > 0$. We define the following sequence of sets in M:

$$\{O^n(M)\}, \qquad O^n(M) = \bigcup_{m \geq 1}^{\infty} F^m(M)$$

and this is a decreasing sequence. Thus $\{\alpha(O^n(M))\}$ is a decreasing sequence of positive numbers. Let $\alpha_\infty(M) = \lim \alpha(O^n(M))$.

DEFINITION 2.7.21.   The mapping $F : M \to M$ is said to be a k-diminishing set if

$$\alpha_\infty(F(M)) \leq k\alpha_\infty(M)$$

whenever $\alpha_\infty(M) > 0$.

DEFINITION 2.7.22.   The mapping F is said to be a diminishing α-set if

$$\alpha_\infty(F(M)) < \alpha_\infty(M)$$

whenever $\alpha_\infty(M) > 0$.

Chapter 3

The Contraction Principle

3.0. INTRODUCTION

The work of Cauchy on differential equations has been fundamental to the concept of existence theorems in Mathematics.

As is well known these are connected with the following topics:

1. The Cauchy–Lipschitz existence and uniqueness theorems;
2. The Cauchy–Picard existence and uniqueness theorems;
3. The Cauchy–Majorant method for analytic equations; and
4. The Cauchy–Poincaré small parameter methods.

It is also known that in fact the first method was known even to L. Euler; A. L. Cauchy (1844) was the first mathematician to give a proof for the existence and uniqueness of the solution of the differential equation

$$dy/dx = f(x, y)$$

$$y(x_0) = y_0.$$

when f is a continuously differentiable function. In 1877, R. Lipschitz simplified Cauchy's proof using what is known today as the 'Lipschitz condition'. A function defined on a subset A of $\mathbf{R}^n$ satisfies the Lipschitz condition if there is a constant K such that

$$|f(x) - f(y)| \leq K\, d(x, y)$$

where d is the metric of $\mathbf{R}^n$,

$$d(x, y) = \left(\sum_1^n (x_i - y_i)^2 \right)^{1/2}, \qquad x = (x_i),\, y = (y_i).$$

Later G. Peano (1890) established a deeper result – the existence of a solution of the equation

$$dy/dx = f(x, y),$$

$$y(x_0) = y_0$$

supposing only the continuity of f. Peano's approach is more related to modern fixed point theory which is used to obtain existence theorems. It is, however, interesting to remark that the method of Cauchy–Picard was in fact used before Cauchy. For example, J. L. Liouville (1838) used the method for heat conduction. In Liouville's paper the iterations are defined and estimates are obtained which are used to prove the uniform convergence of the Neumann series to a solution.

In 1890 Picard applied the method to ordinary differential equations as well as to a class of partial differential equations.

The iterative method was used widely after Picard and some papers dealt with an abstraction of this method. Thus, in 1920, Lawson gave a formulation as an implicit function theorem in abstract spaces: if D is an abstract set and $Y:D+E$, $Z:D+E$ are functions with values in a normed space E and

$$Y(P) = F(Y, Z, P)$$

is a function on D with values in E, Lawson proved that if F satisfies a Lipschitz condition with $k < 1$, uniformly in Z with respect to Y and Y_0, $(P) = F(Y_0, Z_0, P)$ is an identity in P, then there exists a mapping $Y = Y(Z, P)$ satisfying $Y = F(Y, Z, P)$ continuous in Z.

It is important to note that Lawson's result is universal, in the sense that the convergence is proved for any arbitrary initial value. A local version of Lawson's theorem was obtained by Hildebrandt and Graves in 1929.

The formulation of what is known by the name of 'The principle of contraction mapping' or as the 'Banach fixed point theorem' is the following assertion:
'if (X, d) is a complete metric space and $f : X \to X$ has the property that

$$d(f(x), f(y)) \leqq kd(x, y), \qquad k < 1$$

for all $x, y \in X$, then there exists a unique point $x_0 \in X$ such that

$$f(x_0) = x_0$$

and for fixed $x \in X$, the sequence $\{f^n(x)\}$ converges to x_0'.

There exists a great number of attempts to weaken the principle of contraction mapping. We mention here the one proposed by R. Caccioppoli who, in 1930, remarked that it is possible to replace the contraction property by the assumption of that of convergence. More precisely the result of Caccioppoli is as follows:

THEOREM.   *If (X, d) is a complete metric space and $f : X \to X$ has the property* *that*

$$d(f(x), f(y)) \leq \|f\| d(x, y)$$

and if $f^n(x) = f(f^{n-1}(x)), f^n(y) = f(f^{n-1}(y))$ *then the sequence*

$$\{f^n(z)\}$$

converges to a fixed point $z_0 = f(z_0)$ *if* $\sum \|f^n\| < \infty$, *and where*

$$d(f^n(x), f^n(y)) \leq \|f^n\| d(x, y)$$

It is obvious that this reduces to the principle of contraction mapping if $\|f\| < 1$. In the next section we give a precise formulation of many variants of the extension of the principle of contraction mapping.

3.1. THE PRINCIPLE OF CONTRACTION MAPPING IN COMPLETE METRIC SPACES

Suppose that (X, d) is a complete metric space and $f : X \to X$ is any function.

DEFINITION 3.1.1.   The function f is said to satisfy a Lipschitz condition with constant K if

$$d(f(x), f(y)) \leq K d(x, y)$$

holds for all $x, y \in X$. If $K < 1$ then f is called a contraction mapping.

THEOREM 3.1.2   (the Principle of Contraction Mapping). *If f is a contraction mapping on a complete metric space (X, d) then there exists a unique fixed point x_0 of f and if $x^\sim$ is any point in X and $x_n^\sim = f^n(x)$ then*

1. $d(x_n^\sim, x_0) \leq K^n/(1 - K)$,

2. $\lim x_n^\sim = x_0$.

Proof. Let z be an arbitrary point in X and define the sequence

$$z_n = f^n(z).$$

We wish to prove that

$$d(z_{n+1}, z_n) \leq K^n d(z_1, z).$$

Indeed, for arbitrary n, m, we have

$$d(z_{n+1}, z_{m+1}) = d(f(z_n), f(z_m)) \leq K d(z_n, z_m)$$

and this immediately gives us the assertion.

Now, if $n \geq m$, by the triangle inequality,

$$d(z_n, z_m) \leq d(z_m, z_{m+1}) + d(z_{m+1}, z_n)$$
$$\leq d(z_m, z_{m+1}) + d(z_{m+1}, z_{m+2}) + d(z_{m+2}, z_n)$$
$$\leq \cdots \leq d(z_m, z_{m+1}) + \cdots + d(z_{n-1}, z_n)$$
$$= \sum_{i=0}^{n-m} d(z_{m+i}, z_{m+i+1})$$

and applying the above relation we obtain,

$$d(z_m, z_n) \leq K^m(1 + K + K^2 + \cdots + K^{n-m}) \leq K^m \cdot 1/(1 - K).$$

Since $K < 1$ this implies that the sequence $\{z_n\}$ is a Cauchy sequence and since X is complete there exists x_0 such that

$$\lim z_n = x_0.$$

But

$$d(z_{n+1}, f(x_0)) \leq K d(z_n, x_0)$$

and this gives us that $x_0 = f(x_0)$, i.e. x_0 is a fixed point of f.

Suppose now that $x^\sim$ is an arbitrary point in X. In this case we have, for the sequence $\{x_n^\sim\}$ defined by the relation $x_{n+1}^\sim = f(x_n^\sim)$ and where $x_1^\sim = f(x^\sim)$,

$$d(x_{n+1}^\sim, x_0) = d(f(x_n^\sim), f(x_0)) \leq k d(x_n^\sim, x_0)$$

and this gives the relation

$$d(x_{n+1}^\sim, x_0) \leq K^n d(x^\sim, x_0)$$

which implies that

$$\lim x_n^\sim = x_0.$$

We remark that x_0 is the unique fixed point of f. Indeed, if there exists y_0, another fixed point of f, we have

$$d(x_0, y_0) = (f(x_0), f(y_0)) < K d(x_0, y_0)$$

which is a contradiction if $x_0 \neq y_0$.

THEOREM 3.1.3. *If (X, d) is a complete metric space and f is a continuous mapping $f : X \to X$ and, for some integer m, $f^m = f_0 \cdots {}_0 f$ is a contraction mapping then f has a unique fixed point in X.*

Proof. Let x_0 be the unique fixed point of f^m. In this case

$$f(x_0) = f(f^m(x_0)) = f^m(f(x_0))$$

and this gives us that $x_0 = f(x_0)$.

We now give an example to show that the mapping defined on a complete metric space satisfying the relation

$$d(f(x), f(y)) < d(x, y)$$

does not necessarily have fixed points.

Example 3.1.4.   Consider the space $X = \mathbf{R}^+$ with the metric

$$d(x, y) = |x - y|$$

and let f be *defined as*:

$$f(x) = (x^2 + 1)^{1/2}.$$

It is easy to see that this is a continuous mapping satisfying the relation

$$d(f(x), f(y)) < d(x, y)$$

and obviously it has no fixed points.

We now show that contractive mappings defined on incomplete metric spaces fail to have fixed points.

Example 3.1.5.   Let $X = (0, 1]$ with the induced metric from $\mathbf{R}$ and consider f defined as follows:

$$f(x) = x/2$$

which is contractive and has no fixed points.

We give now some examples of contractive mappings.

Example 3.1.6.   Let $f : [a, b] \to [a, b]$ and let f' exist and $\sup |f'(x)| < 1$. Then f is a contraction mapping.

Example 3.1.7.   Consider the space of all functions of the form

$$x(t) = (x_1(t), \ldots, x_n(t))$$

where x_i are real-valued functions and the system of ordinary differential equations

$$dx_i(t)/dt = f_i(t, x_1, \ldots, x_n)$$

which can be written in the form

$$dx/dt = f(t, x)$$

with the initial condition $x(0) = 0$. As is well known this is equivalent to the integral equation

$$x(t) = \int_0^t f(s, x(s)) \, ds \qquad x(0) = 0.$$

We suppose that f is jointly continuous and satisfies the Lipschitz condition

$$|f(t, x) - f(t, y)| < L|x - y|$$

where 'd' is the metric on $\mathbf{R}^n$. The equation can be treated using the following Banach space

$$C_\alpha = \{x(t), x(t) \text{ continuous on } [0, \alpha]\}$$

and the metric is defined as follows:

$$d\tilde{\ }(x(t), y(t)) = \sup \{|x(t) - y(t)|, t \in [0, \alpha]\}.$$

We consider the mapping

$$(Af)(t) = \int_0^t f(s, x(s)) \, ds$$

which leaves C_α invariant. Now since

$$d\tilde{\ }(Ax, Ay) \leqq L\alpha d\tilde{\ }(x, y),$$

for α such that $L\alpha < 1$, we can apply the principle of contraction mapping and we obtain the existence of a solution in the above space.

Remark 3.1.8. We can consider another Banach space and we have the possibility of showing the existence of a solution under more general assumptions.

Consider the Banach space of all continuous functions $x(t)$ on the interval $[0, T]$ with the d metric

$$d\tilde{\ }(x, y) = \max_t e^{-L_1 t} |x(t) - y(t)|$$

where $L_1 > 0$. Using this metric we have, for the operator A defined as above,

$$d^\sim(Ax, Ay) = \max_t |e^{-L_1 t} \int_0^t f(s, x(s)) - f(s, y(s)) \, ds|$$

$$\leq L \max_t \int_0^t e^{L_1(s-t)} e^{-L_1 s} |x(s) - y(s)| ds$$

$$\leq L d^\sim(x, y) \max_t \int_0^t e^{L_1(s-t)} \, ds$$

$$= L/L_1 (1 - e^{-L_1 T}) d^\sim(x, y).$$

For $L = L_1$ we obtain the fact that, with respect to this metric, the mapping A is a contraction mapping and the existence as well as the uniqueness is proved.

Remark 3.1.9. Theorem 3.1.2 asserts the existence and the uniqueness of the solution of the equation $f(x) = x$ in the complete metric space X when f is a contraction mapping. If X is a subspace of a complete metric space Y and there exists a continuous mapping $F : Y \to Y$ such that $f(x) = F(x)$ for all $x \in X$ (i.e., f is the restriction of F to X), then it is possible that the equation $F(y) = y$ may have other solutions.

Indeed, if $X = C_{[0,1]}$ with the metric d defined as

$$d(f, g) = \|f - g\|$$

and the mapping

$$(Af)(t) = \int_0^1 x^2(s) \, ds,$$

considered on the subset of $C_{[0,1]}$ of the following form

$$M = \{f, d(f, 0) \leq k < \tfrac{1}{2}\},$$

has only the solution $x(t) = 0$. In the space X it has also the solution $x(t) = 1$.

3.2. LINEAR OPERATORS AND CONTRACTION MAPPINGS

Let us consider a Banach space X and let $L(X)$ be the set of all bounded linear operators on X. For any element $T \in L(X)$ we denote by r_T its spectral radius, defined as

$$r_T = \lim \|T^n\|^{1/n}.$$

The following result is useful when linear operators are encountered in equations.

THEOREM 3.2.1. *Let* $T \in L(X)$ *and* $\varepsilon > 0$. *Then there exists an equivalent norm*[1] $\|,\|^*$, *such that*

$$\|T\|^* \leq r_T + \varepsilon.$$

Proof. From the definition of the spectral radius it follows that for $\varepsilon > 0$ there exists N_ε such that,

$$\|T^n\| \leq (r_T + \varepsilon)^n, \qquad n \geq N.$$

The norm $\|,\|^*$ is defined now as follows:

$$\|x\|^* = (r_T + \varepsilon)^{N-1}\|x\| + (r_T + \varepsilon)^{N-2}\|Tx\| + \cdots + \|T^{N-1}x\|.$$

From this we obtain

$$(r_T + \varepsilon)^{N-1}\|x\| \leq \|x\|^* \leq (r_T + \varepsilon)^{N-1}\|x\| + \cdots + \|T^{N-1}\|\,\|x\|$$
$$= K\|x\|.$$

Since the fact that $\|,\|^*$ is a norm is obvious from these it is clear that $\|,\|$ and $\|,\|^*$ are equivalent norms.

Now we estimate the norm of T with respect to $\|,\|^*$. We have,

$$\|Tx\|^* = (r_T + \varepsilon)^{N-1}\|Tx\| + (r_T + \varepsilon)^{N-2}\|T^2x\| +$$
$$\cdots + \|T^N x\|$$
$$\leq (r_T + \varepsilon)[(r_T + \varepsilon)^{N-2}\|Tx\| + (r_T + \varepsilon)^{N-3}\|T^2x\| +$$
$$\cdots + (r_T + \varepsilon)\|T^{N-1}x\| + (r_T + \varepsilon)^{N-1}\|x\|]$$
$$= (r_T + \varepsilon)\|x\|^*$$

and the assertion of the theorem is proved.

COROLLARY 3.2.2. *If* $T \in L(X)$ *has the spectral radius* $r_T < 1$, *then there exists an equivalent norm* $\|,\|^*$ *on* X *such that* T *is a contraction mapping*.

3.3. SOME GENERALIZATIONS OF THE CONTRACTION MAPPINGS

There exists a great number of generalizations of the contraction mappings and in what follows we give some of them.

DEFINITION 3.3.1. *If* (X, d) *is a complete metric space and* $f : X \to X$ *then* f

[1] Two norms, $\|,\|_1$ and $\|,\|_2$ on a linear space are called equivalent if there exists m, $M > 0$ such that for all $x \in X$

$$m\|x\|_1 \leq \|x\|_2 \leq M\|x\|_1.$$

is called a generalized contraction mapping in the sense of Krasnoselskii (1964) if

$$d(f(x), f(y)) < \alpha(a, b)d(x,y)$$

for $a < d(x, y) < b$ and $\alpha(a, b) \in [0, 1)$ for $0 < a < b$.

THEOREM 3.3.2.   *If f is a generalized contraction mapping in the sense of Krasnoselskii then there exists a unique fixed point $x_0 \in X$ of f.*
   Proof. Consider the sequence $\{a_n\}$ where

$$a_n = d(x_n, x_{n-1}), \qquad x_1 = f(x_0)$$

and from the properties of f it is clear that $\{a_n\}$ is a nonincreasing sequence. Let $a = \lim a_n$. If $a > 0$ then, for N sufficiently large and for all m, we have

$$a_{N+m} \leq [\alpha(a, a+1)]^m (a+1)$$

and this contradicts the fact that $a > 0$.
   Let $\varepsilon > 0$ be given and choose N such that

$$a_N \leq (\varepsilon/2)[1 - \alpha(\varepsilon/2, \varepsilon)]$$

and we show that f leaves

$$\{x, d(x, x_N) \leq \varepsilon\}$$

invariant. This assertion implies clearly that $\{x_n\}$ is a Cauchy sequence. To prove that the above set is invariant under f, we can see that if $d(x, x_N) \leq \varepsilon/2$ then

$$d(f(x), x_N) \leq d(f(x), f(x_N)) + a_N \leq d(x, x_N) + a_n \leq \varepsilon,$$

and if

$$\varepsilon/2 < d(x, x_N) \leq \varepsilon$$

we obtain

$$d(f(x), f(x_N)) \leq d(f(x), f(x_N)) + a_N \leq \alpha(\varepsilon/2, \varepsilon) + a_N \leq \varepsilon$$

and the invariance is proved. Let $\lim x_n = x_0$, then this is clearly a fixed point for f. The uniqueness is obvious.

   Another class of mappings related to contraction mappings are defined on some special classess of metric spaces.

DEFINITION 3.3.3.   A complete metric space is called ε-chainable if, for any points $a, b \in X$, there eixsts a finite set of points

$$a = x_0, x_1, \ldots, x_{n-1}, x_n = b$$

such that

$$d(x_{i-1}, x_i) \leqq \varepsilon \qquad i = 1, 2, 3, \ldots, n.$$

We define now two classes of mappings.

DEFINITION 3.3.4.   A mapping $f : X \to X$ is called locally contractive if for every $x \in X$ there exists ε_x and λ_x such that for all p, q in

$$(y, d(x, y) \leqq \varepsilon_x\} \qquad \varepsilon_x > 0, \quad \lambda_x \in [0, 1).$$

the relation

$$d(f(p), f(q)) \leqq \lambda_x d(p, q)$$

holds.

DEFINITION 3.3.5.   A mapping $f : X \to X$ is called (ε, λ)-uniformly locally contractive if it is locally contractive and ε and λ do not depend on x.

The following theorem gives a generalization of the contraction mapping principle to a class of mappings on ε-chainable spaces.

THEOREM 3.3.6.   If (X, d) is an ε-chainable space and $f : X \to X$ is an (ε, λ)-uniformly locally contractive mapping then there exists a unique fixed point x_0 of f.
    Proof. Let x be an arbitrary point in X and consider the points x and $f(x)$. We find the ε-chain

$$x = x_0, x_1, \ldots, x_{n-1}, x_n = f(x)$$

such that

$$d(x_{i-1}, x_i) < \varepsilon$$

and thus by the triangle inequality

$$d(x, f(x)) < n \cdot \varepsilon$$

holds. Now we have, obviously,

$$d(f(x_{i-1}), f(x_i)) < \lambda d(x_{i-1}, x_i) < \lambda \cdot \varepsilon$$

and by induction we obtain

$$d(f^m(x_{i-1}), f^m(x_i)) < \lambda^m \cdot \varepsilon.$$

From these we obtain,

$$d(f^m(x), f^{m+1}(x)) < \lambda^m \cdot n \cdot \varepsilon$$

which implies that $\{f^m(x)\}$ is a Cauchy sequence. Indeed, if $n > m$

$$d(f^m(x), f^n(x)) < \sum d(f^{m+i}(x), f^{m+i+1}(x))$$
$$< \lambda^m \cdot n \cdot (1 + \lambda + \lambda^2 + \cdots + \lambda^{n-m})\varepsilon$$
$$< n \cdot \varepsilon \cdot 1/(1 - \lambda)$$

and this proves the assertion. Thus $\lim f^m(x)$ exists and we denote it by x_0. Since f is continuous, we obtain

$$f(x_0) = \lim f(f^m(x)) = \lim f^{m+1}(x) = x_0$$

i.e. x_0 is a fixed point of f. We show now that this is the unique fixed point of f. Indeed, suppose that there exists y_0, another fixed point of f. We find the ε-chain

$$x_0 = x_0, x_1, \ldots, x_{n-1}, x_n = y_0$$

and as above we have

$$d(f(x_0), f(y_0)) < \sum d(f^m(x_{i-1}), f^m(x_i))$$
$$< \lambda^m \cdot n \cdot \varepsilon$$

and for $m \to \infty$ we obtain $d(x_0, y_0) = 0$

The following example illustrates the application of the above theorem.

Example 3.3.7.   Let f be an analytic function in a domain D of the complex plane and suppose that f maps a compact and connected subset C of D into itself. If $|f'(z)| < 1$ for every $z \in C$ then $f(z) = z$ has one and only one solution in C.

Proof. Since f' is continuous on C it follows from the compactness of C that $|f'(z)| < \lambda < 1$ on C. The assertion of the example follows if we show that f is an (ε, λ)-locally contractive mapping of C into C. Let $\{\mathring{S}(z, r)\}$ be an open covering of C with open discs such that $|f'(z)| < \lambda$ in $\mathring{S}(z, 2r)$. Clearly this cover contains a finite subcover, say $\{\mathring{S}(z_i, r_i)\}_1^N$, and if we set $\varepsilon = \min r_i$ we obtain that f is an (ε, λ)-local contraction since any two points of C, z_1 and z_2 with $(d(z_1, z_2) < \varepsilon$ are in some $\mathring{S}(z_i, r_i)$ and thus

$$|f(z_1) - f(z_2)| = \left| \int_{z_1}^{z_2} f'(z) \, dz \right| < \lambda |z_1 - z_2|$$

which proves the assertion.

We give now a theorem on the special spaces.

THEOREM 3.3.8. *If (X, d) is a compact and connected metric space and $f : X \rightarrow X$ is a local contraction mapping then there exists a unique fixed point of f.*

Proof. The proof is in fact an adaptation of the above proof.

First we note that compactness of X implies the existence of K and δ such that if $d(x, y) < \delta$ then $d(f(x), f(y)) \leq Kd(x, y)$. We consider now an arbitrary open covering of X with sets whose diameter is less than δ and let $V_1, V_2, \ldots, V_n$ be a finite subcover.

Since X is a connected set for any points x, y of X there exists a chain of open sets $V_{i_1}, \ldots, V_{i_m}$ such that

$$V_{i_{j-1}} \cap V_{i_j}$$

is nonempty and $x \in V_{i_1}$ and $y \in V_{i_m}$. From these it is clear that $d(x, y) < m \cdot \delta < n \cdot \delta$ and considering the image under f we obtain that $d(f^m(x), f^m(y)) < K^m \cdot n \cdot \delta$. If m is large such that $K^m n < 1$ the restriction of f to $f^m(X)$ is a contraction mapping and thus there exists a fixed point for f. Obviously this is a unique fixed point of f.

The following theorem gives a fixed point result in the case of compact space and local contractions.

THEOREM 3.3.9. *If (X, d) is a compact metric space and $f : X \rightarrow X$ is a local contraction mapping then there exists a unique fixed point of f.*

Proof. We consider the function

$$F(x) = d(f(x), x))$$

and, since X is compact and f continuous, we find that F attains a minimum, say at the point x_0. From the property that f is a locally contractive mapping it follows that x_0 is a fixed point of f. Obviously this is a unique fixed point of f.

Remark 3.3.10. The theorem is valid for a slightly more general class of mapping: f has the property that, for any x and y in X, X a compact metric space:

$$d(f(x), f(y)) < d(x, y).$$

Since the proof is the same as for Theorem 3.3.9 we omit it.

Another interesting generalization of contraction mappings was given by Sehgal [1969].

DEFINITION 3.3.11.   Let (X, d) be a complete metric space and $f : X \to X$ be a continuous mapping. The mapping f is called a local power contraction mapping if there exists a constant $K < 1$ and, for each $x \in X$, there exists an integer $n = n(x)$ such that, for all $y \in X$,

$$d(f^n(x), f^n(y)) \leqq K d(x, y).$$

For this class of mappings we have the following result:

THEOREM 3.3.12.   *If f is a local power contraction mapping then there exists a unique fixed point of f.*

For the proof we need the following lemma:

LEMMA 3.3.13.   *If $f : X \to X$ is a local power contraction mapping then, for each $x \in X$, the number $r(x) = \sup d(x, f^n(x))$ is finite.*
    Proof.   For each $x \in X$ let

$$m(x) = \max \{d(x, f^k(x)), 1 \leqq k \leqq n(x)\}.$$

Now if n is an arbitrary integer then there exists $s > 0$ such that $s \cdot n(x) \leqq n \leqq (s + 1) \cdot n(x)$ and this gives,

$$
\begin{aligned}
d(x, f^n(x)) &\leqq d(f^{n(x)} {}_0 f^{n-m}(x), f^{n(x)}(x)) + \\
&\quad + d(f^{n(x)}(x), x) \\
&\leqq K d(f^{n-m}(x), x) + m(x) \\
&\leqq m(x) + K m(x) + K^2 m(x) + \\
&\quad + \cdots + K^s m(x) \\
&\leqq m(x)/(1 - K)
\end{aligned}
$$

and assertion is proved.

Proof of Theorem 3.3.12. Let x_0 be an arbitrary point in X and let $n_0 = n(x_0)$, $x_1 = f_0^n(x_0)$ and we define inductively a sequence of integers and a sequence of points $\{x_i\}$ in X as follows: $n_i = n(x_i)$ and $x_{i+1} = f^{n}i(x_i)$. First we prove that the sequence $\{x_i\}$ is a Cauchy sequence. For this we estimate $d(x_n, x_{n+1})$. We have,

$$
\begin{aligned}
d(x_n, x_{n+1}) &= d(f^n - 1 {}_0 f^n n(x_{n-1}), f^n n - 1(x_{n-1})) \\
&\leqq K d(f^n n(x_{n-1}), x_{n-1}) \\
&\leqq \cdots \leqq K^n d(f^n n(x_0), x_0)
\end{aligned}
$$

and this implies, obviously, for $n > m$, that

$$d(x_n, x_m) \leqq K^m \cdot 1/(1 - K)$$

which shows that $\{x_i\}$ is a Cauchy sequence. Let $u = \lim x_i$.

We show that u is a fixed point of f. Suppose that this is not true. Then we find a pair of closed neighbourhoods U and V such that $u \in U$ and $f(u) \in V$. Let d_0 be the distance

$$d_0 = \inf\{d(x, y), x \in U, y \in V\} > 0$$

and since f is continuous for n large $f(x_n) \in V$ and also $x_n \in U$.

Now

$$d(x_n, f(x_n)) = d(f^n - 1_0 f^n n(x_{n-1})), f^n n - 1(x_{n-1}))$$
$$\leqq Kd(f(x_{n-1}), x_{n-1}) \leqq \cdots \leqq K^n d(f(x_0), x_0)$$

and for n large this is a contradiction. Thus $f(u) = u$.

Since the uniqueness is obvious, the theorem is proved.

The following example shows that there exists a continuous mapping on a metric space satisfying the condition of the above theorem and no iteration of the mapping is a contraction mapping.

Example 3.3.14.   Let $X = [0,1]$ with the metric

$$d(x, y) = |x - y|$$

and we remark that X is of the form

$$X = \bigcup_1^\infty \{[1/2^n, 1/2^{n-1}]\} \bigcup \{0\}.$$

The mapping $f : X \to X$ is defined as follows:

$$f : [1/2^n, 1/2^{n-1}] \to [1/2^{n+1}, 1/2^n]$$

by the relation

$$f(x) = (n + 2)/(n + 3) \cdot (x - 1/2^n) + 1/2^n$$
$$\text{if } s \in [(3n + 5)/2^{n+1}(n + 2), 1/2^{n-1}]$$
$$= 1/2^{n+1} \quad \text{if } x \in [1/2^n, (3n + 5)/2^{n+1}(n + 2)]$$

and also $f(0) = 0$.

From the definition it follows that f is a nondecreasing continuous function on $[0, 1]$ with values in $[0, 1]$ and the unique fixed point of f is $x = 0$. We show that f satisfies Theorem 3.3.12.

Indeed, if $x \in [1/2^n, 1/2^{n-1}]$ and y is arbitrary, then we have – considering the cases $m \geq n$ and $m < n$ as well as $y \in [1/2^m, 1/2^{m-1}]$ – that f satisfies the inequality

$$|f(x) - f(y)| < (n+3)/(n+4) \cdot |x - y|$$

and thus, for $K = \frac{1}{2}$ for each $x \in X$, $x \in [1/2^n, 1/2^{n-1}]$, $n(x)$ may be $n+3$ and $n(0)$ may be any integer ≥ 1.

Now we show that any iteration of f is not a contraction mapping.

Indeed, if $K < 1$ and with N given we show that there exist x, y in X such that

$$|f^N(x) - f^N(y)| > K|x - y|.$$

Let $n > (NK/(1 - K) - 2)$. Since f is continuous on a compact interval it is uniformly continuous and this assertion is valid for all f^i.

This implies that there exists a $\delta > 0$ such that

$$|x - y| < \delta \to |f^i(x) - f^i(y)| < (n + N + 3)/(n + N + 2)2^{n + N + 1}$$

for $i = 1, 2, 3, \ldots, N$. Let $x = 1/2^{n-1}$ and let y be any point of $[1/2^n, 1/2^{n-1}]$ such that $0 < |x - y| < \delta$ and thus $f^i(x)$ and $f^i(y)$ are in

$$[(3(n + i) + 5)/2^{n + i + 1}(n + i + 2), 1/2^{n + i - 1}].$$

This gives us that

$$|f(x) - f(y)| = (n+2)/(n+3) \cdot |x - y|$$

$$|f^2(x) - f^2(y)| = (n+2)/(n+4) \cdot |x - y|$$

$$\text{---}$$

$$|f^N(x) - f^N(y)| = (n+2)/(n + N + 2) \cdot |x - y| > k|x - y|$$

and the assertion is proved.

Another interesting and useful generalization of the contraction mappings was proposed by E. Rakotch (1962) and is as follows:

DEFINITION 3.3.15. A mapping $T : M \to M$, where M is a complete metric space is called a Rakotch contraction if there exists a decreasing function $\alpha(t)$ with $\alpha(t) < 1$ and such that, for all $x, y \in M$,

$$d(Tx, Ty) < \alpha(d(x, y))d(x, y).$$

A more general class was considered by Boyd and Wong [1969] and is as follows:

DEFINITION 3.3.16. A mapping $T : M \to M$ is called contractive (in the sense of Boyd and Wong) if there exists a function on the range of the metric d such that:

1. $\varphi(t) < t,$

2. $d(Tx, Ty) < \varphi(d(x, y)),$

3. φ is upper semicontinuous from the right.

Remark 3.3.17. For $\varphi(t) = \alpha(t)t$ with $\alpha(t) < 1$ this is exactly the class of Rakotch. Later we give an example to show that the class of Boyd and Wong is strictly larger than the class of Rakotch.

Let M be a metric space and P be the range of d, the metric of M. Also we suppose that M is complete.

THEOREM 3.3.18. Let $T : M \to M$ satisfy the Boyd–Wong conditions for $t \in \bar{P} \backslash 0$. In this case T has a unique fixed point in M.

Proof. For any x fixed in M we define the sequence of real numbers $\{a_n\}$

$$a_n = d(T^n x, T^{n-1} x)$$

and it is clear that

$$a_{n+1} \leqq a_n$$

for all n. Thus the sequence is convergent and we let $a = \lim a_n$.

If $a > 0$ we obtain a contradiction. Indeed,

$$a_{n+1} \leqq \varphi(a_n)$$

and thus

$$a \leqq \varphi(a)$$

which is in contradiction with the property of φ. Thus $a_n \to 0$ and we prove now that $\{T^n_x\}$ is a Cauchy sequence. Suppose that this is not so. Then we find $\varepsilon > 0$ such that, for two sequences of integers $\{m(k)\}$ and $\{n(k)\}$, $m(k) \geqq n(k) \geqq k$, we have the relation

$$d_k = d(T^m x, T^n x) \geqq \varepsilon, \qquad k = 1, 2, 3, \ldots .$$

Of course we can assume that $d(T^m x, T^{m-1} x) \leqq \varepsilon$ and thus we have

$$d_k < d(T^m x, T^{m-1} x) + d(T^{m-1} x, T^n x) \leqq a_m + \varepsilon \leqq a_k + \varepsilon$$

which implies that

$$\lim_k d_k = 0.$$

Since

$$d_k = d(T^m x, T^n x) \leq d(T^m x, T^{m+1} x) + d(T^{m+1} x, T^{n+1} x) +$$
$$+ d(T^{n+1} x, T^n x)$$
$$\leq 2a_k + \varphi(d(T^m x, T^n x)) = 2a_k + \varphi(d_k)$$

and for $k \to \infty$ we obtain

$$\varepsilon \leq \varphi(\varepsilon)$$

which is a contradiction.

Thus $\{T^n x\}$ is a Cauchy sequence for any x and it is obvious that $\lim T^n x = x^*$ is a fixed point of T. Also it is clear that it is the unique point with this property.

Another generalization of the notion of contraction mapping is the following, which contains as particular cases many of the classes of mappings considered above:

DEFINITION 3.3.19.   Let (X, d) be a complete metric space and $T : X \to X$ be a continuous mapping. Then we say that T is convex of order 2 if there exist $a, b \in \mathbf{R}^+$ such that

1. $d(T^2 x, T^2 y) \leq ad(x, y) + bd(Tx, Ty)$,

2. $a + b < 1$.

This class of mappings can be further generalized as in the following definition:

DEFINITION 3.3.20.   A continuous mapping $T : X \to X$ is called a convex contraction of order n if there exist $a_0, \ldots, a_{n-1}$ in $\mathbf{R}^+$ such that

1. $d(T^n x, T^n y) \leq \sum_{0}^{n-1} a_i d(T^i x, T^i y)$

2. $\sum_{0}^{n-1} a_i < 1$.

For results about these classes of mappings see V. Istrățescu (1979 a).

The following generalization of the Contraction Principle was found by I. Ekeland (1972) and rediscovered independently by J. Caristi (1976). The proof of this generalization is that of A. Brønstedt (1976).

THEOREM 3.3.21.   *Let (X, d) be a complete metric space and $f : X \to X$ be a*

continuous mapping. Suppose that there exists a lower semicontinuous function
$\varphi : X \to \mathbf{R}^+$ *such that for all* $x \in X$,

$$d(x, f(x)) < \varphi(x) - \varphi(f(x)).$$

In this case f has a fixed point in X.

Before proving the theorem we remark that any contraction $T : X \to X$,
$d(Tx, Ty) < kd(x, y)$ satisfies the condition of the theorem. Indeed, in this
case we can choose φ so that

$$\varphi(x) = 1/(1 - k) d(x, f(x)).$$

Now we give the proof of Theorem 3.3.21. We define a partial order on X
by the following rule:

$$x \leq y \quad \text{iff} \quad d(x, y) \leq \varphi(x) - \varphi(y).$$

Let x_0 be an arbitrary but fixed point in X. We can apply Zorn's Lemma
to obtain a maximal totally ordered subset M in X containing x_0. Let
$\{x_\alpha\}_{\alpha \in I}$ where I is totally ordered and let

$$x_\alpha \leq x_\beta \quad \text{iff} \quad \alpha \leq \beta.$$

Since $(\varphi(x_\alpha))$ is a decreasing set in $\mathbf{R}^+$ we choose $r > 0$ such that

$$\varphi(x_\alpha) \to r, \quad \alpha \uparrow .$$

Let $\varepsilon > 0$ and $\alpha_0 \in I$ such that

$$\alpha \geq \alpha_0 \to r \leq \varphi(x_\alpha) \leq r + \varepsilon.$$

Let $\beta \geq \alpha \geq \alpha_0$ and thus

$$d(x_\alpha, x_\beta) \leq \varphi(x_\alpha) - \varphi(x_\beta) \leq \varepsilon$$

which gives us that (x_α) is a Cauchy net in X. Since X is complete we can
consider that $x = \lim x_\alpha$. But φ is a lower semicontinuous function and
thus we have that $\varphi(x) \leq r$. For $\beta \uparrow$ we obtain, from the above inequality,

$$d(x_\alpha, x) \leq \varphi(x_\alpha) - \varphi(x)$$

which gives us

$$x_\alpha \leq x.$$

Now M is maximal and thus $x \in M$. But we have

$$d(x, f(x)) \leq \varphi(x) - \varphi(f(x)),$$

i.e.

$$x_\alpha \leqq x \leqq f(x),$$

and again by maximality $f(x) \leqq x$. Thus $x = f(x)$.

For interesting applications of this theorem we refer to Caristi (1976), and Kirk (1976).

In Definition 3.3.4, a localization of the contraction mapping condition was defined (Rakotch, 1962). We define now, following R. D. Holmes (1976), another localization of the contraction mapping condition.

DEFINITION 3.3.22.   Let (X, d) be a complete metric space and $f : X \to X$ be a mapping. The mapping f is said to be a local radial contraction if there exists $k \in (0, 1)$ with the following property: for each $x \in X$ there exists a neighbourhood $V_x \ni x$ such that

$$d(f(x), f(y)) \leqq k d(x, y)$$

for all $y \in V_x$.

Equivalently, f is a local radical contraction if, for each $x \in X$, there exists $\varepsilon > 0$ such that for any $y \in X$, $d(x, y) < \varepsilon_x$ $d(f(x), f(y)) \leqq k \cdot d(x, y)$.

For results concerning fixed points for such a class of mappings we refer to R. D. Holmes, and T. Hu and W. A. Kirk (1979).

We consider now a class of mappings related to the contraction mappings as well as to the local power contraction class of mappings introduced by Sehgal (see Definition 3.3.11).

Our class of mappings is defined as follows :

DEFINITION 3.3.23   (V. Istrăţescu, 1980). Let (X, d) be a complete metric space and $f : X \to X$. The mapping f is said to have the local power diminishing diameter property if there exists $k \in (0, 1)$ such that, for any bounded set M in X such that $f(M) \subseteq M$, there exists an integer $n = n(M)$ such that $d(f^n(M)) \leqq k d(M)$.

LEMMA 3.3.24.   *Every local power contraction mapping has the local power diminishing diameter property.*

Proof. Let $h : X \to X$ be a local power contraction mapping and thus there exists $k \in (0, 1)$ such that, for each $x \in X$, there exists an integer $n = n(x)$ such that for all $y \in X$, $d(f^n(x), f^n(y)) \leqq k d(x, y)$. Let us now consider M to be a bounded set in X such that $f(M) \subseteq M$. Let x be arbitrary in M and define the following sequence of points of M :

$$x_1 = f^{n_1}(x), \quad n_1 = n(x),$$
$$x_2 = f^{n_2}(x_1), \quad n_2 = n(x_1), \ldots,$$
$$x_m = f^{n_m}(x_{m-1}), \quad n_m = n(x_{m-1}), \ldots.$$

Then we have, for z arbitrary in M and Y arbitrary in M,

$$d(f^{n_1 + \cdots + n_m}(x), f^{n_1 + \cdots + n_m}(z))$$
$$\leq d(f^{n_1 + \cdots + n_m}(x), x_m) +$$
$$\quad + d(x_m, f^{n_1 + \cdots + n_m}(z))$$
$$\leq k(d(f^{n_1 + \cdots + n_{m-1}}(x), x_{m-1}) +$$
$$\quad + d(x_{m-1}, f^{n_1 + \cdots + n_{m-1}}(z))$$
$$\leq \cdots \leq k^m(d(y, x) + d(x, z))$$
$$\leq 2 \cdot k^m d(M)$$

and if we choose m such that $2 \cdot k^m < 1$ we obtain the assertion of the Lemma.

We suppose now that the metric space X is bounded, i.e. its diameter is bounded. Then we have

THEOREM 3.3.25. *If $f: X \to X$ is a mapping with the local power diminishing diameter property, then f has a unique fixed point in X.*
Proof. It is obvious that the set of fixed points of f contains at most one point and thus we prove only the existence of a fixed point.
To this end we consider the following sequence of sets:

$$\bar{X} \supseteq \overline{f(X)} \supseteq \overline{f^2(X)} \supseteq \overline{f^3(X)} \supseteq \cdots \supseteq \overline{f^n(X)} \supseteq \overline{f^{n+1}(X)} \supseteq \cdots.$$

and we show that this sequence has the property that $d(\overline{f^n(X)}) \to 0$.
Indeed, we consider the following sequence of sets:

$$X_1 = f^{n_1}(X), \quad n_1 = n(X),$$
$$X_2 = f^{n_2}(X_1), \quad n_2 = n(X_1), \ldots,$$

and we remark that each of the X_i has the closure in the sequence considered above. But we have $d(X_m) \leq k^m d(X)$. Since $k \in (0, 1)$ we get that $d(\bar{X}_m) \to 0$. This implies that our assertion about the diameters of $\overline{f^n(X)}$ is true. Then $\bigcap \overline{f^n(X)} = (x_0)$. But from the above relation we obtain that, for any $x \in X$, $(f^n(x))$ converges to the point x_0. This implies that the set $(x_0, f(x_0), f^2(x_0), \ldots.)$ is closed and invariant for f. Since the image of this set under f is exactly itself we see that it reduces to the point x_0 which is a fixed point for f.

3.4. HILBERT'S PROJECTIVE METRIC AND MAPPINGS OF CONTRACTIVE TYPE

In his studies on foundations of geometry D. Hilbert (1895) introduced a metric which is interesting in its own right but has also applications to analysis, as was proved by G. Birkhoff (1957).

For example the famous theorem of Perron–Frobenius on non-negative matrices as well as Jentzch's theorem in the theory of integral operators can be proved using the contraction principle with respect to the Hilbert metric defined on a Banach space.

In what follows, following mainly the paper of Bushell (1973), we give an account of the Hilbert metric and some of its applications.

DEFINITION 3.4.1.   A set K in a Banach space E is called a cone if the following properties hold:

1. $K + K \subset K$,

2. $\lambda K \subset K$ for all $\lambda > 0$,

3. $K \cap (-K) = \{0\}$

and we call K solid if $\overset{\circ}{K}$ is nonempty.

In what follows we consider only solid cones.

If K is a cone in a Banach space E then we can define an 'order relation' as follows:
$$x \geqq y \quad \text{if } x - y \in K$$
and
$$x > y \quad \text{if } x - y \in \overset{\circ}{K}.$$

A cone K is Archimedean if $nx \leqq y$ for all $n = 1, 2, 3, \ldots$ implies that $x \leqq 0$. For a cone K we set $K^+ = K \backslash \{0\}$.

DEFINITION 3.4.2.   Let K be a cone and $x, y \in K^+$. The numbers $m(x, y)$ and $M(x, y)$ are defined as follows:

1. $m(x, y) = \sup \{\alpha, \alpha y \leqq x\}$

2. $M(x, y) = \inf \{\beta, x \leqq \beta y\}$

and $M(x, y) = \infty$ if the set is empty.

The following proposition gives the relation between these numbers.

PROPOSITION 3.4.3. *For any* $x, y \in K^+$ *the following relations hold*:

$$m(x, y)y \leq x \leq M(x, y)y$$

Proof. First we remark that in any closed cone the partial order defined above is Archimedean. Now, from the definition of $M(x, y)$ it follows that, for each n,

$$x \leq [M(x, y) + 1/n]y,$$

and from this we obtain

$$n[x - M(x, y)y] \leq y,$$

which gives further;

$$x - M(x, y) \leq 0.$$

From the definition of $m(x, y)$ and again using the fact that the order is Archimedean we obtain the relation

$$x - m(x, y)y \geq 0.$$

Thus, for all $x, y \in K^+$,

$$0 \leq m(x, y) \leq M(x, y) \leq \infty.$$

Indeed, in the contrary case we have for some $x, y \in K^+$

$$m(x, y) > M(x, y).$$

In this case we obtain

$$[M(x, y) - m(x, y)]y = [M(x, y)y - x] + [x - m(x, y)y] \geq 0$$

and

$$-[M(x, y) - m(x, y)]y \geq 0,$$

which gives us that

$$[M(x, y) - m(x, y)]y = 0$$

and, since

$$M(x, y) - m(x, y) \geq 0,$$

$y = 0$, which is a contradiction since $y \in K^+$.

Now the definition of the Hilbert metric is given as follows:

DEFINITION 3.4.4.   For any $x, y \in K^+$ the Hilbert metric $d(x, y)$ is defined
as

$$d(x, y) = \log [M(x, y)/m(x, y)].$$

The properties which connect this metric with the Banach space are
given in

THEOREM 3.4.5.   *On $\overset{\circ}{K}, d(,)$ is a pseodometric, and $d(,)$ is a metric on
$\overset{\circ}{K} \cap S(0, 1)$.*
   Proof. Since $d(x, y) > M(x, y)$ it follows that $d(x, y) > 0$. Also $d(x, y) = 0$
iff $x = y\lambda$ for some $\lambda > 0$. We prove now that for all $x, y \in \overset{\circ}{K} d(x, y)$ is finite.
Indeed, choose $\varepsilon > 0$ such that $S(x, \varepsilon) \subset K, S(y, \varepsilon) \subset K$.
   In this case we have

$$x - (\varepsilon/\|y\|)y > 0$$

$$y - (\varepsilon/\|x\|)x > 0$$

which implies that

$$0 < (\varepsilon/\|y\|) \leqq m(x, y) \leqq M(x, y) \leqq (\|x\|/\varepsilon) < \infty.$$

Since

$$x \leqq M(x, z)y \leqq M(x, y)M(y, z)z$$

then

$$M(x, z) \leqq M(x, y)M(y, z)$$

and similarly

$$m(x, z) \geqq m(x, y)m(y, z).$$

From these it follows that

$$d(x, z) \leqq d(x, y) + d(y, z).$$

The relation

$$d(x, y) = d(y, x)$$

is obvious and thus the assertion is proved.

Also, it is easy to see that the following assertion is true:

PROPOSITION 3.4.6. *For any* $x, y \in \overset{\circ}{K}$ *and* $a, b \in \mathbf{R}^+$

$$d(ax, by) = d(x, y).$$

We give now an example.

Example 3.4.7. Given $\mathbf{R}^n$ *consider the set*

$$K = \{x = (x_1, x_2, \ldots, x_n), x_i \geq 0\}.$$

Then it is obvious that this is a solid cone and also closed, and, thus the above results can be applied.

Now if $y = (y_1, \ldots, y_n)$ is also in K then

$$m(x, y) = \max_i \{x_i/y_i\}$$

$$M(x, y) = \inf_i \{x_i/y_i\}$$

and we obtain the Hilbert metric which has the form

$$d(x, y) = \log \max_{i,j} \{x_i y_j / y_i x_j\}.$$

Other properties of the numbers $m(x, y)$ and $M(x, y)$ are given in

PROPOSITION 3.4.8. *For any* $x, y \in \overset{\circ}{K}$ *and* $a, b > 0$,

1. $M(ax + by, y) = aM(x, y) + b,$

2. $m(ax + by, y) = am(x, y) + b,$

3. $m(x, y) \cdot M(x, y) = 1.$

Proof. We have, from the definition of $m(x, y)$ and $M(x, y)$,

$$m(x, y) \leq M(x, y) \quad \text{and} \quad x \geq 0$$

which implies that

$$0 \leq m(x, y) \leq M(x, y)$$

and also

$$M(-x, y) = \inf\{a, -x < ay\}$$
$$= \inf\{a, -ay < x\}$$
$$= -\sup\{b, by < x\}$$

or

$$M((-x, y)) = -m(x, y), \qquad m(-x, y) = -M(x, y).$$

Also

$$M((x + y), z) \leqq M(x, z) + M(z, y)$$
$$m((x + y), z) \geqq m(x, z) + m(z, y)$$

and from these we conclude the above assertions.

We now give some properties of the mappings defined on the cone K.

DEFINITION 3.4.9. A map $T : K \to K$ is called non-negative and, if $T : \mathring{K} \to \mathring{K}$, we say that T is positive.

An important class of mappings is considered in

DEFINITION 3.4.10. A positive mapping T is called positive homogeneous of degree p in $\mathring{K}$ if, for all $x \in \mathring{K}$ and $a > 0$,

$$T(ax) = a^p T(x).$$

For any positive mapping T we consider the so called 'projective diameter' which is defined as follows:

DEFINITION 3.4.11. If $T : \mathring{K} \to \mathring{K}$ then the projective diameter $\Delta(T)$ of T is the number

$$\Delta(T) = \sup \{d(Tx, Ty), x, y \in \mathring{K}\}.$$

DEFINITION 3.4.12. If T is positive then the 'contraction ratio $k(T)$' is defined as

$$k(T) = \inf \{a, d(Tx, Ty) \leqq a d(x, y), x, y \in \mathring{K}\}$$

and the oscillation of x and y is the number

$$\operatorname{osc}(x, y) = M(x, y) - m(x, y).$$

The oscillation of T when T is positive, $\operatorname{osc} T$, is defined as

$$\operatorname{osc} T = \inf \{a, \operatorname{osc}(Tx, Ty) \leqq a \operatorname{osc}(x, y), x, y \in \mathring{K}\}.$$

For applications the following class of mappings is very important:

DEFINITION 3.4.13. A mapping $f : E \to E$ is called monotone increasing if $x \leqq y$ implies $f(x) \leqq f(y)$.

THEOREM 3.4.14. *If $T : E \to E$ is monotone increasing and positive homogeneous of degree p on $\mathring{K}$ then $k(T) \leqq p$.*
Proof. Since by Proposition 3.4.3.

$$m(x, y)y \leqq x \leqq M(x, y)y$$

and T being monotone increasing and homogeneous of degree p,

$$m(x, y)^p Ty \leq Tx \leq M(x, y)^p Ty$$

and thus

$$m(x, y)^p \leq m(Tx, Ty) \leq M(Tx, Ty) \leq (M(x, y)^p)$$

which obviously implies the assertion of the theorem.

The following theorem of Birkhoff is of fundamental importance for applications.

THEOREM 3.4.15. *If* $T : E \rightarrow E$ *is a positive linear mapping then*

$$k(T) = \tanh \tfrac{1}{4}\Delta(T).$$

For the proof we refer to the paper by G. Birkhoff (1957) and to that of P. Bushell (1972) for a simpler proof.

For applications the following result is useful.

THEOREM 3.4.16. *Let* $f : E \rightarrow E$ *satisfy one of the following conditions*:

1. F *is monotone increasing and positive homogeneous of degree* $p \in (0, 1)$ *in* $\mathring{K}$,

2. f *is a positive linear mapping with finite projective diameter.*

Then, in the case when 1 *is verified and the metric space* $\mathring{K} \cap S(0, 1)$ *is complete under the Hilbert metric, there exists a unique* $x \in \mathring{K}$ *such that* $Tx = x$; *in the case when* 2 *is verified then there exists a unique positive element* $x \tilde{} \in \mathring{K} \cap S(0, 1)$ *such that* $Tx = x$.

Proof. We consider the map

$$T\tilde{}x = Tx/\|Tx\|$$

which is a composition of a strict contraction with an isometry.

From the Banach theorem (the contraction principle) it follows that there exists an $x\tilde{} \in E$ such that $T\tilde{}x\tilde{} = x\tilde{}$. In this case the element $x = \|Tx\tilde{}\|^{1/(1-p)}x\tilde{}$ satisfies the assertions of the theorem.

It is now clear that an interesting problem is as follows: if E is a Banach space as given above, then it is possible to construct new Hilbert projective metrics? The following result give an answer to this problem by showing a construction of a family of Hilbert projective metrics and pointing out that the Hilbert projective metric gives the best contraction ratio.

THEOREM 3.4.17.   *Let F be a differentiable real-valued function in* $[0, \infty)$ *such that*:

1. $F(t) \geq 0$ *for* $t \in (0, \infty)$ *and* $F(0) = 0$,

2. $F'(t) \geq 0$ *for* $t \in (0, \infty)$ *and* $F'(0) > 0$,

3. $F(t + s) \leq F(t)F(s)$.

In this case

$$d_F(x, y) = F(\log [M(x, y)/m(x, y)])$$

is a pseudometric on $\mathring{K}$ *and if* K_F *is the associated contraction ratio then*

$$K_F^{(A)} \geq K(A)$$

where $K(A)$ *is the contraction ratio for the Hilbert projective metric.*

   Proof. It is not difficult to see, using the properties of $m(x, y)$ and $M(x, y)$ that d_F is a pseudometric on $\mathring{K}$. From the definition of the contraction ratio, if $x, y \in \mathring{K}$ and $\varepsilon > 0$ then we have

$$K_F(A) \geq \frac{F(\log ((\varepsilon M(Ax, Ay) + 1)/\varepsilon m(Ax, Ay) + 1))}{F(\log ((\varepsilon M(x, y) + 1)/\varepsilon m(x, y) + 1))}$$

which for $\varepsilon \to 0$, by Hospital's rule, gives

$$K_F(A) \geq (\text{osc}(Ax/Ay)/\text{osc}(x/y))$$

and thus

$$K_F(A) = N(A) = k(A).$$

THEOREM 3.4.18.   *Let* $X = \mathbf{R}^n$ *and* $K = (x \in \mathbf{R}^n, x = (x_1, x_2, \ldots, x_n), x_i \geq 0,$ $i = 1, 2, \ldots, n)$. *Then* E *is a complete metric space.*

   Proof. Let x, y be in E and then clearly we have

$$m(x/y) \leq 1 \leq M(x/y)$$

which implies that

$$\|x - y\| = \left(\sum_1^n (x_i - y_i)^2\right)^{1/2} \leq \left(\sum_1^n (M(x/y) - m(x/y))^2 y_i^2\right)^{1/2}$$

$$\leq (M(x/y) - m(x/y)) = (\exp(d(x, y)) - 1)m(x/y).$$

But

$$M(x/y) = \max(1 + (x_i - y_i)/y_i, 1 \leq i \leq n) \leq 1 + \|x - y\|/m(y/u)$$

where $u = (1, 1, \ldots, 1) \in \mathbf{R}^n$.

In a similar way if

$$\|x - y\| < m(y/u)$$

we get

$$m(x/y) \geq 1 - \|x - y\|/m(y/u)$$

and thus if $\|x - y\| < m(y/u)$ we obtain

$$\|x - y\| \geq m(y/u) \tanh\left(\tfrac{1}{2}d(x, y)\right)$$

and the completeness follows.

In a similar way we can prove the completeness of E in the case when $X = C_{[0,1]}$ and $K = (f, f \in C_{[0,1]}, f \geq 0)$. Here u is the function $u(t) = 1$ for all $t \in [0, 1]$. Bushell (1973) has proved the completeness of E when $X = L(\mathbf{R}^n)$ and K is the cone of all real positive semidefinite symmetric matrices. For other results concerning these problems we refer the reader to the papers by Bushell and by Birkhoff.

We now give some applications. Suppose that $\mathbf{A}$ is an $n \times n$ matrix with non-negative entries and suppose also that $\mathbf{A}$ is undecomposable, i.e. there does not exist a permutation matrix $\mathbf{P}$ such that

$$\mathbf{PAP'} = \begin{pmatrix} \mathbf{B} & \mathbf{0} \\ \mathbf{C} & \mathbf{D} \end{pmatrix}$$

where $\mathbf{B}, \mathbf{D}$ are square submatrices. In this case there exists an integer $m > 1$ such that $\mathbf{A}^m = \mathbf{B}$ has positive entries. Also $\mathbf{A}^m$ is a positive linear mapping in the interior of the positive cone in $\mathbf{T}^n$ and

$$\Delta(\mathbf{A}^m) = \max\{\log[b_{ij}b_{pq}/b_{iq}b_{pj}]\} < \infty$$

and from the above results we have that $\mathbf{A}$ has a unique positive eigenvector.

Consider now a matrix $\mathbf{A} = \{a_{ij}\}, 1 \leq i, j \leq n$ with non-negative entries and at least one positive element in each row. Then, for any $p \in (0, 1)$, there exists a unique positive solution $x = (x_1, \ldots, x_n)$ of

$$x_i = \sum_j a_{ij} x_j^p.$$

Indeed we can apply the above results to

$$F(x) = \left\{\sum_{j=1}^{n} a_{ij} x_j^p\right\}$$

and the assertion follows.

In the space C_I where $I = [0, 1]$ we consider the operator

$$Tx(t) = \int_0^1 K(s, t)x(s)\,ds$$

with the kernel K a positive and continuous function on I^2 which satisfies the inequality

$$0 < a < K(s, t) < b.$$

Then there exists a unique positive eigenfunction for T.
For more details we refer to publications by Birkhoff and by Bushell.

Remark 3.4.19.   It appears to be of considerable interest to extend the Hilbert projective metric to other classes of mappings which were defined in the precedings sections.

3.5. APPROXIMATE ITERATION

Suppose that $T: X \to X$ is a contraction mapping defined on a complete metric space and

$$d(Tx, Ty) \leq kd(x, y)$$

for all $x, y \in X$ and $k \in (0, 1)$. If x_0 is an arbitrary element in X and x^* is the fixed point of T then

$$d(x_k, x^*) \leq k/(1 - k)d(x_k, x_{k-1})$$

where $x_{k+1} = Tx_k$ for $k = 0, 1, 2, \ldots$.

Indeed,

$$
\begin{aligned}
d(x_k, x^*) < d(x_k, Tx^*) &\leq (x_k, Tx_k) + d(Tx_k, Tx^*) \\
&= d(Tx_{k-1}, Tx^*) + d(Tx_k, Tx^*) \\
&\leq kd(x_{k-1}, x^*) + kd(x_k, x^*)
\end{aligned}
$$

which gives further

$$(1 - k)d(x_k, x^*) \leq kd(x_{k-1}, x_k)$$

and this is equivalent to our assertion.

Thus, if T is a contraction mapping for which we know the number k or a majorant of it, then we can compute the actual error $d(x_m, x^*)$ after m steps in terms of the last step.

However, in practice, because of rounding or discretization in the function T, an approximate sequence is used in place of $\{x_m\}$. Thus the problem becomes that of relating these sequences. The basic result in this direction was obtained by A. Ostrovski (1964) and is as follows:

THEOREM 3.5.1. *Let* $T : X \to X$ *be a contraction mapping of a complete metric space and*

$$d(Tx, Ty) \leq kd(x, y)$$

for all $x, y \in X$ *and* $k \in (0, 1)$. *Let* x_0 *be an arbitrary point in* X *and* $x_{k+1} = Tx_k$ *for* $k = 0, 1, 2, 3, \ldots$ *and* $\{y_m\}$ *be a sequence in* X *and*

$$\varepsilon_m = d(y_{m+1}, Ty_m).$$

Then

$$d(x^*, y_{m+1}) \leq (1 - k)^{-1}[\varepsilon_m + kd(y_{m+1}, y_m)] = a_m,$$

$$d(x^*, y_{m+1}) \leq d(x_{m+1}, y_{m+1}) + k^m d(x_0, y_0) + \sum_{i=0}^{m} k^{m-i}\varepsilon_i = b_m$$

and

$$\lim y_m = x^*$$

iff $\lim \varepsilon_m = 0$.

Proof. The first assertion follows from

$$d(y_{m+1}, x^*) \leq d(y_{m+1}, Ty_m) + d(Ty_m, Ty_{m+1}) +$$
$$+ d(Ty_{m+1}, x^*)$$
$$\leq \varepsilon_m + kd(y_m, y_{m+1}) + kd(y_{m+1}, x^*),$$

and the next relation follows from

$$d(x_{m+1}, y_{m+1}) \leq d(Tx_m, Ty_m) + d(Ty_m, y_{m+1})$$
$$\leq kd(x_m, y_m) + \varepsilon_m$$
$$\leq \sum k^{m-1}\varepsilon_i + k^{m+1}d(x_0, y_0)$$

and

$$d(y_{m+1}, x^*) \leq d(y_{m+1}, x_{m+1}) + d(x_{m+1}, x^*).$$

Suppose now that $\lim \varepsilon_m = 0$. Let $\varepsilon > 0$ and, for $m \geq N_\varepsilon$, we have $\varepsilon_m \leq \varepsilon$. In this case we obtain, for $r_m = \sum k^{m-i} \varepsilon_i$,

$$r_m \leqq k^{m-N_\varepsilon} r_{N_\varepsilon} + \sum_{j=N_\varepsilon+1} k^{m-i} \varepsilon_i \leqq k^{m-N_\varepsilon} r_{N_\varepsilon} + \varepsilon(k^{N_\varepsilon+1}/(1-k))$$

which implies that

$$\lim r_m = 0$$

and this gives

$$\lim y_m = x^*.$$

Conversely, if $\lim y_m = x^*$ then we have

$$0 \leqq \varepsilon_m \leqq d(Ty_m, y_{m+1}) \leqq d(Ty_m, x^*) + d(x^*, y_{m+1})$$
$$\leqq kd(y_m, x^*) + d(x^*, y_{m+1})$$

which implies that

$$\lim \varepsilon_m = 0.$$

Remark 3.5.2.   The theorem obviously admits extensions to more general classes of mappings. In particular it is true for mappings having as iterate a contraction mapping and of course, the mapping is continuous.

Another useful result in the computation of fixed points for contraction mappings is the following theorem of Urabe (1956).

THEOREM 3.5.3.   *Suppose that D is a subset in $\mathbf{R}^n$, that D is closed, and that $T : D \to D$ is a contraction mapping. Suppose that there exists a mapping $S : D \to D$ such that, for all $x \in D$,*

$$\|Tx - Sx\| \leqq \varepsilon$$

and also for some $y_0 \in D$,

$$S = \{y, \|y - Sy_0\| \leqq r\}$$

is in D and where

$$r = [1/(1-k)](\|k(Sy_0 - y_0)\| + 2\varepsilon).$$

Then the sequence

$$y_{k+1} = Sy_k, \qquad k = 0, 1, 2, \ldots$$

is in S and

$$\|y_{m+1} - x^*\| \leqq k/(1-k) \|y_m - y_{m+1}\| + \varepsilon/(1-k)$$

where k is the constant in the definition of the contraction and x^* is its fixed point.

Proof. The assertion follows from the relation

$$\|Sx - Sy\| \leq \|Sx - Tx\| + \|Tx - Ty\| + \|Ty - Sy\|$$
$$\leq k\|x - y\| + 2\varepsilon.$$

3.6. A CONVERSE OF THE CONTRACTION PRINCIPLE

Various generalizations of the contraction principle replace the global hypothesis – that the mapping is a contraction – with various local hypotheses. It is possible to show that, after a suitable change of the metric, the mapping is actually a contraction mapping.

The first result of this type seems to be that of C. Bessaga (1959) and his result is as follows:

THEOREM 3.6.1. *Suppose that S is an arbitrary abstract set and that $f : S \to S$ is a mapping with the property that each iteration f^n has a unique fixed point in S. Then, for each $s \in (0, 1)$, there exists a metric d_s on S such that (S, d_s) is a complete metric space and f satisfies the inequality*

$$d_s(f(x), f(y)) \leq s d_s(x, y).$$

Another interesting result was obtained by L. Janos (1967) and is as follows:

THEOREM 3.6.2. *Suppose that (X, d) is a compact metric space and $f : X \to X$ is a continuous mapping with the property that*

$$\bigcap_0^\infty f^n(X) = \{x_0\}$$

then for each $s \in (0, 1)$ there exists a metric d_s such that (X, d_s) is a compact space and f is a contraction mapping with respect to d_s.

Generalizations of this result as well as of related results were obtained by Meyers (1970a, 1970b), Leader (1977), Rosenholtz (1976), etc.

In what follows we give some of these results. Also we note that a similar problem for the case of linear operators on Banach spaces was treated in Section 3.2.

First we give the proof of Theorem 3.6.2. since it contains the idea which has been used to prove more general results.

The proof is presented in a series of Lemmas.

LEMMA 3.6.3.   *If f is a mapping as in Theorem 3.6.2. then there exists a new metric $d*$ such that*

1.   *D and $d*$ are equivalent,*

2.   *$d*(f(x), f(y)) \leq d*(x, y)$ for all $x, y \in X$.*

We recall that two metrics are called equivalent if they generate the same topology.

Proof. Let us define the metric $d*$ by

$$d*(x, y) = \max \{d(f^n(x), f^n(y)), \quad n = 0, 1, \dots \}$$

and it is obvious that this is a metric and, of course, that it satisfies condition 2.

Since $d*(x, y) > d(x, y)$ it follows that any Cauchy sequence with respect to $d*$ is a Cauchy sequence with respect to d. We show now that any Cauchy sequence with respect to d is also a Cauchy sequence with respect to $d*$. Consider (x_n) and, since X is compact, we can suppose that $x_n^d \to x$. We now show that $x_n^{d*} \to x$. Suppose that this is not so. Then there exists an $\varepsilon > 0$ such that $d*(x_{n_k}, x) \geq \varepsilon$ and we can suppose, without loss of generality, that this is just the sequence $\{x_n\}$. We define now a sequence $\{k_n\}$ of integers such that

$$d*(x_n, x) = d(f^k n(x_n), f^k n(x))$$

and if $\{k_n\}$ is bounded then there exists some k which is repeated infinitely often and thus, using the continuity of f in the topology generated by d

$$d(f^k(x_n), f^k(x)) \geq \varepsilon,$$

which is a contradiction.

If $\{k_n\}$ is unbounded then this contradicts the property

$$\bigcap f^n(X) = \{x_0\}$$

and the Lemma is proved.

LEMMA 3.6.4.   *If f is as in Theorem 3.6.2. then for each $s \in (0, 1)$ there exists a metric d_s equivalent to d possessing the property that*

$$d_s(f(x), f(y)) \leq s d_s(x, y).$$

Proof. We consider the sequence of sets

$$H_0 = X, \qquad H_1 = f(X), \dots, H_n = f^n(X), \dots$$

and define the functions

$$n(x) = \max \{n, x \in H_n\}$$

and with this the function

$$n(x, y) = \min \{n(x), n(y)\}$$

and obviously we have the relation

$$n(f(x), f(y)) \geq n(x, y) + 1.$$

We then define

$$d_s(x, y) = s^{n(x, y)} d^*(x, y)$$

and it is easy to see that this satisfies the property

$$d_s(f(x), f(y)) < s d_s(x, y)$$

and it is possible that this is not a metric.

We can easily modify this to obtain a metric, however, as follows:
For any $x, y \in X$ we consider a chain

$$C_{x,y} = \{x = x_0, x_1, \ldots, x_n = y\}$$

and put

$$D_s(C_{x,y}) = \sum d_s(x_{i-1}, x_i).$$

The metric which satisfies all of our requirements is defined as follows:

$$d_s^*(x, y) = \inf \{D_s(C_{x,y})\}$$

where the inf is taken over all possible chains.

We show now that this is a metric. Clearly

1. $\qquad d_s^*(x, y) = d_s^*(y, x)$

2. $\qquad d_s^*(x, x) = 0$

and the triangle inequality follows from the fact that the union of chains $C_{x,y}$ with chains $C_{y,z}$ is a subset of the set of chains $X_{x,z}$.

We show now that, for all $x \neq y$, we have $d_s^*(x, y) > 0$.

We may assume, without loss of generality, that $n(x) < n(y)$.

Now if $y = x_0$ then

$$d_s^*(x, x_0) \geq s^{n(y)} \min \{d^*(x, y) d^*(y, H_{n(y)+1})\}$$

which is positive. If $y \neq x_0$ then every chain $C_{x, y}$ lies entirely in the set

$H_{n(y)+1}$ or it does not, hence

$$d_s^*(x, y) \geq s^{n(y)} \min \{d^*(x, y), d^*(y, H_{n(y)+1})\}$$

and the assertion that d_s^* is a metric is proved.

We now prove that these metrics are equivalent and we remark that f is a contraction mapping with respect to d_s^*.

From the definition of d_s^* it is clear that

$$d_s^* \leq d_s \leq d^*$$

and for the assertion it is sufficient to show that every Cauchy sequence $\{x_n\}$ with respect to d_s^* is a Cauchy sequence with respect to d^*.

Since X is a compact space with respect to d^* we can choose a subsequence, which we suppose to be $\{x_n\}$, convergent to a point y. We note that $x_n \to x$ with respect to d_s^*.

Since it is clear that $x_n \to y$ in d_s^* we have a contradiction and the assertion is proved.

Since

$$D_s(f(C_{x,y})) \leq s D_s(C_{x,y})$$

this gives us, by taking inf, that f is contraction mapping with respect to d_s^*.

From this theorem we obtain the following result:

THEOREM 3.6.5.   *Let (X, d) be a compact metric space and $F : X \to X$ be a continuous mapping satisfying the property that $\{x_0\}$ is the only non-empty invariant set in X. Then there exists an equivalent metric d^* on X such that (X, d^*) is a compact space and f is a contraction with respect to d^*.*

Proof. First we remark that the sequence of compact sets $(f^n(X))$ is decreasing and since X is compact it has a non-empty intersection. Let $M = \bigcap f^n(X)$. By Theorem 3.5.2. it suffices to prove that $f(M) = M$.

It is obvious that $f(M) \subseteq M$ and we prove the converse inclusion. Let $x \in M$ be arbitrary and consider the sequence of sets

$$G_n = f^{-1}(x) \cap f^n(X)$$

which are non-empty and compact. Their intersection is also non-empty,

$$\bigcap G_n = f^{-1}(x) \cap M$$

and this gives us that $x \in f(M)$.

The next theorem shows a new way to obtain Theorem 3.6.5.

THEOREM 3.6.6. *If (X, d) is a complete metric space and ε-chainable then for any $F : X \to X$ which is an (ε, λ)-uniformly locally contractive mapping there exists an equivalent metric D^* on X such that f is a contraction mapping with respect to this metric.*

Proof. Clearly f is a continuous mapping. For any $x \in X$ and $y \in X$ we define

$$d_\lambda(x, y) = \inf \{D(C_{x,y})\}$$

where $C_{x,y}$ is an ε-chain. It is not difficult to show that this is a metric which is equivalent to d and with respect to this metric f is a contraction.

We recall that $f : X \to X$ is called locally contractive if for each point $x \in X$ there exists a neighbourhood $x \in U$ so that for any point x, z of U

$$d(f(x), f(z)) < d(x, z).$$

THEOREM 3.6.7. *If X is a compact and connected metric space and $F : X \to X$ is a locally contractive mapping then there exists an equivalent metric on X, say d^* such that f with respect to this metric is a contraction mapping.*

Proof. For any x and y in X we define the number

$$d^*(x, y) = d(x, y) + d(f(x), f(y)) + \cdots + d(f^n(x), f^n(y)) + \cdots$$

which, as is easy to see, is well defined since the series converges.

It is also obvious that this generates a metric.

We show now that it is equivalent to d. For this it is sufficient to show that the identity $I : (X, d) \to (X, d^*)$
is a homeomorphism.

Since (X, d) is compact it is clear that it suffices to show continuity. Let $\varepsilon > 0$ and, if $d(x, y) < a$, where

$$a = \min \{k, (1 - M)\varepsilon\},$$

this implies that $d(x, y) < k$. Here we recall that f is said to be a local contraction if $d(x, y) < a$, which implies that $d(f(x), f(y)) < M d(x, y)$ and $M < 1$.

Further we obtain

$$\begin{aligned} d^*(x, y) &= d(x, y) + d(f(x), f(y)) + \cdots + \\ &\leq d(x, y) + M d(x, y) + M^2 d(x, y) + \cdots + \\ &= (1 - M)^{-1} d(x, y) \leq \varepsilon. \end{aligned}$$

Now it is easy to see that if $f : X \to X$ is a local contractive mapping with respect to the metric d then it is also locally contractive with respect to $d*$ with the same constants.

Also it is clear from the definition of $d*$ that f is contractive with respect to $d*$. We show now that f is a contraction mapping with respect to $d*$. Indeed, suppose that this is not so.

In this case we find the sequences $\{x_n\}$ and $\{y_n\}$ in X such that

$$d*(f(x_n), f(y_n)) > (1 - 1/n)d*(x_n, y_n).$$

Since X is compact we may suppose, without loss of generality, that these sequences are convergent and we let

$$\lim x_n = x, \qquad \lim y_n = y.$$

We have two cases to consider.

Case 1. $x \neq y$ and, using the fact that f is continuous, we obtain

$$d*(f(x), f(y)) \geq d*(x, y)$$

which is clearly a contradiction.

Case 2. $x = y$. Since f is a local contraction we find a neighbourhood U of x ($= y$) and a number $M < 1$ such that, for all a, b in this set U,

$$d*(f(a), f(b)) \leq Md*(a, b).$$

Since $x_n \to x$ and $y_n \to y = x$ for n large, x_n and y_n are in U and

$$(1 - 1/n)d*(x_n, y_n) \leq d*(f(x_n), f(y_n)) \leq Md*(x_n, y_n)$$

which implies that

$$1 - 1/n \leq M$$

and this is a contradiction.

The condition that X be a compact space mentioned in Theorem 3.6.2. has been shown to be unnecessary by Meyers. The result is as follows:

THEOREM 3.6.8.   *Let (X, d) be a complete metric space and $f : X \to X$ be a continuous mapping staisfying the following properties:*

1.   *f has a unique fixed point x_0,*

2.   *for each $x \in X$, $f^n(x) \to x_0$,*

3.   *there exists a neighbourhood U of x_0 such that, for any neighbourhood V of x_0, there exists an N such that for all $n > N$ $f^n(U) \subset V$.*

Then there exists an equivalent metric on X such that f is a contraction mapping with respect to this metric.

Another characterization was obtained by S. Leader (1977) and is as follows:

THEOREM 3.6.9.   *Let (X, d) be a complete metric space and $f : X \to X$ be a continuous mapping. Then there exists an equivalent metric d^* on X such that f is a contraction mapping with respect to this metric iff:*

1.   *for each $x \in X, f^n(x) \to x_0$, the fixed point of f,*

2.   *the convergence of $f^n(x) \to x_0$ is uniform in some neighbourhood of x_0.*

For details about these as well as for some applications we refer to the papers of Opoitsev (1976), Meyers (1967) and Leader (1977).

3.7. SOME APPLICATIONS OF THE CONTRACTION PRINCIPLE

The contraction principle has a great many applications and they are scattered throughout almost all branches of mathematics.

In what follows we give some applications to illustrate the power and the way of applying the contraction principle.

A. *An Existence Theorem for Differential Equations*

Let us suppose that we have the differential equation

$$dy/dx = f(x, y) \qquad y(x_0) = y_0.$$

and we suppose that f is a continuous function of (x, y) in a domain D in the complex plane. We suppose also that f satisfies the Lipschitz condition with respect to y uniformly in x, i.e.,

$$|f(x, y_1) - f(x, y_2)| \leqq K|y_1 - y_2|.$$

It is not difficult to see that the above equation is equivalent to the following integral equation

$$y(x) = \int_{x_0}^{x} f(s, y(s))\, ds + y_0$$

where the unknown function is $y = y(x)$. This equation suggests the definition of a mapping which, under some conditions and in a suitable space, is a contraction mapping.

In fact, we can prove the following assertion : *if f is as given above and h is a positive number such that hK < 1, then there exists a function y˜ defined on* $[x_0, x_0 + h]$ *such that*

1. $dy˜/dx = f(x, y˜(x))$,

2. $y˜(x_0) = y_0$.

Proof. For this we consider the Banach space of all continuous real-valued functions on the interval $[x_0, x_0 + h] = I$ with the supnorm,

$$g \in C_I \rightarrow \|g\| = \sup \{|g(s)|, s \in I\}$$

and we consider the mapping

$$T : C_I \rightarrow C_I$$

defined by

$$Tg(x) = \int_{x_0}^{x} f(s, g(s)) \, ds + y_0.$$

Since for any g and $g˜$ we have,

$$\|Tg - Tg˜\| = \sup_{x} \left| \int_{x_0}^{x} [f(s, g(s)) - f(s, g˜(s))] \, ds \right|$$

$$\leq Kh\|g - g˜\|$$

and thus T is a contraction mapping. Thus we find a fixed point say $y˜ \in C_I$ and this obviously is a solution of the equation.

B. *A Proof of the Square Roots Lemma in Banach *-Algebras*

For the notion related to Banach algebras we refer to Ch. Rickart (1960).

The following proof of the square root lemma does not require the continuity of the involution.

In fact we prove the following assertion: let $\mathscr{A}$ be a Banach *-algebra and take $a \in \mathscr{A}$ with the property that $a^* = a$ (i.e. a is a Hermitian element) and the spectral radius r_a of a satisfies the inequality $r_a < 1$. Then there exists a unique $x \in \mathscr{A}$ such that

1. $2x - x^2 = a$,

2. $x^* = x$.

First we prove that, given $a \in \mathcal{A}$ such that $r_a < 1$, there exists a unique element $x \in \mathcal{A}$ such that

1. $2x - x^2 = a,$

2. $r_x < 1.$

Indeed, by Theorem 3.2.1. – which can be proved for Banach algebras in the same way – we may choose a norm $\|,\|_+$ such that

$$\|a\|_+ < 1$$

and we now consider the Banach algebra $\mathcal{A}_a$, which is the last closed subalgebra of $\mathcal{A}$ containing a, and we define the set M_a by

$$M_a = \{y, \ y \in \mathcal{A}_a, \ \|y\|_+ < 1\}$$

Now we define a mapping on $\mathcal{A}_a$ by the relation

$$Tx = \tfrac{1}{2}(a + x^2)$$

and it is easy to see that M_a is an invariant set for T. Since

$$\|Tx - Ty\|_+ = \tfrac{1}{2}\{\|x^2 - y^2\|_+$$
$$= \tfrac{1}{2}\{\|(x + y)(x - y)\|_+\}$$
$$\leq \tfrac{1}{2}\|x + y\|_+ \|x - y\|_+$$
$$\leq a\|x - y\|_+$$

T is a contraction mapping. The existence of x having the required properties is proved. We show now that x is unique. Indeed, in the contrary case we have another element y which satisfies the same conditions,

1. $2y - y^2 = a,$

2. $r_y < 1.$

Now, since y commutes with a and x is a limit of polynomials in a it follows that y commutes with x. Thus $r_{x+y} < 2$ and we can choose a new equivalent norm, say $\|,\|^*$ such that $\|x + y\|^* < 2$; we have further,

$$\|x - y\|^* = \|\tfrac{1}{2}(a + x^2) - \tfrac{1}{2}(a + y^2)\|^*$$
$$= \tfrac{1}{2}\|x^2 - y^2\|^* \leq \tfrac{1}{2}\|x + y\|^* \|x - y\|^*$$

which gives us that $\|x - y\|^* = 0$.

It remains to prove that $x^* = x$. Since $\sigma(x)^* = \sigma(x^*)$ it follows that $r_{x^*} < 1$.

Also

$$a = (2x - x^2)^* = 2x^* - x^{*2}.$$

and by the uniqueness property, we have $x = x^*$.

For instructive and interesting applications we refer also to Graffi (1951) (1954), Hayden and Suffridge (1976), Earle and Hamilton (1970).

Chapter 4

Brouwer's Fixed Point Theorem

4.0. INTRODUCTION

Perhaps the most important result in the fixed point theory is the famous theorem of Brouwer which says that the closed unit sphere of E_n has the fixed point property, i.e. if S_n denotes this sphere and $f: S_n \to S_n$ is any continuous function then there exists a point $x_0 \in S_n$ such that $f(x_0) = x_0$.

This result, published by L. E. J. Brouwer in 1910, was previously known to H. Poincaré in an equivalent form. In 1886, Poincaré proved the following result: if $f: E_n \to E_n$ is any continuous function with the property that, for some $r > 0$ and any $\alpha > 0$,

$$f(x) + \alpha x \neq 0 \qquad \|x\| = r,$$

then there exists a point x_0, $\|x_0\| \leq r$ such that $f(x_0) = x_0$.

Now it is known that this assertion is equivalent to the Brouwer fixed point theorem. Another interesting fact is that the Poincaré theorem was also rediscovered by P. Bohl in 1904.

There exist many proofs of the Brouwer fixed point theorem.

Some of them are very short but use results from algebraic topology. In what follows we give some proofs. For the reader who prefers an analytical proof, this is given as Theorem 4.1.5. The reader who is familiar with the differential forms will find a proof of Brouwer's fixed point theorem using them given as Theorem 4.2.1. The reader who prefers an elementary proof, using only basic properties of polynomials and differentiable functions, will find this in Section 4.5.

In recent years attention has been directed to finding methods for computing fixed points. Here the well-known and important paper of H. Scarf (1967) was very inspiring. We present Scarf's proof which gives also a new proof of Brouwer's fixed point theorem. Some related results are also discussed.

4.1. THE FIXED POINT PROPERTY

Let X be a topological space.

DEFINITION 4.1.1.   The space X is said to have the 'fixed point property' (abbreviated as f.p.p.) if, for any continuous function $f : X \rightarrow X$, there exists $x \in X$ such that $f(x_0) = x_0$.

LEMMA 4.1.2.   *If X, Y are topological spaces and*

$$h : X \rightarrow Y$$

is a homeomorphism, then if X has the f.p.p. Y also has the f.p.p.
    Proof. Since the proof is simple we omit it.

An interesting and important way to obtain spaces with the f.p.p. is to use the so-called retracts.

DEFINITION 4.1.3.   Let X and X_1 be topological spaces. Then X_1 is called a retract of X if

1. $X_1 \subset X$,

2.   There exists a continuous function

$$r : X \rightarrow X_1$$

such that $r(x) = x$ for all $x \in X_1$.

The function r is called a retraction.

THEOREM 4.1.4.   *If X has the f.p.p. and X_1 is any retract of X then X_1 has the f.p.p.*
    Proof. Let r be the retraction of X onto X_1 and $f : X_1 \rightarrow X_1$ be any continuous function. We can define a function $g : X \rightarrow X$ by the relation

$$g(x) = f(r(x))$$

and since X has the f.p.p. and obviously g is continuous (and has values in X_1) it follows that there exists $x_0 \in X$ such that $g(x_0) = x_0$. Obviously $x_0 \in X_1$ and thus $r(x_0) = x_0$ and this gives the theorem.

The connection between Brouwer's fixed point theorem and retractions is given in the following

THEOREM 4.1.5.   *The Brouwer fixed point theorem is equivalent to the following assertion: there exists no retraction which is continuously differentiable of*

$$\{x, \|x\| \leq 1, x \in E_n\}$$

on the boundary

$\{x, \|x\| = 1, x \in E_n\}$.

Proof. Suppose that the Brouwer theorem is true. Suppose also that there exists a retraction $r(.)$ which is also continuously differentiable. We define,

$$r_1(x) = -r(x)$$

which is a continuous function on $\{x, \|x\| \le 1\}$ with values in the same set. It is obvious that r_1 has no fixed points. This contradicts the Brouwer theorem.

Conversely, suppose that there does not exist a retraction r which is continuously differentiable for the above pair of spaces. We show that any function which is continuously differentiable on $\{x, \|x\| \le 1\}$ with values in $\{x, \|x\| = 1\}$ has a fixed point. Indeed, in the contrary case, if f is such a function then for any x, $\|x\| \le 1$, $f(x) \ne x$ and we can consider the line through x and $f(x)$ which intersects $\{x, \|x\| = 1\}$ in a point, say $g(x)$. The function

$$x \to g(x)$$

is well defined and of course $g(x) = x$ for all x, $\|x\| = 1$. We prove now that g is continuous differentiable. Indeed, since the line through x and $f(x)$ has the form

$$\alpha(x)x + (1 - \alpha(x))f(x).$$

To prove the assertion it is necessary to prove the continuous differentiability of $\alpha(x)$. Since

$$\langle \alpha(x)x + (1 - \alpha(x)f(x), \alpha(x)x + (1 - \alpha(x))f(x) \rangle = 1$$

for all x, $\|x\| = 1$ we have that

$$\alpha(x)^2 \|x\|^2 + 2\alpha(x)(1 - \alpha(x))\langle x, f(x) \rangle +$$

$$+ (1 - \alpha(x))^2 \|f(x)\|^2 = 1$$

which gives us that $\alpha(x)$ is the solution of an equation the coefficients of which are continuously differentiable. Then it follows, by the well-known formula, that the solution has the same property. Thus the fixed point exists. Then the Weierstrass–Stone theorem gives the assertion for all continuous functions. Indeed, we find a sequence of continuous differentiable functions, say $\{f_n\}$ such that $\{f_n\}$ converges uniformly to f. Choose x_n such that $x_n = f(x_n)$. We can suppose, without loss of generality, that the sequence $\{x_n\}$ is convergent to a point x_0 which is obviously a fixed point for f. For an elementary proof of noncontractibility of the circle, as well as for some generalizations, see R. F. Brown (1974).

4.2. BROUWER'S FIXED POINT THEOREM. EQUIVALENT FORMULATIONS

We give now the Brouwer theorem.

THEOREM 4.2.1.   *Let* $f : \{x, \|x\| \leq 1\} \to \{x, \|x\| \leq 1\} = \bar{S}_n(0, 1) = the$ *unit ball in an n-dimensional space with f continuous. Then f has a fixed point in* $\bar{S}_n(0, 1)$.

Proof. According to Theorem 4.1.5. it is sufficient to show that a retraction which is continuous differentiable of $\{x, \|x\| \leq 1\}$ on $\{x, \|x\| = 1\}$ does not exist.

Suppose on the contrary that there exists such a retraction say f. We consider the differential form

$$\alpha = x_1 dx_2 \cdots dx_n$$

and since

$$\int\limits_{\{x,\, \|x\| = 1\}} \alpha = \int\limits_{\{x,\, \|x\| = 1\}} f^*\alpha = \int\limits_{\{x,\, \|x\| \leq 1\}} df^*\alpha$$

$$= \int\limits_{\{x,\, \|x\| \leq 1\}} f^* d\alpha = 0$$

we have a contradiction, since

$$\int\limits_{\{x,\, \|x\| = 1\}} \alpha = \int\limits_{\{x,\, \|x\| \leq 1\}} d\alpha = \mathrm{vol}\{x, \|x\| \leq 1\} \neq 0$$

and the assertion is proved.

Another useful theorem, equivalent to the Brouwer fixed point theorem is the following assertion, first proved by Henri Poincaré in 1886 and rediscovered by P. Bohl in 1904.

THEOREM 4.2.2.   *Let $f : E_n \to E_n$ be a continuous mapping and suppose that, for some $r > 0$ and all $\lambda > 0$,*

$$f(u) + \lambda u \neq 0$$

for any u, $\|u\| = r$. Then there exists a point u_0, $\|u_0\| < r$ such that $f(u_0) = 0$.

Proof. We define the mapping

$$g(u) = -f(u)/\|f(u)\| \cdot r$$

on $\{x, \|x\| \leq r\}$ and it is clear that this mapping is continuous and $\|g(u)\| = r$. Thus by Brouwer's fixed point theorem there exists a point, say $u\tilde{}$ such that $g(u\tilde{}) = u\tilde{}$ and this gives us that

$$f(u\tilde{})r + \|f(u\tilde{})\|u\tilde{} = 0$$

and this contradicts the property of f. Thus there exists a point u_0 such that $f(u_0) = 0$.

Thus we see that the Brouwer fixed point theorem implies the assertion of Theorem 4.2.2.

We now prove the converse. We suppose that the assertion of Theorem 4.2.2. is true and we show that this implies Brouwer's fixed point theorem.

We consider the function

$$g(x) = x - f(x)$$

which has the property that, for all x, $\|x\| \leq 1$, $g(x) \neq 0$. We take $r = 1$ and we show that for all $\lambda > 0$

$$g(x) + \lambda x \neq 0.$$

Let

$$\delta = \inf\{\|x - f(x)\|, \|x\| \leq 1\}.$$

Clearly we can suppose that $\delta > 0$ since in the contrary case we easily obtain the Brouwer fixed point theorem. We define the mapping

$$g\tilde{}(x) = (x - f(x))/\delta$$

and we prove that for all $\lambda > 0$, $g\tilde{}(x) + \lambda x \neq 0$ if $\|x\| = 1$. Indeed, the assertion is obviously true for $\lambda \in (0, 1)$. Thus we can suppose that $\lambda \geq 1$. Also it is obvious that we can consider only the case when $\|x\| = 1$ and $\lambda = \|x - f(x)\| = \delta$. In the contrary case, i.e. if for some x_0, $\|x_0\| = 1$, $\|x_0 - f(x_0)\| = \delta$ and $g\tilde{}(x_0) + \delta x_0 = 0$ we obtain

$$\langle x_0, g\tilde{}(x_0) \rangle + \delta = 1 - \langle x_0, g\tilde{}(x_0) \rangle + \delta$$

and thus

$$-2\delta = 2(1 - \langle x_0, g\tilde{}(x_0) \rangle).$$

But

$$\delta^2 = -2\delta - 1 + \|g\check{\ }(x_0)\|^2$$

and since

$$-1 + \|g\check{\ }(x)\| \leqq 0$$

and this gives us $\delta^2 \leqq 0$. This contradiction proves the assertion. Thus for all $\lambda > 0$, $g\check{\ }(x) + \lambda x \neq 0$ with $\|x\| = 1$. From the hypothesis, there exists an x_0 such that $g\check{\ }(x_0) = 0$ and this obviously implies the Brouwer fixed point theorem.

We give now another equivalent form of Brouwer fixed point theorem. This form is known to be useful in the theory of differential equations.

THEOREM 4.2.3.   *Let $f_i : D_n \to R$ be continuous, and let it satisfy the following inequalities:*

1. $f_i(x_1, \ldots, x_{i-1}, a_i, x_{i+1}, \ldots, x_n) \geqq 0$

2. $f_i(x_1, \ldots, x_{i-1}, b_i, x_{i+1}, \ldots, x_n) \leqq 0$

for $i = 1, 2, 3, \ldots, n$ where

$$D_n = \{x = (x_1, \ldots, x_n) \quad a_i \leqq x_i \leqq b_i\}.$$

Then there exists a point $x_0 \in D_n$ such that $f_i(x_0) = 0$, $i = 1, 2, 3, \ldots, n$.

Proof. Let m_i and M_i be the minimum and the maximum of f_i on D_n and let δ_i', δ_i'' denote, respectively, the distances of the sets of points in D_n at which $f_i < 0$, $f_i > 0$, respectively, from the hyperplanes $x_i = a_i$ and $x_i = b_i$. Since f_i is continuous and $\delta_i' > 0$, $\delta_i'' > 0$ we find n numbers ε_i such that

$$0 < \varepsilon_i \leqq -\delta_i'/m_i$$

$$0 < \varepsilon_i \leqq \delta_i''/M_i.$$

We define now the functions

$$g_i(x_1, \ldots, x_n) = x_i + \varepsilon_i f_i(x_1, \ldots, x_n)$$

and we remark that at the point $(x_1, \ldots, x_n)$ we have

$$a_i \leqq x_i < g_i(x_1, \ldots, x_n) \leqq b_i$$

if $f(x_1, \ldots, x_n) = 0$; at the points where $f_i > 0$ or $f_i < 0$ we have also

$$a_i \leqq x_i + \varepsilon_i f_i(x_1, \ldots, x_n) < b_i - \delta_1'' + \varepsilon_i M_i < b_i,$$

respectively,

$$b_i \geqq x_i + \varepsilon_i f_i(x_1, \ldots, x_n) > \delta'_1 + a_i + \varepsilon_i m_i > a_i.$$

Thus, by the Brouwer fixed point theorem there exists a point $x^\sim = (x_1^\sim, x_2^\sim, \ldots, x_n^\sim)$ such that

$$g_i(x^\sim) = x^\sim$$

and since $\varepsilon_i > 0$ we obtain

$$f_i(x^\sim) = 0.$$

For the converse, we consider the functions

$$g_i^*(x) = f_i(x_1, \ldots, x_n) - x_i$$

and by hypothesis there exists a point $x^\sim = (x_1, \ldots, x_n)$ such that

$$g_i^*(x^\sim) = 0$$

and this implies the assertion of the theorem.

As a consequence of the Brouwer fixed point theorem we note the following result of Abian and Brown. See also C. Meyer (1964/65).

COROLLARY 4.2.4. *Let $f : \{x, \|x\| \leqq 1\} \to E_n$ be a continuous mapping having the property that for all x, $\|x\| = 1$, $\|f(x)\| \leqq 1$.*

In this case there exists x_0, $\|x_0\| \leqq 1$ such that $f(x_0) = x_0$.

Proof. Let $L_{x, f(x)}$ be the line through x and $f(x)$. Suppose that $x \neq f(x)$ for all x, $\|x\| = 1$. We define the function

$$x \to g(x)$$

where $g(x)$ represents the point on $L_{x, f(x)}$ of norm 1. It is not difficult to show that g is a continuous function and that for all x, $\|x\| = 1$, $g(x) = x$, i.e. g is a retraction mapping. This is clearly a contradiction and the corollary is proved.

As a consequence of the Brouwer fixed point theorem we can prove the following generalization:

THEOREM 4.2.5. *If C is a compact convex set in a finite dimensional space and $f : C \to C$ is any continuous mapping then there exists a point $x^\sim \in C$ such that $f(x^\sim) = x^\sim$.*

In other words, any compact and convex set in a finite dimensional space has the fixed point property (f.p.p.).

Proof. We make, if necessary, a translation and a homothety, so that

$$C \subset \{x, \|x\| \leqq 1\}.$$

If P_C denotes the Minkowski functional for this set, then clearly C is homeomorphic with $\{x, \|x\| \leqq 1\}$ through the mapping

$$x \to x \cdot \|x\| / p_C(x)$$

and by Lemma 4.1.2. the assertion follows.

In what follows we wish to present a proof of the Brouwer fixed point theorem using some notions from the theory of smooth manifolds. For details about manifolds we refer to the excellent books by Milnor (1965), G. de Rham (1955). For the reader's convenience we repeat here some notions.

Let $\mathbf{R}^m$ be the real m-dimensional space and U be an open set in $\mathbf{R}^m$. A function

$$f : U \to V$$

where V is an open set in $\mathbf{R}^n$ is called smooth if

$$\frac{\partial^m f}{\partial x_{i_1} \cdots \partial x_{i_m}}$$

exist and are continuous. If we have an arbitrary set $M \subset \mathbf{R}^m$ and a function $f : M \to \mathbf{R}^n$ then we say that f is smooth at x if there exists a neighbourhood $U_x \ni x$ such that f coincides with a smooth $g : U_x \to \mathbf{R}^n$ on the set $M \cap U_x$. We say that $f : M \to \mathbf{R}^n$ is smooth if it is smooth at every point $x \in M$.

DEFINITION 4.2.6.   If $M \to N$ then we say that f is a diffeomorphism iff f is smooth and also f^{-1} is smooth.

The notion of a smooth manifold is defined as follows:

DEFINITION 4.2.7.   A subset $M \subset \mathbf{R}^m$ is called a smooth manifold of dimension n if for each $x \in M$ there exists a neighbourhood $x \in U_x$ which is diffeomorphic with an open subset of $\mathbf{R}^n$.

If M is a smooth manifold then

$$g : U \to W \cap M$$

is called the parametrization of $W \cap M$ and g^{-1} is called the system of coordinates on $W \cap M$.

The smooth manifold of dimension zero are those sets M for which, for each $x \in M$, there exists a neighbourhood W such that $W \cap M = \{x\}$.

Now we define the derivative and the tangent space of smooth maps and smooth manifolds.

Let M and N be two smooth manifolds and $f : M \rightarrow N$ be a function.

First we consider the case when $M = U$ where U is an open set in $\mathbf{R}^m$. The derivative of f at x is then defined as

$$\mathrm{d}f_x(h) = \lim (f(x + th) - f(x))/t$$

and it is clear that this is a mapping (linear)

$$\mathrm{d}f_x : \mathbf{R}^m \rightarrow \mathbf{R}^n.$$

The well known Chain rule is true for the derivative defined as above. Also, if

$$I : U \rightarrow U$$

is the identity map then $\mathrm{d}f_x$ is the identity map and moreover, if f is a linear map then $\mathrm{d}f_x = f$.

We note the following important fact which is, however, easy to prove.

PROPOSITION 4.2.8.   *If $f : U \rightarrow V$ is a diffeomorphism and $U \subset \mathbf{R}^m$, $V \subset \mathbf{R}^n$ then $m = n$.*

The notion of tangent space is defined as follows: suppose that we have a smooth manifold M in $\mathbf{R}^m$ and let

$$g : U \rightarrow M$$

be a parametrization of a neighbourhood $g(U)$ of x in M, $g(u) = x$.

In this case

$$\mathrm{d}g_u : \mathbf{R}^n \rightarrow \mathbf{R}^m$$

has sense and we set TM_x as the image $\mathrm{d}g_u(\mathbf{R}^n)$ of $\mathrm{d}g_u$. It is not difficult to see that the object so defined is independent of the parametrization used.

Now if we have arbitrary manifolds M and N and $f : M \rightarrow N$ is a smooth map then the derivative

$$\mathrm{d}f_x : TM_x \rightarrow TN_y, \qquad y = f(x)$$

is defined using neighbourhoods. Since f is a smooth map we find an open set W containing x and a smooth map

$$f^{\sim} : W \rightarrow \mathbf{R}^n$$

such that

$$f^\sim/W \cap M = f/W \cap M.$$

Then $df_x(z)$ is defined as $df_x^\sim(z)$ for all $z \in TM_x$.

This also proves that this is indeed a 'well defined' notion as well as the validity of the 'chain rule' and the property of diffeomorphism.

We consider now two smooth manifolds M and N and let $f : M \to N$ be a smooth map. We suppose first that dim $M =$ dim N.

DEFINITION 4.2.9.   A point $x \in M$ is called regular if the derivative df_x is nonsingular.

In this case, by the inverse function theorem, there exists a neighbourhood of x which is diffeomorphic with an open set in N.

The point $y \in N$ is called a regular value of f if $f^{-1}(y)$ contains only regular points. A point x is called singular if df_x is singular.

Let M be a compact smooth manifold and $f : M \to N$. In this case for any regular value y, $f^{-1}(y)$ is a finite set.

Indeed, $f^{-1}(y)$ is a compact set and it is discrete since f is one-to-one in a neighbourhood of each $x \in f^{-1}(y)$

It is clear now that for any $f : M \to N$, M compact and smooth, we can define an integer $\# f^{-1}(y)$ as the number of points in $f^{-1}(y)$ and

$$y \to \# f^{-1}(y)$$

is locally constant.

Suppose now that $f : M \to N$ is a smooth map and M, N are two smooth manifolds of dimension m (respectively n) with $m > n$.

Also let C be the set of all $x \in M$ for which

$$df_x : TM_x \to TN_{f_x}$$

has the rank less than n. The set C is called the set of all critical points of f and $f(C)$ is the set of all critical values.

Of course $N - f(C)$ is the set of all regular values of f.

Now, since M can be covered by a countable collection of neighbourhoods, each diffeomorphic with an open subset of $\mathbf{R}^n$, it follows that the set of regular values of a smooth map is every where a dense set in N.

PROPOSITION 4.2.10.   Let $f : M \to N$ be a smooth map, and let M, N be smooth manifolds of dim $M = m \geqq n =$ dim N. For any regular value $y \in N$ the set $f^{-1}(y)$ is a smooth manifold of dimension $m - n$.

Proof. Let $x \in f^{-1}(y)$ and since y is a regular value, df_x maps TM_x onto

TM_{f_x} which gives us that the null space df_x is of dimension $m - n$. Let $M \subset \mathbf{R}^k$ and let $L : \mathbf{R}^k \to \mathbf{R}^{m-n}$ be nonsingular on null $d(f_x)TM_x \subset \mathbf{R}^k$. We define

$$F : M \to N \cap \mathbf{R}^{m-n}$$

by

$$F(x) = (f(z), L(z))$$

which has the derivative

$$dF_x(z) = (df_x(z), L(z))$$

and it is nonsingular. This implies that F maps some neighbourhood U of x diffeomorphically onto a neighbourhood of $(y, L(x))$, say V.

Also since F maps

$$f^{-1}(y) \cap U \to (y \times \mathbf{R}^{m-n}) \cap V$$

diffeomorphically and onto, $f^{-1}(y)$ is a smooth manifold of dimension $m - n$.

Example 4.2.11. Consider $\mathbf{R}^m$ and $\mathbf{R}$ and let

$$f : \mathbf{R}^m \to \mathbf{R}$$

be defined by the relation

$$f(x) = x_1^2 + \cdots + x_m^2$$

and from the form of f it is obvious that it is a smooth map.

In this case every strictly positive number has a regular value for f. From Proposition 4.2.10. we obtain that the unit sphere $S^{m-1} = f^{-1}(1)$ is a smooth manifold of dimension $m - 1$.

Consider now a manifold M which is supposed to be smooth and contains the manifold M_1. In this case for each $x \in M_1$, TM_{1x} is subspace of TM_x. We can consider the orthogonal complement and this subspace is called the space of all normal vectors to M at x. It is clear that the dimension of this space is $m - m_1$. where $m = \dim M, m_1 = \dim M_1$.

As an example of the above results, for each regular value, $f^{-1}(y)$ is a smooth manifold in M for any smooth map f.

LEMMA 4.2.12. *Null* (df_x) *is in the tangent space* $TM'_{1x} \subseteq TM_x$.

Proof. Since df_x maps the subspace TM'_{1x} into zero, counting the dimension, it is clear that df_x is an isomorphism between the space of normal vectors and TN_y.

The following results will be essential for the new proof of Brouwer's fixed point theorem.

Let

$$H^m = (x = (x_1, x_2, \ldots, x_m), x_m \geq 0)$$

and the boundary of this set is $\mathbf{R}^{m-1} \times 0$, i.e. a hyperplane in $\mathbf{R}^m$.

DEFINITION 4.2.13.   A subset M in $\mathbf{R}^k$ is called a smooth m-manifold with boundary if for each $x \in M$ there exists a neighbourhood U of x such that $U \cap M$ is diffeomorphic to an open subset V of H^m. The boundary ∂M is the set of all points which correspond to H^m under such diffeomorphism.

From these it is clear that ∂M and $M - \partial M$ are smooth manifolds of dimensions $m - 1$ and m respectively.

For every $x \in M$ we define the tangent space TM_x which is an m-dimensional space.

If M is a manifold without boundary and $F: M \to \mathbf{R}$ is a smooth map having 0 as a regular value then $M_1 = (x, x \in M, g(x) \geq 0$ is clearly a manifold with the boundary equal to $(x, F(x) = 0)$.

THEOREM 4.2.14.   *Let M, N be two manifolds and let M have a boundary and* dim $M = m$, dim $N = n$. *If $f: M \to N$ is a smooth map such that for $y \in N$ there exists a regular value for f and for $f/\partial M$, then $f^{-1}(y)$ is a smooth $(m - n)$-manifold with boundary.*

Proof. Clearly it is sufficient to consider only the case when $M = H^m$ and $N = \mathbf{R}^n$. Let $f: H^m \to \mathbf{R}^n$ be a smooth map with the properties stated in the theorem. If x' is in $f^{-1}(y)$ and is an interior point then the assertion of the theorem is clear.

It remains to consider the case when x' is a boundary point.

In this case we consider another map g defined on a neighbourhood of x' and equal to f on the part of the neighbourhood situated in H^m. We can suppose that g so defined has no critical points and thus that $g^{-1}(y)$ is a smooth manifold of dimension $m - n$.

Define

$$p: g^{-1}(y) \to \mathbf{R}$$

by

$$p(x_1, \ldots, x_m) = x_m$$

and we show that p has 0 as a regular value. From the above results about the tangent space, the tangent space of $g^{-1}(y)$ at some point in $p^{-1}(0)$ is equal to the null space of the dg at the respective point. Since f/H^m is regular we get that this space is not exactly $\mathbf{R}^{m-1} \times 0$ and thus the set $g^{-1}(y) \cap H^m = f^{-1}(y) \cap U$, U is the neighbourhood where g is defined, is a smooth manifold. Clearly the boundary is equal to $p^{-1}(0)$.

THEOREM 4.2.15. *For any manifold M with boundary there is no smooth map* $f : M \to M$ *having M pointwise invariant.*

Proof. Suppose that such a mapping exists and let $y \in M$ be a regular value of f. From the hypothesis we get that y is a regular value of the identity map (on M). In this case $f^{-1}(y)$ is a smooth manifold of dimension 1 and the boundary is the point y. Since $f^{-1}(y)$ is compact then it is known that the only manifolds with dimension 1 which are compact are in fact finite disjoint unions of circles and segments and thus the boundary has an even number of points. This is in contradiction to the fact that the boundary is exactly (y). Thus such a function f does not exist.

THEOREM 4.2.16 (Brouwer). *Let* $f : \bar{S}_n(0, 1) \to \bar{S}_n(0, 1)$ *be any continuous mapping. Then f has a fixed point in* $\bar{S}_n(0, 1)$.

Proof. An approximation argument shows that we can suppose, without loss of generality, that f is smooth.

Suppose that f has no fixed points and thus for any x in $\bar{S}_n(0, 1)$, $f(x) \neq x$. We consider the following $g : \bar{S}_n(0, 1) \to S^{n-1}$ defined as follows: for each x, $g(x)$ is the point on S^{n-1} which is nearer to x and is on the line generated by the points x and $f(x)$. Since f has no fixed points it is clear that this is again a smooth map. Clearly g has S^{n-1} pointwise invariant and thus g does not exist and this implies that f has a fixed point in $\bar{S}_n(0, 1)$.

Remark 4.2.17. The above proof was based essentially on Theorem 4.2.15. The proof of Theorem 4.2.15. is due to M. Hirsch (1963) and plays an important role in the algorithms for fixed points in Brouwer's theorem. See also Section 4.8.

4.3. ROBBINS COMPLEMENTS OF BROUWER'S THEOREM

Let $\bar{S}_n(0, 1) = \{x, \|x\| \leq 1\}$ where $x \in \mathbf{R}^n$ and the norm on $\mathbf{R}^n$ is

$$x \to \|x\| = \left(\sum_{1}^{n} |x_i|^2 \right)^{1/2}.$$

If $f : \bar{S}_n(0, 1) \to \bar{S}_n(0, 1)$ is a continuous mapping then there exists at least one point $x \in \bar{S}_n(0, 1)$ such that $f(x) = x$. Let

$$F(f) = \{x, f(x) = x\}$$

be the fixed point set of the mapping f. From the continuity of f it follows that the set $F(f)$ is closed. It is natural to ask about the nature of the sets $F(f)$ for f arbitrary continuous mapping. The following theorem of H. Robbins (1967) gives the answer to this problem.

THEOREM 4.3.1. *If F is any closed nonempty set in $\bar{S}_n(0, 1)$ then there exists a continuous mapping f,*

$$f : \bar{S}_n(0, 1) \rightarrow \bar{S}_n(0, 1)$$

such that $F(f) = F$.

Proof. Let F be closed and nonempty in $\bar{S}_n(0, 1)$. We define the mapping f as follows: for each $x \in \bar{S}_n(0, 1)$ we consider the number

$$d(x, F) = \inf \{ \|x - y\|, \, y \in F \}$$

and f has the form

$$f(x) = \begin{cases} x + d(x, F)(x - x_0)/(\|x - x_0\|) & x \neq x_0 \\ x_0 & x = x_0. \end{cases}$$

From the properties of the norm and

$$x \rightarrow d(x, F)$$

it follows that f is continuous. Since $d(x, F) = 0$ iff $x \in F$ we have that $F(f) = F$. In the above formula for f, x_0 is an arbitrary point in $S_n(0, 1)$.

It is quite natural to ask if the above assertion is true when we consider some special classes of mappings. We mention the following negative result of Robbins.

THEOREM 4.3.2. *For any n odd there exists a closed nonempty set F such that it is not a fixed set for any homeomorphism of $\bar{S}_n(0, 1)$.*

For a proof as well as for other results and problems concerning these results we refer to Robbins' paper.

Remark 4.3.3. For the case of compact mappings or more general mappings, see the next chapter.

4.4. THE BORSUK–ULAM THEOREM

One of the most interesting generalizations of Brouwer's fixed point theorem is the so-called Borsuk–Ulam theorem and Borsuk's theorem about the antipodal points.

Let $S_n = (x = (x_1, x_2, \ldots, x_{n+1}),$ some $x_i = \pm 1)$
and choose a mapping

$$\alpha : S_n \to S_n$$

defined by

$$\alpha(x) = -x,$$

then two points corresponding under $\alpha(,)$ are called antipodal.

The famous Borsuk–Ulam theorem is the following assertion:

THEOREM 4.4.1. *Let $f : S_n \to \mathbf{R}^n$ be a continuous mapping. Then the equation*

$$f(x) = f(-x)$$

has a solution.

Borsuk's theorem about antipodal points is the following assertion:

THEOREM 4.4.2. *Let $B_{n+1} = (x, x \in \mathbf{R}^{n+1}, \|x\| \leq 1)$. Then there is no continuous mapping*

$$f : B_{n+1} \to S_n$$

such that for all $x \in S_n$, $f(x) = -f(-x)$.

In what follows we give the proof of these results. The proof of the Borsuk–Ulam theorem which we give is that of Meyerson–Wright (1979) and the proof of Borsuk's theorem using the Borsuk–Ulam theorem is more or less standard.

The following result is the main tool in proving these theorems.

THEOREM 4.4.3. *Let $f : S_n \to \mathbf{R}^n$ be a piecewise linear map. Then there exists an x_0 such that $f(x_0) = f(-x_0)$.*

Proof. We recall that if

$$(p_1, p_2, \ldots, p_{m+1})$$

are points in $\mathbf{R}^n$ then we say that these $(m+1)$-points are in the general position if no three of them lie on a line, no four lie in a plane, $\ldots$, no $(s+1)$-points lie in a linear space of dimension less than s. Then if $(p_1, \ldots, p_{m+1})$ are points in a general position, the set

$$\operatorname{conv}(p_1, \ldots, p_{m+1}) = \left(\sum_1^{m+1} a_i p_i, a_i \geq 0, \sum a_i = 1 \right)$$

is called the m-simplex, and the points $p_1, \ldots, p_{m+1}$ are called the vertices of the simplex and the simplex is denoted by

$$\sigma^m(p_1, \ldots, p_{m+1}).$$

If $\sigma^m(p_1, \ldots, p_{m+1})$ is a simplex then any set of points

$$(p_{i_1}, \ldots, p_{i_{r+1}})$$

$r < m$, can generate a simplex, since these points are in a general position and this simplex is called the r-dimensional face of the simplex $\sigma^m(p_1, \ldots, p_{m+1})$.

Now, if $\sigma^m(p_1, \ldots, p_{m+1})$ is any simplex then a point $x \in \sigma^m$ is characterized by the numbers $(a_1, \ldots, a_{m+1})$,

$$x = \sum a_i p_i, \qquad a_i \geq 0, \qquad \sum a_i = 1.$$

These numbers are called the barycentric coordinates of x. Suppose that we have a simplex $\sigma^m(p_1, \ldots, p_{m+1})$ and the set of faces

$$\sigma^1, \sigma^2, \ldots, \sigma^{r+1}$$

with the property that

$$\sigma^j \quad \text{is a face of } \sigma^{j+1} =, \qquad j = 1, 2, \ldots, r.$$

We consider the points $b_{\sigma^{j+1}}$ in σ^j with the barycentric coordinates

$$(1/(j+1), \ldots, 1/(j+1))$$

which is called the barycentre of σ^{j+1}. Thus the points

$$(b_{\sigma^1}, \ldots, b_{\sigma^{r+1}}).$$

which are (as is easy to see) in general position, permits us to consider the simplex

$$\sigma^r(b_{\sigma^1} \cdots b_{\sigma^{r+1}}).$$

We define now the so called barycentric division of the simplex $\sigma^m(p_1, \ldots, p_{m+1})$ as the collection of all simplexes which are obtained as for the simplex

$$\sigma^r(b_{\sigma^1}, \ldots, b_{\sigma^{r+1}}).$$

A triangulation T of the sphere in $\mathbf{R}^{n+1}$ is any finite collection of simplexes $C = (\sigma_q^p)$ with the following properties:

1. $\bigcup \sigma_q^p$ is the sphere,

2.   if $\sigma \in C$, σ_1 is a face of σ, then $\sigma_1 \in C$,

3.   if σ, $\sigma' \in C$ then $\sigma \cap \sigma'$

is empty or a face of both σ and σ'.

Now we are ready to prove the piecewise version of the Borsuk–Ulam theorem.

Since f is piecewise linear, it is linear on each simplex of a triangulation T of the set S_n and we consider the subdivision $T \cap \alpha T$ of T into convex cells

$$(\sigma \cap \alpha\sigma, \ \sigma \in T, \ \alpha\sigma \text{ is the image of } \sigma \text{ under } \alpha).$$

We remark that f is linear on each cell of $T \cap \alpha T$ and from the definition of $T \cap \alpha T$ it is obvious that it is invariant under α (since $\alpha^2 = I$). Now we can consider a subdivision of $S_n \times [0, 1]$ by crossing each cell of $T \cap \alpha T$ with $[0, 1]$.

We remark that these are again convex and further we can divide these new cells without adding new vertices to obtain a triangulation T_1 which is invariant under the mapping

$$(x, t) \to (-x, t).$$

To this end we can proceed as follows: first we order the pair of vertices $(v, H(v))$ (where $H(x, t) = (-x, t)$). Suppose now that we have subdivided the $(r - 1)$-skeleton and consider C to be an r-cell and, if v is the vertex of C from the first vertex pair meeting C, we use this v to cone on the faces of C not containing v.

Consider a to be a point in $\mathbf{R}^n$ such that

$$(a, 1, 1)$$

is in the interior of an n-simplex of T_1. We then define

$$G_0 : S_n \to \mathbf{R}^n$$

$$G_1 : S_n \to \mathbf{R}^n$$

by the relations

$$G_0(x) = f(x) - f(-x),$$

$$G_1(z, s) = z - sa$$

for all $(z, s) \in \mathbf{R}^n$. We extend piecewise linearly to

$$G : S_n \times [0, 1] \to \mathbf{R}^n$$

using T_1 and we remark that we have the relations

$$G(x, t) = -G(-x, t)$$
$$G_1^{-1}(0) = ((a, 1), (-a, -1)).$$

Now if $G(\sigma - S_n \times 0)$ does not contain 0 for some $(n-1)$-simplex of T_1 then in the adjustments of G we take $a' = a$; in the contrary case we modify G as follows:

1. $\qquad G/S_n \times 0$ remains unchanged,

2. $\qquad$ on each pair $((z, s, 1), (-z, -s, 1))$ of vertices of T_1 we define G piecewise linearly using T_1 such that

$\qquad 2_1.$ $\quad G(z, s, 1) = G(-z, -s, 1)$ for all $(z, s, 1) \in S_n \times [0, 1]$,

$\qquad 2_2.$ $\quad$ for some $a' \in \mathbf{R}^n$, $G_1^{-1}(0) = ((a', 1), (-a', -1))$,

$\qquad 2_3.$ $\quad$ no $G(\sigma - S_n \times 0)$ contains 0 for any $\sigma \in T_1$ of dim $\leq n-1$.

Now 2_1 is clear from the definition, 2_2 follows from making small changes in G and, for 2_3, suppose we have adjustments at v and Tv contradicting 2_3 while satisfying 2_2. In this case any adjustment of $G(v)$ out of the plane containing $G(T_1 v)$ has the property that $T_1 v$ satisfies 2_3.

Now the proof of the piecewise linear version of the Borsuk–Ulam theorem is as follows: if C is the component of $G^{-1}(0) - (S_n - 0)$ containing $(p, 1, 1)$ then it is a polygonal arc which has its endpoints either in $S_n \times 0$ or $S_n \times 1$. Now since $G(x, t) = -G(-x, t)$, $H(C)$ is the component of $G(0) - (S_n - 0)$ containing $(-a', -1, 1)$.

We have either

$$C = H(C)$$

or

$$C \cap H(C) = \varnothing$$

We have in fact $C \cap H(C) = \varnothing$ since in the contrary case C is a closed arc and H would have a fixed point, as a consequence of the piecewise linear version of intermediate value theorem.

Thus $\mathrm{Cl}(C)$ must be an arc connecting $S_n \times 1$ to $S_n \times 0$ and thus $\mathrm{Cl}(C) \backslash S_n \times 0$ is the solution.

Thus from this form of the solution we have an indication of the algorithm to find the solution: this consists, in fact, of moving along the polygonal arc in $G^{-1}(0)$ from $S_n \times 1$ to $S_n \times 0$.

We remark that this is a homotopy-type algorithm to find a solution.

From the piecewise linear version of the Borsuk–Ulam theorem we readily obtain the Borsuk–Ulam theorem.

THEOREM 4.4.4. *Let $f: S_n \to \mathbf{R}$ be any continuous map. Then there exists a point x_0 such that $f(x_0) = f(-x_0)$.*

Proof. Let (T_k) be a sequence of triangulations of S_n with the mesh $T_k \leqq 1/k$. We define

$$f^k : S_n \to \mathbf{R}^n$$

by $f^k(x) = f(x)$ at the vertices of the triangulation T_k and extends f^k linearly. Clearly $f^k \to f$ uniformly. Let $x_k \in S_n$ such that

$$f^k(x_k) = f^k(-x_k)$$

and we suppose, without loss of generality, that the sequence (x_k) is convergent to a point x'. Then from the continuity of f and the uniform convergence of (f^k) to f we easily obtain that x' satisfies the relation

$$f(x) = f(-x)$$

and the theorem is proved.

COROLLARY 4.4.5. *There is no continuous mapping of S_n into S_{n-1} which maps antipodal points of S_n into antipodal points of S_{n-1}.*

Proof. Suppose that such a map exists:

$$f: S_n \to S_{n-1}, \qquad f(x) = -f(-x)$$

for each $x \in S_n$. We can consider this map as a map on S_n with values in $\mathbf{R}^n$ and thus we find a point with the property that

$$f(x) = f(-x)$$

and this is a contradiction and the Corollary is proved.

COROLLARY 4.4.6 (Borsuk's antipodal theorem). *There is no continuous mapping*

$$f: B_{n+1} \to S_n, \qquad f(x) = -f(-x), \quad x \in S_n.$$

Proof. Suppose on the contrary that such a mapping exists. If

$$f = (f_1, \ldots, f_{n+1}); \qquad f_i : B_{n+1} \to \mathbf{R}$$

is this mapping then we have

$$\sum_{1}^{n+1} |f_i(x)| = 1 \qquad x \in B_{n+1}$$

$$f_i(x) = -f_i(x) \qquad x \in S_n.$$

Let r be the retraction of $\mathbf{R}^{n+2}$ onto B_{n+1} and

$$g_i(x) = \begin{cases} f_i(rx) & x \in B_{n+1}^{+} = (x, x \in B_{n+1}, x_{n+1} \geqq 0) \\ -f_i(r(-x)) & x \in B_{n+1}^{-} = (x, x \in B_{n+1}, x_{n+1} \leqq 0). \end{cases}$$

Then clearly these are continuous functions and

$$\sum_{1}^{n+1} g_i(x) = 1 \qquad x \in S_{n+1}$$

$$g_i(x) = -g_i(-x) \qquad x \in S_n$$

which implies that

$$g = (g_1, \ldots, g_n)$$

is a mapping on S_{n+1} with values in S_n which maps antipodal points into antipodal points. This contradicts the result of the above Corollary.

COROLLARY 4.4.7 (The Domain Invariance Theorem). *Let $f : U \to \mathbf{R}^n$ be a continuous function on an open set in $\mathbf{R}^n$ which is supposed to be locally injective (i.e. for each $x \in U$ there exists a neighbourhood V_x such that f/V_x is injective). Then f is an open map.*

Since we prove a more general theorem in the case of Banach space we give no proof of this important result.

4.5. AN ELEMENTARY PROOF OF BROUWER'S THEOREM

We wish to present now an elementary proof of Brouwer's fixed point theorem, in the sense that the proof requires only the properties of continuous functions and properties of compact subsets of $\mathbf{R}^n$. The proof we give is that of Kuga (1974) and this uses the generalization to n dimension of Bernstein's polynomials to prove an approximation theorem. See also 12.5.

As is well known, for any continuous function on $[0, 1]$ with real values the sequence of polynomials $\{B_n(f)\}$, defined as follows:

$$B_n(f)(x) = \sum_{k=0}^{n} f(k/n)x^k(1-x)^{n-k};$$

converges uniformly to f on $[0, 1]$.

We consider now the set S_{n-1} defined as follows:

$$S_{n-1} = \{x = (x_1, x_2, \ldots, x_n), \sum_{i=1}^{n} x_i = 1, x_i \geq 0\}$$

which is obviously a compact and convex subset in $\mathbf{R}^n$. We remark also that $S_1 = [0, 1]$.

To define the Bernstein's polynomials on S_{n-1} we consider the set N_n of all n-dimensional nonnegative integers of which the elements add up to v, i.e.,

$$N_n = \{(a_1, a_2, \ldots, a_\mu)\}$$

where

$$a_i = (a_{i1}, a_{i2}, \ldots, a_{in}),$$
$$\mu = n(n+1)\cdots(n+v-1)/n!$$

and $\sum_{j=1}^{n} a_{ij} = v$.

We define now,

$$g(x, a_i) = v!/a_{i1}!\cdots a_{in}! \cdot x_1^{a_{i1}}\cdots x_n^{a_{in}}.$$

It is easy to see that the functions $g(x, a_i)$ have the following properties:

(1) $$\sum_{a_i} g(x, a_i) = 1 \quad \text{for all } x \in S_{n-1},$$

(2) $$vx_j = \sum_{a_i} g(x, a_i)a_{ij}, \quad j = 1, 2, \ldots n, \quad x \in S_{n-1},$$

(3) $$v(v-1)(x_j)^2 = \sum_{a_i} a_{ij}(a_{ij} - 1)g(x, a_i),$$
$$j = 1, 2, 3, \ldots, n, \quad x \in S_{n-1}.$$

The first relation represents the expansion of $(\sum_{i=1}^{n} x_i)^v$ restricted to S_{n-1}. From this, by differentiation, we can obtain the second and the third relation. As we have noted above, the Bernstein polynomials were introduced by S. Bernstein to give a simple and constructive proof of the Weierstrass approximation theorem (i.e. the space generated by polynomials is dense in $C_{[0,1]}$). Now we show that the polynomials

$$\sum_{a_i \in N_n} f(a_i/v)g(x, a_i) = B_v(f)$$

are dense in $C_{S_{n-1}}$.

This means the following:

THEOREM 4.5.1.   *Let f be a continuous function on S_{n-1} with values in S_{n-1} then for any $\varepsilon > 0$ there exists $B_v(f)$ such that*

$$\|f(x) - B_v(f)\| \leq 2\varepsilon.$$

Proof. Since S_{n-1} is a compact set and f is a continuous function for $\varepsilon > 0$ there exists $\delta > 0$ such that if $\|x - y\| \leq \delta$ then $\|f(x) - f(y)\| \leq \varepsilon$ for all $x, y \in S_{n-1}$. Also, since f is continuous and S_{n-1} compact, there exists a $K > 0$ such that for all $x \in S_{n-1}$,

$$\|f(x)\| \leq K.$$

Now we use the above relation (1) to obtain that

$$\|f(x) - B_v(f)\|$$

is equal to

$$\| \sum_{a_i \in N_n} (f(x) - f(a_i/v))g(x, a_i)\|.$$

We shall evaluate the sum by partitioning N_n into two subsets N_1 and N_2 defined as follows:

$$N_1 = \{a_i, \|x - a_i/v\| \leq \delta\}$$

and

$$N_2 = \{a_i, \|x - a_i/v\| > \delta\}.$$

In this case we have,

$$\| \sum_{a_i \in N_1} [f(x) - f(a_i/v)]g(x, a_i)\|$$
$$\leq \varepsilon \sum g(x, a_i) = \varepsilon.$$

and

$$\left\| \sum_{a_i \in N_2} [f(x) - f(a_i/v)] g(x, a_i) \right\| \leq 2K \sum_{a_i \in N_2} g(x, a_i)$$

$$\leq 2K \sum_{a_i \in N_2} \left[\sum_{j=1}^{n} (x_i - a_{ij}/v)^2 \right] / \delta^2 \cdot g(x, a_i)$$

$$\leq 2K/\delta^2 v^2 \sum_{a_i \in N} \sum_{j=1}^{n} (vx_j - a_{ij})^2 g(x, a_i)$$

$$= 2K \left[v^2 \sum_{j=1}^{n} (x_j)^2 - 2v \sum_{j=1}^{n} \sum_{a_i \in N} a_{ij} g(x, a_i) + \right.$$

$$\left. + \sum_{j=1}^{n} \sum_{a_i \in N} (a_{ij})^2 g(x, a_i) \right]$$

$$\leq 2K \left[1 - \sum_{j=1}^{n} (x_j)^2 \right] / v \delta^2 \leq \varepsilon$$

if we take

$$v \geq 2K(1 - 1/n)/\delta^2 \varepsilon$$

and this proves our assertion.

We consider now the functions $\{B_v(f)\}$ which are obviously continuous and have values in S_{n-1}. Suppose now that Brouwer's theorem is true for such functions. Then we find the points $\{x_v\}$ in S_{n-1} such that

$$B_v(f)(x_v) = x_v.$$

Since S_{n-1} is a compact set we find a subsequence of $\{x_v\}$ which converges, say to $x_0 \in S_{n-1}$. In this case we can clearly see that x_0 is a fixed point of f. Thus to prove the Brouwer's theorem we can restrict our considerations only to the functions $B_v(f)$.

At this point we prove the assertion using an induction argument.

It is obvious that in the case of S_1 the assertion is true, in fact it is true for a more general class of mappings, namely, the class of functions with the property that

$$F(x) = x - f(x)$$

has the Darboux property.

Let

$$B_v(f) = \{B_{v,i}(f)\}$$

and if $B_{v,n}(f)(0, 0, \ldots, 0, 1) = (0, 0, \ldots, 0, 1)$ then clearly the point $(0, 0, 0, \ldots, 0, 1)$ is a fixed point for $B_v(f)$. Thus we can assume that

$$B_{v,n}(f)(0, 0, 0, \ldots, 0, 1) < 1.$$

Let $\xi = (\xi_1, \ldots, \xi_{n-1}) \in S_{n-2}$
and set

$$B_{v,n}(f)(\xi, x_n) = x_n = x_n(1 + x_n)^{v-1}$$

which has v roots (as an equation in x_n). Also these roots are continuous functions on ξ. Now, if z is any complex number, $z = x + iy$ then we write

$$x = \operatorname{Re} z, \qquad y = \operatorname{Im} z$$

and we set

$$\rho(\xi, t) = \max_{1 < i < v} [\operatorname{Re} z_i(\xi) - t|\operatorname{Im} z_i(\xi)|]$$

which is a continuous function on $S_{n-2} \times \mathbf{R}^+$. From the fact that $B_{v,n}(f)$ $(0, 0, 0, \ldots, 1) < 1$ and from the intermediate value theorem we get $\rho(\xi, t) \geqq 0$.

We define now the following functions,

$$h_i(\xi, t) = B_{v,i}(f)(\xi/(1 + \rho(\xi, t)), \rho(\xi, t)/(1 + \rho(\xi, t))) - \lambda \xi_i$$

for $i = 1, 2, 3, \ldots, n-1$ and where

$$\lambda = \sum_{j=1}^{n-1} \xi_j B_{v,i}(f)(\xi/(1 + \rho(\xi, t)), \rho(\xi, t)/(1 + \rho(\xi, t))) \cdot$$

$$\cdot \left[\sum_{j=1}^{n-1} (\xi_j)^2 \right]^{-1}.$$

The mapping $\{h_i\}$ satisfies the so-called 'Walras' law', i.e.

$$\sum_{i=1}^{n-1} \xi_i h_i(\xi, t) = 0$$

for all $\xi \in S_{n-2}$. Since the verification of this identity is simple we omit it.

We consider the following functions, $\{d_i(\xi, t)\}$, defined as follows:

$$d_i(\xi, t) = \xi_i + \max(h_i(\xi, t), 0) \Big/ \left(1 + \sum_{j=1}^{n-1} \max(h_i(\xi, t), 0) \right)$$

for $i = 1, 2, 3, \ldots, n-1$.

For any $t \in \mathbf{R}^+$ the mapping

$$\xi \to (d_i(\xi, t))$$

is continuous and maps S_{n-2} into itself. Therefore by the induction hypothesis, we find a fixed point ξ^t, i.e.,

$$(d_i(\xi^t, t)) = (\xi_i^t), \qquad i = 1, 2, 3, \ldots, n-1.$$

This implies, according to the definition of d_i's functions that,

$$\max\{h_i(\xi^t, t), 0\} = \xi_i^t \sum_{j=1}^{n-1} \max\{h_j(\xi^t, t), 0),$$

$$i = 1, 2, 3, \ldots, n-1.$$

Now using the fact that the functions $\{h_i\}$ satisfy the Walras law we get,

$$\sum_{i=1}^{n-1} [\max(h_i(\xi^t, t), 0)]^2 = \sum_{i=1}^{n-1} h_i(\xi^t, t)\xi_i^t$$

$$\sum_{j=1}^{n-1} \max(h_j(\xi^t, t), 0) = 0$$

This clearly implies that

$$h_i(\xi^t, t) \leqq 0$$

for all $i = 1, 2, 3, \ldots, n-1$.
Also from the fact that the functions $\{h_i\}$ satisfy the Walras law, we have

$$h_i(\xi^t, t) = 0 \quad \text{if } \xi_i^t > 0.$$

But the functions $B_{v,i}(f)$ are nonnegative and the last two relations lead to

$$B_{v,i}(f)(\xi^t/(1 + \rho)(\xi^t, t)), \rho(\xi^t, t)/(1 + \rho(\xi^t, t)) = \lambda^t \xi_i^t$$

for $i = 1, 2, 3, \ldots, n-1$. Of course λ^t is the λ at the point $\xi = \xi^t$.

Let $\{t_n\}$ be a nonnegative sequence such that

$$\lim t_n = \infty$$

and the corresponding fixed points denoted by ξ^{t_n}. Since these are in S_{n-2} which is a compact set we can assume without loss of generality that the sequence $\{\xi^{t_n}\}$ converges to a point denoted by $\xi^* \in S_{n-2}$.

Now there exists an index v such that $\rho(\xi^{t_n}, t_n)$ is attained infinitely many times by

$$\text{Re } z_v(\xi^{t_n}) - t_n |\text{Im } z_v(\xi^{t_n})|$$

and again, taking a subsequence of $\{t_n\}$, say $\{t_{n_l}\}$ we set

$$\rho(\xi^{t_{n_l}}, t_{n_l}) = \operatorname{Re} z_v(\xi^{t_{n_l}}) - t_{n_l}|\operatorname{Im} z_v(\xi^{t_{n_l}})|$$

for all t_{n_l}.

Now, since $\rho(\xi, t) \geqq 0$ and $\operatorname{Re} z_v(\xi)$ is bounded on S_{n-1}, then this inequality implies,

$$\operatorname{Im} z_v(\xi^*) = 0$$

because the sequence $\{t_{n_l}\}$ converges to ∞.
    But

$$B_{v,n}(\xi, x_n) = x_n(1 + x_n)^{v-1}$$

has a real nonnegative solution for any $\xi \in S_{n-2}$ we find another index $v\tilde{\ }$ such that

$$\operatorname{Re} z_{v\tilde{\ }}(\xi^{t_{n_l}}) = \max_{1 \leqq l \leqq v,\ \operatorname{Im} z_1(\xi t_{n_l}) = 0} \operatorname{Re} z_l(\xi^{t_{n_l}})$$

holds infinitely many times.
    Taking now a subsequence of $\{\xi^{t_{n_l}}\}$, say $\{\xi^{t_{n_r}})$ we may put

$$\operatorname{Re} z_v(\xi^{t_{n_r}}) \geqq \operatorname{Re} z_v(\xi^{t_{n_r}}) - t_{n_r}|\operatorname{Im} z_v(\xi^{t_{n_r}})|$$
$$\geqq \operatorname{Re} z_{v\tilde{\ }}(\xi^{t_{n_r}})$$

for all t_{n_r}. This implies, as above, that

$$\lim t_{n_r}|\operatorname{Im} z_v(\xi^{t_{n_r}})| = 0$$

Now taking limits, we obtain

$$B_{v,n}(f)(\xi^*/(1 + \operatorname{Re} z_v(\xi^*)), \operatorname{Re} z_v(\xi^*)/(1 + \operatorname{Re} z_v(\xi^*)) = \lambda^*\xi_i^*$$

and

$$B_{v,n}(f)(\xi^*, \operatorname{Re} z_v(\xi^*)) = \operatorname{Re} z_v(\xi^*)[1 + \operatorname{Re} z_v(\xi^*)]^{v-1}$$

for $i = 1, 2, 3, \ldots, n-1$ and where

$$\lambda^* = \lim_{r \to \infty} \lambda^{t_{n_r}}.$$

Since $B_{v,n}(f)$ is homogeneous of degree v with respect to (ξ, x_n) we have,

$$B_{v,n}(f)(\xi^*/(1 + \operatorname{Re} z_v(\xi^*)), \operatorname{Re} z_v(\xi^*)/(1 + \operatorname{Re} z_v(\xi^*))$$
$$= \operatorname{Re} z_v(\xi^*)/(1 + \operatorname{Re} z_v(\xi^*)).$$

Since

$$\sum_{a_i \in N} g(x, a_i) = 1$$

from the last two relations we obtain,

$$\lambda^* = 1/(1 + \operatorname{Re} z_v(\xi^*))$$

and thus

$$x^v = (\xi^*/(1 + \operatorname{Re} z_v(\xi^*)), \operatorname{Re} z_v(\xi^*)/(1 + \operatorname{Re} z_v(\xi^*)))$$

is a fixed point of $B_{v,n}$.

This clearly gives the assertion of the theorem of Brouwer.

4.6. SOME EXAMPLES

First we give an example of spaces which have the fixed point property, using the Brouwer fixed point theorem. Next we mention an example of interesting spaces without the fixed point property.

Example 4.6.1. The $\sin 1/x$ circle has the fixed point property.

First we remark that we can give an entirely elementary proof of this theorem (i.e. we use only the notion of continuity and the properties of continuous functions). Indeed, we remark that what we need is the Brouwer theorem for the case of a set of the form $[a, b]$. In this case the Brouwer theorem can be proved in an elementary fashion. This proof is given in

THEOREM 4.6.2. *Let $f : [a, b] \to [a, b]$ be a function such that $g(t_0) = f(t_0) - t_0$ defined on $[a, b]$ has the Darboux property. Then f has a fixed point in $[a, b]$.*

Proof. We can suppose, without loss of generality, that $[a, b]$ is the segment $[0, 1]$. The function g has values in the points $t = 0$ and $t = 1$ as follows:

$$g(0) = f(0) \geqq 0, \qquad g(1) = f(1) - 1 \leqq 0$$

and, since g has the Darboux property, we find a $t_0 \in [0, 1]$ such that

$$g(t_0) = 0$$

and thus t_0 is a fixed point of f.

For the proof of Theorem 4.6.1. we recall that the $\sin 1/x$ circle is the closure of the graph of the function $y = \sin 1/x$ for $0 < x \leqq 1/\pi$ and an arc from $p_1 = (0, 1)$ to $p_2 = (1/\pi, 0)$. We denote the point $(0, -1)$ as p_1.

For the proof of the fact that the sin $1/x$ circle has the fixed point property we use, following R. H. Bing (1969), the dead-end method.

In what follows, t denotes the dog and $f(t)$ denotes the rabit, the direction along the arc in the sin $1/x$ circle from t to $f(t)$ is the direction of the chase, since we consider the dog chasing the rabbit. Suppose that the dog starts at the point p_1, and we remark that the rabbit is in the front of the dog, i.e. t is on the arc $[p_1, f(t)]$ of the sin $1/x$ circle, and if the dog moves slightly beyond p_1 towards p_2 and, because f is a continuous function, when t moves past p_3 and then back along the sin $1/x$ graph to near $[p_1, p_2]$, $f(t)$ stays in front of t. Since each point in $[p_1, p_2]$ is a limit of a sequence of points $\{t_i\}$ in the sin $1/x$ graph and since each t_i is ahead $f(t_i)$, it follows that $\{f(t_i)\}$ approaches $[p_1, p_2]$. This implies that $t_0 = \lim t_i$ has the property that $f(t_0) \in [p_1, p_2]$. Thus $f\{[p_1, p_2]\} \subset [p_1, p_2]$.

In this case by Theorem 4.3.2, f has a fixed point in $[p_1, p_2]$.

An interesting example of a space without the fixed point property was found by S. Kinoshita. The space has the name 'a can-with-a-roll-of-toilet-paper' and is described mathematically as follows: a cylindrical can with a bottom but no top plus the part of the solid cylinder which lies above a spiral on the base converging to the edge of the base.

The result of Kinoshita (1953) (which we give without proof) is as follows:

THEOREM 4.6.3.   *The 'can-with-a-roll-of-toilet-paper' is a contractible space and does not have the fixed point property.*

4.7. SOME APPLICATIONS OF BROUWER'S FIXED POINT THEOREM

The Brouwer fixed point theorem has many important applications. These are scattered through the literature. In what follows we present two applications. The first refers to the well known theorem concerning the existence of roots for complex polynomials, the so-called Fundamental Theorem of Algebra. We show that the theorem can be proved using the Brouwer fixed point theorem. The second application refers to the existence of equilibrium points in an economy.

For further information about this we refer to the excellent book of J. T. Schwartz (1961). Other very good sources are the books or lecture notes of M. Hazewinkel (1972), H. Nikaido (1968) and K. J. Arrow and F. J. Hahn (1971).

First we give the Fundamental Theorem of Algebra.

THEOREM 4.7.1.   *Let $p(z) = a_0 + a_1 z + \cdots + a_n z^n$ be a complex polynomial. Then there exists a z_0 such that $p(z_0) = 0$.*

Proof. (B. H. Arnold, 1949). We can suppose, without loss of generality, that $a_n = 1$. Let $z = re^{i\theta}$, $0 \le \theta < 2\pi$ and

$$R = 2 + |a_0| + \cdots + |a_{n-1}|.$$

We now define a function g on the complex plane as follows:

$$g(z) = \begin{cases} z - f(z)/(Re^{i(n-1)\theta} & |z| \le 1 \\ z - f(z)/(Rz^{n-1}) & |z| \ge 1. \end{cases}$$

From the form of g it is clear that it is a continuous function. We consider now the set

$$C = (z, |z| \le R)$$

which is obviously a compact and convex set in the plane.

We now show that C is an invariant set for g. Indeed, suppose that $|z| \le 1$ and thus,

$$|g(z)| \le |z| + |f(z)|/R \le 1 + (1 + |a_0| + \cdots + |a_{n-1}|)/R$$

$$\le 1 + 1 = 2 < R.$$

Suppose now that $|z| \ge 1$. Then we have,

$$|g(z)| \le /z - z/R - (a_0 + \cdots + a_{n-1}z^{n-1})/(Rz^{n-1})|$$

$$\le (R-1) + (|a_0| + \cdots + |a_{n-1}|)/R$$

$$\le R - 1 + (R-2)/R \le R$$

and thus C is an invariant set for g. Let z_0 be a fixed point of g. It is clear that this satisfies the relation

$$p(z_0) = 0.$$

We consider now the application to equilibrium points in an economy. In what follows we refer briefly to the well-known Leontieff models.

We suppose that we have N producers each producing a product. Suppose that X_i represents the total output of producer P_i and x_{ij} represents the total amount of the product of producer P_i which is consumed by producer P_j. Define

$$Y_i = X_i - \sum_{j=1}^{N} x_{ij}$$

and these numbers can be interpreted as the difference between the total output of producer P_i and the amount consumed by the producers $P_1, \ldots, P_N$.

The number Y_i is known as 'final demand' of the commodity i.

The closed Leontieff model assumes that $Y_i = 0$ for all $i = 1, 2, 3, \ldots N$. The numbers

$$a_{ij} = x_{ij}/X_j$$

are called the 'production coefficients'. In the linear Leontieff model the numbers a_{ij} are supposed to be constants and independent of X_i. In this case we have

$$(I - A)X = Y$$

where $A = (a_{ij})$, $X = (X_1, \ldots, X_N)$, $Y = (Y_1, \ldots, Y_N)$, and the problem is the following one: given Y, find X.

However, to draw a closer parallel to real economic systems, the assumption that a_{ij} are constants is changed to the following one: a_{ij} are continuous positive functions.

If $f_{ij}(x)$ represents the amount of the capital that the producer P_i spends on the commodity G_j produced by P_j when his income is x, then $f_{ij}(x)$ can be interpreted as 'demand functions' and of course $f_{ii}(x) = 0$ for all $i = 1, 2, 3, \ldots, N$.

Now, if each producer spends his income by buying goods from the other producers, then we must have the relation

$$(*) \qquad x = \sum_{j=1}^{N} f_{ij}(x)$$

and since the economic law asserts that the income x_i of the producer P_i is determined in such a way that the total amount of each commodity sold by a producer must be equal to the total amount of that commodity bought by the other producers we must have – in mathematical terms – the relation

$$(**) \qquad x_j = \sum_{i=1}^{N} f_{ij}(x_i).$$

Now since the functions $f_{ij}(,)$ are supposed to be nonlinear and continuous we can obtain the existence of a point x, $x = (x_1, \ldots, x_N)$ satisfying the relation $(**)$.

We have, in fact, the following theorem:

THEOREM 4.7.2.   *Suppose that $f_{ij}(,)$ are continuous positive functions satisfying* (*). *Then there exist* $x = (x_1, \ldots, x_N)$ *satisfying* (**).

Proof. We consider the set C defined as follows:

$$C = (y = (y_1, \ldots, y_N), \qquad y_i \geqq 0, \qquad \sum_1^N y_i = 1)$$

which is obviously closed and convex. Since it is bounded, it is in fact compact. We define $g : C \to C$ by the relation

$$(g(y))_j = \sum_{i=1}^N f_{ij}(y)$$

which has C as an invariant set. Then by Brouwer's fixed point theorem, g has a fixed point which clearly satisfies the conditions of the theorem.

4.8.   THE COMPUTATION OF FIXED POINTS. SCARF'S THEOREM

The fixed point algorithms can be regarded as one of the major breakthroughs in computational methods, giving us for the first time a numerical algorithm for approximating, in some sense, the fixed points for some classes of mappings.

The first algorithm was proposed by H. E. Scarf in 1967 and since then many algorithms have been proposed having as a basis different proofs of Brouwer's theorem. In what follows we present Scarf's algorithm and we note that the algorithms proposed in connection with Brouwer's theorem can be classified as being based on combinatorial proofs of Brouwer's theorem and on differential topology proofs of Brouwer's theorem, as well as on first- second- and third-generations of algorithms, depending on how they utilize the information obtained from earlier computer runs:

1.   the first generation: do not allow inclusion of information obtained from earlier computer runs;

2.   second generation: allowed, but only in a restart procedure;

3.   third generation: allow a continuous refinement of the grid size.

Since many subsequent proofs and algorithms use the ideas in Scarf's proof, we first give Scarf's result. Next we present Scarf's algorithm as well as an application to economics. Some other algorithms are mentioned.

The basic idea of Scarf is to perform a systematic search of a sequence of points which are to be examined. This point of view was further developed by Harold Kuhn. For other developments we refer to Eaves (1976).

Let us consider the simplex

$$S = (x = (x_1, x_2, \ldots, x_n), \ x_i \geq 0, \ \sum x_i = 1)$$

and let

$$A_k = (p^1, \ldots, p^n, \ldots, p^k)$$

be a finite set of vectors in $\mathbf{R}^n$. We suppose that

$$p^1 = (0, M_1, \ldots, M_n)$$
$$p^2 = (M_2, 0, M_2, \ldots, M_n)$$

$$p^{n-1} = (M_{n-1}, \ldots, M_{n-1}, 0, M_{n-1})$$
$$p^n = (M_n, \ldots, M_n, 0)$$

with

$$M_1 > M_2 > \cdots > M_n > 1.$$

DEFINITION 4.8.1.   A set of n vectors in A_k,

$$(p^{i1}, \ldots, p^{in})$$

is called a primitive set if there are no vectors $p^j \in A_k$ such that

$$p_1^j > \min(p_1^{i_1}) \qquad 1 \leq k \leq n$$
$$p_2^j > \min(p_2^{i_k})$$

$$p_n^j > \min(p_n^{i_k}).$$

DEFINITION 4.8.2.   Let $(p^{i_1}, \ldots, p^{i_n})$ be a primitive set in A_k. Then the subsimplex of S defined by

$$(x = (x_1, x_2, \ldots, x_n), \ x_i > \min(p_i^{i_k}))$$

is called a primitive subsimplex.

The following assumption will be of crucial importance.

NONDEGENERACY ASSUMPTION 4.8.3.   If p^i, p^j are two different vectors in A_k then they do not have the same ith coordinate for any i.

From this assumption it follows that each n-bounding face parallel to one of the coordinate hyperplanes contains only one element of the primitive set, namely, the vector in the primitive set for which the corresponding coordinate is smallest.

Now we give the combinatorial lemma of Scarf.

LEMMA 4.8.3. *Suppose that $(p^{j_1}, \ldots, p^{j_n})$ is a primitive set and let p^{j_r} be a fixed element of this set. Then there exists a unique element p^j in A_k, $p^j \neq p^{j_r}$ such that*

$$(p^{j_1}, \ldots, p^{j_{r-1}}, p^j, p^{j_{r+1}}, \ldots, p^{j_n})$$

is a primitive set and this is true with one exception, namely when $(n-1)$-elements $p^{j_k}(k \neq r)$ are selected from the first n vectors of A_k. In this last case no replacement is possible to obtain a new primitive set.

Proof. We can construct this element p^j as follows: first we may assume that

$$p_i^{j_i} = \min_k (p_i^{j_k})$$

and thus p^{j_i} lie on the face of the primitive subsimplex on which the ith coordinate is constant. Suppose that p^{j_1} is being removed and consider $p^{j_{i^*}}$ to be that vector in the primitive set with the second-smallest value of its first coordinate. In this case j_{i^*} is greater than n unless the exceptional case already mentioned arises. We move now the face containing p^{j_i} parallel to itself, lowering the ith coordinate until we first intersect a vector p^j in A_k having the property that

$$p_i^j > p_i^{j_i} \qquad i \neq 1, i^*$$

$$p_1^j > p_1^{j_{i^*}}$$

or the face of S with $x_i = 0$; thus we have a vector which replaces p_1^j. We remark that the above construction works in all cases but the vectors $(p^{j_i})(i \neq 1)$ are all in the first n vectors of A_k.

The set obtained in this way is obviously a primitive set. Now suppose that

$$(p^s, p^{j_2}, \ldots, p^{j_n})$$

is a primitive set and then we get that, for $i \neq 1, i^*$, p^{j_i} are on that bounding face of the new primitive subsimplex whose ith coordinate is constant, since, in the contrary case, there exists i_0 such that $p^{j_{i_0}}$ are not on bounding faces of the new primitive subcomplex and this is a contradiction. Then we have that for each $i \neq 1, i^*$,

$$p_i^{j_i} = \min(p_1^s, p_i^{j_2}, \ldots, p_i^{j_n}).$$

Now it remains to choose the coordinates 1 and i^*. Obviously, the vector p^s is either on the face with constant first coordinate and $p^{j_{i^*}}$ on that with i^*th coordinate constant or vice versa. Indeed, if the first does not appear, then $p^{j_{i^*}}$ would be on the face of the subsimplex with the ith coordinate constant for $i = 2, \ldots, n$. Now if $(p^s, p^{j_1}, \ldots, p^{j_n})$ is a primitive set and $p^s \neq p^{j_1}$ then the vector p^s is on the face with ith coordinate constant and $p_{j_{i^*}}$ is on the face with the first coordinate constant. Indeed, if this is not so, if $p_1^{j_1} < p_1^s$ (we suppose that $p^s \neq p^{j_1}$) the old simplex has p^s in its interior and if the contrary inequality holds, p^{j_1} is in the interior of the new simplex. Then we must have $p_1^s = p_1^{j_1}$ and the nondegeneracy assumption implies that $s = j_1$ and thus we are at the initial conditions.

If $(p^s, p^{j_1}, \ldots, p^{j_n})$ is a primitive set and $p^s \neq p^{j_1}$ then p^s is the vector p^j considered above. Indeed, since $p^{j_{i^*}}$ is on the face with ith coordinate constant $(i \neq 1, i^*)$ and $p^{j_{i^*}}$ lie on the face with the first coordinate constant, then $p^s = p^j$.

Now from the form of vectors in the exceptional case we get that no replacement in possible and thus the vector p^j is found.

We now show that we can determine a primitive set with all vectors indexed differently.

If A_k is a primitive set, then each element of A_k is associated with an index which is a number in $(1, 2, \ldots, n)$. The index is arbitrary, except for the first n vectors in the set, and we require that the first element has the index 1, the second has the index 2, etc. Now the algorithm which permits us to obtain a primitive set with all members indexed differently is as follows: we begin with a primitive set with all members except only two being indexed differently. We consider the vectors

$$(p^2, \ldots, p^n, p^{j^*})$$

with p^{j^*} such that the first coordinate is maximal beyond the first n vectors. Then

$$\min(p_i^{j^*}, p_i^2, \ldots, p_i^n) = \begin{cases} p_i^j & i = 1 \\ 0 & i > 1 \end{cases}$$

and $(p^2, p^3, \ldots, p^n, p^{j^*})$ is a primitive set. Now we have two cases to consider:

Case 1. p^{j^*} is associated with the index 1,
Case 2. p^{j^*} is associated with one of the indices $2, \ldots, n$.

In the first case we have a primitive set with all members differently indexed. Thus we are concerned with the following primitive sets:

(a) 1 is not an index for vectors in the set,

(b) all vectors are indexed differently, except two vectors which have the same index.

We then take one vector from those having the same index and remove it from our primitive set. In this case we then obtain either a set with the same properties or a solution. In the case when we are not at the initial primitive set, then one of the two vectors existing in the primitive set with the same index will have been introduced to arrive at the current position and the algorithm proceeds to dismiss the other vector from the pair.

Thus we have at each stage two possible removals from our primitive set: (1) the first, to get the current position, (2) since there is only the vector $p^j (2 \leq j \leq n)$ which can be removed from the primitive set to obtain another primitive set, the other possibility is to remove p^{j^*} and, in this case, we can apply the Lemma 4.8.3. We continue in this way until we find a primitive set with all members indexed differently and since there exists only a finite number of primitive sets we have a finite number of steps.

Now, we can never return to a previous primitive set since, if the first returns to a primitive set which is not the initial one then we have three ways to emerge from that primitive set, and if the first return is to the initial primitive set, then we have two ways to emerge from that initial primitive set.

Thus we have

THEOREM 4.8.4. *There exists a primitive set with all members indexed differently.*

As a corollary we obtain Brouwer's fixed point theorem.

COROLLARY 4.8.5. (Brouwer fixed point theorem). *If $f : B_n \to B_n$ is continuous and $B_n = (x, x \in \mathbf{R}^n, \|x\| \leq 1)$ then f has a fixed point in B_n.*

Proof. We chose a sequence of primitive sets A_k such that for $k < \infty$, the maximum diameter of a primitive subsimplex tends to zero and thus we find a sequence of primitive subsimplices converging to a point, say x', and from the continuity of f we have

$$f(x')_i \geq x'_i$$

and thus $x' = f(x')$.

Following Scarf we give some indications about the computational techniques. Of course, a basic problem is to select the vectors in the primitive set A_k, which is a primitive set of n vectors (n is the dimension of the space $\mathbf{R}^n$ in which the vectors are situated). From this primitive set one vector is eliminated and it is replaced by another vector. Of course when the primitive set is of the form

$$(k_1/D, k_2/D, \ldots, k_n/D)$$

where k_i are positive integers and $k_1 + \cdots + k_n = D$ the search for vectors can be simpler. Indeed, in this case, p^j will be either the vector given by the formulas

$$p_i^j = a_i + 1/D \qquad i \neq i*$$
$$p_{i*}^j = 1 - \sum_{i \neq i*} (a_i + 1/D)$$

or one of the first n members of A_k.

The vector $a = (a_i)$ is calculated examining all vectors in A_k with

$$p_i^j > a_i \qquad i \neq i*$$

and selecting that vector with the largest value of p_{i*}^j. It is possible that the A_k obtained in this way does not satisfy the nondegeneracy assumption. Scarf's method of avoiding this difficulty consist in constructing at each step of the algorithm the matrix

$$\begin{Bmatrix} 0 & M_2 & M_3 \cdots & M_n & p_1^{n+1} \cdots & p_1^l \\ M_1 & 0 & M_3 & M_n & \cdot & \cdot \\ M_1 & M_2 & 0 & M_n & \cdot & \cdot \\ \cdot & & M_3 & \vdots & \cdot & \cdot \\ \cdot & & & M_n & \cdot & \cdot \\ M_1 & M_2 & \cdots 0 & & p_n^{n+1} \cdots & p_n^l \end{Bmatrix}$$

and, if the matrix has two columns with identical elements in the ith row, the first is assumed to be larger, and if there exists a vector in the matrix having an entry in the ith row identical to a vector not in the matrix, the former is supposed to be larger. It is not difficult to see that this method gives a finite algorithm. The determination of p^j is then reduced to a search over vectors which have been used in previous steps and the rest of the vectors in A_k are found by one calculation.

The algorithm continues and terminates with a primitive set with all members differently indexed. For the approximation of the fixed point of f any point of the subsimplex of the primitive set obtained can be taken as an approximate fixed point.

For certain estimates of error bounds for approximate fixed points we refer to the very interesting paper of Ch. Bowman and S. Karamardian (1976).

We give now some indications about the applicability of the above results to equilibrium prices in a general economic model of exchange. See also Section 4.7.

Suppose that we have n commodities in an economy and m economic agents. If the prices are

$$\pi = (\pi_1, \pi_2, \ldots, \pi_n)$$

for the n commodities and the l agents respond by the excess demand

$$g_1^l(\pi), g_2^l(\pi), \ldots, g_n^l(\pi)$$

in which $g_1^l(\pi)$ represents the net increase in the commodity j desired by the l agent at the price π. We make the following assumptions about the excess demand functions $g_j^l(,)$:

1. $g_j^l(,)$ is homogeneous of degree $0, j = 1, 2, \ldots n$.

2.   for each j

$$\sum_{i=1}^{n} \pi_i g_i^j = 0, \qquad j = 1, 2, \ldots n.$$

3. $g_j^l(,)$ are continuous functions on the simplex
$$S = (x, x = (x_1, \ldots, x_n) x_i \geqq 0, \sum x_i = 1).$$

We define now the market excess demand for the commodity j

$$g_j(\pi) = \sum_{l=1}^{m} g_i^l(\pi)$$

and we say that a vector of prices is in equilibrium if

$$g_j(\pi) \leqq 0$$

and is actually zero if the price associated with that commodity is strictly positive. As in Section 4.7. we can conclude that there exists at least one equilibrium price.

The mapping

$$f : S \to S$$

is defined by

$$f = (f_1, f_2, \ldots, f_n)$$

where

$$f_j(\pi) = (\pi_j + s \max(0, g_j(\pi)) / (1 + s \sum_k \max(0, g_k(\pi)))$$

and s is a small positive constant. Then from our hypotheses using Brouwer's fixed point theorem we conclude that there exists at least one point π^* such that $f(\pi^*) = \pi^*$.

We show now that this is an equilibrium price vector.

Suppose that

$$\sum_k \max(0, g_k(\pi^*)) > 0$$

and thus, $f(\pi^*) = \pi^*$ gives

$$\pi_j^* + {}_s \max(0, g_j(\pi^*)) = C\pi_j^*$$

with $C > 1$. This gives that

$$g_j(\pi^*) > 0$$

for each j. But by assumption 2

$$\sum \pi_j^* g_j(\pi^*) = 0$$

and, since π^* is in S, this is a contradiction.

Thus we must have

$$\sum_k \max(0, g_k(\pi^*)) = 0$$

and this implies that

$$g_j(\pi^*) \leq 0$$

for $j = 1, 2, \ldots, n$. Assumption 2 implies once again that

$$g_j(\pi^*) = 0$$

if $\pi_j^* > 0$. Thus π_j^* is an equilibrium price vector. To find the fixed point, the labelling of the vector x will be done as follows: the index j is associated with the vector χ if

$$f_j(\chi) \geqq \chi_j$$

or

$$\max(0, g_j(\chi)) \geqq \chi_j \sum \max(0, g_k(\chi)).$$

For another very interesting algorithm for computing fixed points of functions with continuous derivatives, based on Hirsch's proof of the noncontractibility of the cell onto its boundary (see Section 4.1), we refer to the paper of Kellog, Li and Yorke (1976).

For further information see Eaves (1976), Garcia (1976) Kojima (1978), Forster (1978), the book edited by Karamardian (1977).

Chapter 5

Schauder's Fixed point Theorem and some Generalizations

5.0. INTRODUCTION

L. E. J. Brouwer proved his famous theorem in 1912. In 1922 J. W. Alexander, under the impression that it was proved for homeomorphisms only, gave a new proof. Also, G. D. Birkhoff and O. D. Kellogg, labouring under the same impression, gave a new proof and, what is more important, they obtain extensions to a more general setting, namely, that any compact convex set in the spaces $L^2_{[0,1]}$ and $\mathbf{C}^n_{[0,1]}$ has the fixed point property for continuous mappings.

In 1926 S. Lefschetz gave an extension of Brouwer's theorem to orientable n-manifolds without boundary, using what is called now the Lefschetz number $\Lambda(f)$, and also extended this to the case of n-manifolds with boundary. This result was further extended to finite polyhedra by H. Hopf in 1929.

This direction of research uses as a tool the theory of homology and we remark that there exists another approach to the extension of Brouwer's theorem suggested by the paper of Birkhoff and Kellogg.

This direction has as an essential characteristic an assumption about convexity and compactness. Thus in 1927 J. Schauder extended the Birkhoff-Kellogg theorem to metric linear spaces having a linear base, and in 1930 he obtained the result that every compact convex set in a Banach space has the fixed point property for continuous mappings, as well as that every weakly compact convex set in a separable Banach space has the fixed point property for weakly continuous mappings. An improvement of the last assertion was obtained by M. Krein and V. Smulian in 1940: if C is a closed convex set in a Banach space and $f : C \to C$ is weakly continuous with the property that $f(C)$ is separable and the weak closure of $f(C)$ is weakly compact then f has a fixed point in C.

Thus if X is a Banach space and f is a mapping such that:

1. $f : X \to X$ is continuous, $f(X)$ is bounded and the image of each bounded set has a compact closure then there exists in X a fixed point of f.

2.   If C is a closed convex set of X and $f:C\to C$ is continuous and the closure of $f(C)$ is compact then f has a fixed point in C.

3.   If C is a compact convex set in X and $f:C\to C$ is continuous then f has a fixed point in C.

Assertion 1 was extended to locally convex spaces by A. N. Tychonoff in 1935 and assertion 2 by M. Hukuhara in 1950 to the same setting.

An interesting extension of both assertions 1 and 2 was obtained by F. Browder (1959), under some conditions for the iterations of the mapping. The result is as follows:

(a). If $f:X\to X$ is continuous and such that, for some integer m, $f^m(X)$ is bounded and the image of each bounded set has a compact closure, then f has a fixed point in X.

(b). If C and C_1 are open convex sets in X, C_0 is a closed convex set and the following relation holds

$$C_0 \subset C_1 \subset C$$

and $f:C\to X$ is continuous with the property that $f(C)$ is compact, and if, for a positive integer m, f^m is well defined on C_1, $\bigcup_{i=0}^{m} f^i(C_0) \subset C_1$; while if $f^m(C_1) \subset C_0$, then f has a fixed point in C_0.

A new and important step in extending the Schauder theorem to more general classes of mappings was made by G. Darbo in 1955.

Using what now is called the Kuratowski measure of noncompactness Darbo defines a new class of mappings, the so-called α-set contractions, which contains the compact mappings as well as the contraction mappings, and proves a fixed point theorem for this class of mappings.

This very interesting paper has escaped the attention of the mathematical world for a long time. In 1967 Sadovski, using a new measure of non-compactness, proved a generalization of the Schauder Fixed Point Theorem for mappings which are now called condensing or densifying.

The idea in Darbo's paper was further developed and now there exists a great variety of generalizations of this extension of Schauder's theorem. Also we may mention that the Tychonoff theorem was generalized using a suitable notion of measure of non-compactness in locally convex spaces.

In what follows we give an account of these results as well as of some applications. Among them we mention the beautiful result of Lomonosov in the problem of invariant subspaces: if T is an element in $L(X)$ and commutes with a nonzero compact operator then there exists a nontrivial hyperinvariant subspace for T. (A closed subspace of X, say X_1, is called hyperinvariant for an operator S in $L(X)$ if X_1 is invariant for all operators commuting with S).

5.1. THE SCHAUDER FIXED POINT THEOREM

We give now the Schauder fixed point theorem and some variants.

Let X be a Banach space.

DEFINITION 5.1.1. A continuous mapping $f : M \to M$, where M is a subset of X, is called compact if for each bounded sequence $\{x_n\} \subset M$, the sequence $\{f(x_n)\}$ has a convergent subsequence.

Our purpose in what follows is to prove the following results which are known as Schauder's Fixed Point Theorems. We give them here because in applications one form or another is used in the sense the sometimes it is easy to prove that the mapping is compact and sometimes that the set invariant for the mapping is compact.

THEOREM 5.1.2. If C is a bounded closed convex set in X and $f : C \to C$ is a compact mapping then there exists a point $x_0 \in C$ such that $f(x_0) = x_0$.

THEOREM 5.1.3. If C is a compact convex set in X and $f : C \to C$ is a continuous mapping then there exists $x_0 \in C$ such that $f(x_0) = x_0$.

For the proof we need some results about sequences of mappings.

LEMMA 5.1.4. If $\{f_n\}$ is a sequence of compact mappings converging uniformly on a bounded set M of X to a mapping f_0 then $\overline{\{f_0(x), x \in M\}}$ is a compact set and f_0 is a continuous mappings.

Proof. The fact that f_0 is a continuous function is obvious from the uniform convergence. We prove now that $\overline{\{f_0(x), x \in M\}}$ is a compact set. Since $f_n \to f_0$ uniformly on M for any $\varepsilon > 0$, given there exists N_ε such that for all $n \geq N_\varepsilon$, then

$$\| f_n(x) - f_0(x) \| \leq \varepsilon/2$$

holds for all $x \in M$. Now the set $\{f_n(x), x \in M\}$ is compact and thus for $\varepsilon/2$ there exists a finite number of elements, say $x_1, \ldots, x_m$, such that for any $x \in M$ there exists an index m_i such that

$$\| f_n(x) - f_n(x_{m_i}) \| \leq \varepsilon/2.$$

and in this case we obtain

$$\|f_0(x) - f_n(x_{m_i})\| = \|f_0(x) - f_n(x) + f_n(x) - f_n(x_{m_i})\|$$
$$\leqq \|f_0(x) - f_n(x)\| + \|f_n(x) - f_n(x_{m_i})\|$$
$$\leqq \varepsilon/2 + \varepsilon/2 = \varepsilon.$$

From Hausdorff's characterization of compact sets the assertion of the Lemma follows.

The following Lemma gives the main tool in proving the Schauder fixed point theorem.

LEMMA 5.1.5. *Let* $f : M \to M$, *and let* M *be a bounded set in* X *and* f *be a compact mapping. In this case there exists a sequence of mappings* $\{f_n\}$ *such that* :

1. $f_n \to f$ *uniformly on* M,

2. $\{f_n(M)\}$ *generates a finite-dimensional subspace of* X.

Proof. Since f is a compact mapping the set $\overline{\{f(x), x \in M\}}$ is a compact set. Let $\varepsilon_n \to 0$ and $\varepsilon_n > 0$. From the compactness, for each n we find a finite number of elements, say $x_1^n, \ldots, x_m^n$, such that for each $x \in M$ there exists x_i^n such that

$$\|Tx - x_i^n\| \leqq \varepsilon_n.$$

We define now the mapping g_n on TM in this way: for $x \in \overline{TM}$ let

$$T_n x = \left(\sum_1^m \alpha_i^{(n)}(x) x_i^n \right) \Big/ \left(\sum_1^m \alpha_i^{(n)}(x) \right)$$

and where

$$\alpha_i^{(n)}(x) = \begin{cases} \varepsilon_n - \|x - x_i^n\| & \text{if } \|x - x_i^n\| \leqq \varepsilon_n \\ 0 & \text{in the contrary case.} \end{cases}$$

From the definition of the mapping T_n it is clear that the convex hull of the set $\{x_1^n, x_2^n, \ldots, x_m^n\}$ is invariant for T_n.

Also it is clear that for each x there exists at least one i_0 such that $\alpha_{i_0}^n(x) \neq 0$ and the functions

$$x \to \alpha_i^n(x)$$

are continuous functions. From this it follows that

$$x \to \alpha_i^n(x) \bigg/ \left(\sum_1^m \alpha_i^n(x) \right)$$

is also continuous. Now we have,

$$\|x - T_n x\| = \left\| x - \left(\sum_1^m \alpha_i^n(x) x_i^n \right) \bigg/ \sum_1^m \alpha_i^n \right\|$$

$$= \left\| \left(\sum_1^m \alpha_i^n(x)(x - x_i^n) \right) \bigg/ \sum_1^m \alpha_i^n \right\|$$

$$\leq \varepsilon_n \left(\sum_1^m \alpha_i^n(x) \bigg/ \sum_1^m \alpha_i^n(x) \right) = \varepsilon_n$$

since if $\|x - x_i^n\| > \varepsilon_n$ then $\alpha_i^n(x) = 0$.

To prove the assertion of the Lemma we define the mappings $T_n^\sim$ as follows:

$$T_n^\sim x = \begin{cases} T_n(Tx) & \text{for } x \in M, \\ 0 & \text{in the contrary case.} \end{cases}$$

From the above estimate we have that $T_n^\sim$ converges uniformly on M to Tx and from the definition of T_n the second property is also satisfied.

LEMMA 5.1.6.   *Let M be a compact and convex set in a Banach space X and also let $T : M \to M$ be a continuous mapping. In this case there exists a sequence of mappings $\{T_n\}$ with the following properties:*

1. $T_n \to T$ *uniformly on* M,

2. $\{ T_n M]$ *is contained in a finite-dimensional space.*

Proof. Let $\varepsilon_n \to 0$ be a sequence of positive numbers. Since M is a compact set, for each n there exists a finite system of elements, say $\{x_1^n, x_2^n, \ldots, x_m^n\}$ such that for each $x \in M$ there exists an i with the property that

$$\|x - x_i^n\| < \varepsilon_n.$$

We define now the numbers μ_i^n as follows:

$$\mu_i^n(x) = \begin{cases} \|x - x_i^n\| & \text{if } \|x - x_i^n\| \leq \varepsilon_n/2 \\ \varepsilon_n - \|x - x_i^n\| & \text{if } \varepsilon_n/2 < \|x - x_i^n\| \leq \varepsilon_n \\ 0 & \text{in the contrary case.} \end{cases}$$

As above we can remark that for each x there exists at least one i such that $\mu_i^n > 0$ and that the functions

$$x \to \mu_i^n(x)$$

are continuous.

Now we can define the mappings T_n as follows:

$$T_n x = \sum_1^m \mu_i^n(x) x_i^n \Big/ \sum_1^m \mu_i^n(x),$$

and it is clear that these mappings have the following properties:

1. T_n is continuous,

2. for each x, $T_n x$ is in the convex hull of $\{x_1^n, x_2^n, \ldots, x_m^n\}$,

3. for each $x \in M$, $\|x - T_n x\| \leqq \varepsilon_n$,

4. the convex hull of $\{x_1^n, x_2^n, \ldots, x_m^n\}$ is invariant for T_n.

The assertion of the lemma follows from this, if we take

$$T_n^\sim(x) = \begin{cases} T_n x & \text{for } x \in M, \\ 0 & \text{in the contrary case.} \end{cases}$$

Also we need the following result, given as

LEMMA 5.1.7. *If $\{T_n\}$ is a sequence of compact operators converging uniformly on a set E to a mapping T, then the following assertion holds:*

$$E^\sim = \overline{\bigcup T_n E} \quad \text{with } T_n E = \{T_n x, x \in E\}$$

is a compact set.

Proof. Since $T_n \to T$ uniformly on E, for any $\varepsilon > 0$ there exists an integer N_ε such that for all $n \geqq N_\varepsilon$,

$$\|Tx - T_n x\| \leqq \varepsilon/2$$

for all $x \in E$.

Since the sets $\overline{T_0 E}, \ldots, \overline{T_{N_\varepsilon} E}$ are compact sets it follows that $\bigcup_0^{N_\varepsilon} \overline{T_i E}$ is also a compact set. Thus we find a finite set, say $x_0, x_1, \ldots, x_m$ such that if $T_j x \varepsilon \bigcup \overline{T_i E}$ then there exists an i such that

$$\|T_j x - x_i\| < \varepsilon/2.$$

In this case $\{x_1, x_2, \ldots, x_m\}$ is an ε-net for $\bigcup_0^\infty \overline{T_i E}$ and from Hausdorff's theorem the assertion of the Lemma follows.

Now we are ready to prove the Theorems 5.1.2. and 5.1.3.

Proof of Theorem 5.1.2. Let $\varepsilon_n \to 0$ be a sequence of positive numbers. From Lemma 1.5 there exists a sequence of continuous mappings $\{T_n\}$ and a convex and compact set E_n (the convex hull of $\{x_1^n, x_2^n, \ldots x_m^n\}$) such that

1. $\qquad T_n E_n \subset E_n,$

2. $\qquad \|T_n x - x\| \leqq \varepsilon_n.$

By Brouwer's fixed point theorem for each n there exists a fixed point x_n of T_n. From Lemma 1.7 it follows that this sequence is contained in a compact set. In this case we can suppose, without loss of generality, that $x_n \to x_{\tilde{0}}$. We show now that this is a fixed point for f. Indeed,

$$\|f(x_{\tilde{0}}) - x_{\tilde{0}}\| = \|f(x_{\tilde{0}}) - T_n x_{\tilde{0}} + T_n x_{\tilde{0}} - T_n x_n + T_n x_n - x_{\tilde{0}}\|$$
$$\leqq \|f(x_{\tilde{0}}) - T_n(x_{\tilde{0}})\| + \|T_n x_{\tilde{0}} - x_n\| + \|x_n - x_{\tilde{0}}\|$$
$$\leqq \varepsilon_n + \|T_n x_{\tilde{0}} - T_n x_n\| + \|x_n - x_{\tilde{0}}\|$$

and for $n \to \infty$ we obtain that

$$\|f(x_{\tilde{0}}) - x_{\tilde{0}}\| = 0$$

and the assertion of the theorem is proved.

Proof of Theorem 5.1.3. As above, we take a sequence of positive numbers $\{\varepsilon_n\}$, $\varepsilon_n \to 0$ and by Lemma 1.6 we find a sequence of mappings $\{T_n\}$ such that,

1. $\qquad T_n E_n \subset E_n, \qquad n = 1, 2, 3, \ldots,$

2. $\qquad \|T_n x - Tx\| \leqq \varepsilon_n, \qquad n = 1, 2, 3, \ldots,$

3. $\qquad E_n$ is a compact and convex set invariant for T_n.

Thus we can apply the Brouwer fixed point theorem and we obtain that for each n there exists x_n such that $T_n x_n = x_n$.

Now we can simply prove that there exists a fixed point for f as follows: since $\{x_n\}$ is a subset of a compact set, we can suppose, without loss of generality, that $\lim x_n$ exists and is equal to a point x_0. Since,

$$\|f(x_0) - x_0\| = \|f(x_0) - T_n x_0 + T_n x_0 - T_n x_n +$$
$$+ T_n x_n - T_n x_n - x_0\|$$
$$\leqq \|f(x_0) - T_n x_0\| + \|T_n x_0 - T_n x_n\| +$$
$$+ \|x_n - x_0\|$$

and for $n \to \infty$ this gives that

$$\| f(x_0) - x_0 \| = 0$$

and the assertion of the theorem is proved.

The following extension of the Schauder fixed point theorem was obtained by G. Jones (1963) and has applications in the theory of periodic systems of functional-differential equations.

We denote by h the mapping defined on bounded sets of a linear space which associates with each such subset its closed convex hull.

The extension of the Schauder fixed point theorem is as follows:

THEOREM 5.1.8 (G. Jones). *Let C be a closed and convex bounded set in a Banach space X and $f : C \to C$ be a continuous mapping such that, for some integer n, $((fh)^n C)$ is contained in a compact set.*

In this case f has a fixed point in C.

Proof. Since C is closed and convex we have that $(fh)C \subset C$ and thus $(fh)^n C$ is contained in $(fh)^{n-1} C$. Thus

$$f(h((fh)^n C)) \subset (fh)^n C \subset h(fh)^n C$$

and thus f has $(fh)^n C$ as an invariant set. From the Schauder fixed point theorem it follows that f has a fixed point in $(fh)^n C$ and thus in C.

For a generalization of this result see Section 5.2.

5.2. DARBO'S GENERALIZATION OF SCHAUDER'S FIXED POINT THEOREM

As we know, the Schauder fixed point theorem uses in an essential way some compactness condition about the mapping (this can, of course, be about the mapping itself or about the domain of the definition), and an interesting and important problem appears to be to extend the assertion of the existence of fixed points in the Schauder theorem to mappings without strong relation with compactness.

The first result in this direction appears to have been given by M. Volpato in 1953. However the first essential step in extending the Schauder fixed point theorem was made by G. Darbo in 1955.

The main idea is to use the measure of noncompactness of Kuratowsi (defined in 1939) and to define classes of mappings using this number, called now the Kuratowski measure of noncompactness.

Using this measure, as well as other measures of noncompactness, several classes of mappings were defined and, at the same time, various extensions of the Schauder theorem were obtained.

In what follows we give some of these extensions. See also Section 2.7.

First we begin with some definitions of various classes of mappings as well as some simple examples of mappings in these classes. We also make the remark that we define these mappings only for Kuratowski's measure of noncompactness, but that the definitions have sense also for other measures of noncompactness.

DEFINITION 5.2.1.   Let X be a complete metric space and $f : X \to X$ be a continuous mapping. The mapping f is called an α-set contraction if there exists $k \in [0, 1)$ such that, for all bounded noncompact subsets of X, the following relation holds

$$\alpha(f(A)) \leq k\alpha(A).$$

DEFINITION 5.2.2.   The continuous mapping $f : X \to X$ is called condensing if, for any bounded noncompact subset in X,

$$\alpha(f(A)) < \alpha(A).$$

It is obvious that every α-set contraction is a condensing mapping. The following example shows that the converse is not true.

Example 5.2.3.   We give now an example of a condensing mapping which is not an α-set contraction for any $k \in [0, 1)$.

For this purpose we consider a strictly decreasing function $\varphi : [0, 1] \to \mathbf{R}^+$ with the property that $\varphi(0) = 1$. Let B_1 be the unit ball with center in 0 of an infinite dimensional Banach space.

We now define the following mapping:

$$f : B_1 \to B_1,$$

by the relation $f(x) = \varphi(\|x\|)x$.

First we show that this is not an α-set contraction for any $k \in [0, 1)$. Indeed, for this we consider the sets $B_r = \{x, \|x\| \leq r\}$ which have the property that $f(B_r) \supset B_{\varphi(r)r}$ and thus

$$\alpha(f(B_r)) = 2\varphi(r)r$$

and, since

$$\alpha(B_r) = 2r,$$

it follows that f is at best an α-set contraction with $k = \varphi(r)$.

Since $\varphi(r) \to 1$ for $r \to 0$ it follows that f cannot be an α-set contraction for any $k \in [0, 1)$.

We now show that f is a condensing mapping. Let A be a subset and $A \subset B$, with $\alpha(A) > 0$.

Since for any bounded set S, $S \subset B_1$,

$$f(S) \subset \overline{\text{conv}} \{S, 0\}.$$

we obtain

$$\alpha(f(S)) \leq \alpha(S \cup \{0\}) = \alpha(S).$$

Now if S is in B_1 and is noncompact, i.e. $\alpha(S) = d > 0$ we take $r < d/2$ and define two sets:

$$S_1 = S \cap B_r, \qquad S_2 = S \cap B_r'$$

where B_r' is the complement of B_r. Clearly we have

$$f(S) = f(S_1) \cup f(S_2)$$

and since f is a 1-set contraction we get,

$$\alpha(f(S_1)) \leq 2r < d = \alpha(S).$$

The function φ is supposed to be strictly decreasing and thus, for x, $\|x\| \geq r$,

$$f(S_2) \subset (sa, 0 \leq s \leq \varphi(r), a \in S_2) \subseteq \overline{\text{conv}\,(\varphi(r)S_2)}$$

which implies that

$$\alpha(f(S_2)) \leq \varphi(r)\alpha(S) < \alpha(S).$$

But from the property

$$\alpha(A \cup B) = \max\,(\alpha(A),\, \alpha(B)),$$

we obtain

$$\alpha(f(S)) \leq \max\,(\alpha(f(S_1)),\, \alpha(f(S_2))) < \alpha(S)$$

i.e. f is a condensing map.

DEFINITION 5.2.4. Let X be a metric space and $f : X \to X$ be a continuous α-set contraction. The index α_f of f is the infimum of all $s > 0$ such that f has the property that, for any noncompact set S in M,

$$\alpha(f(S)) \leq s\alpha(S).$$

Concerning this number associated with α-set contractions we can give the following properties first enunciated by G. Darbo:

LEMMA 5.2.5.   *Let X be a metric space and f, g be two α-set contractions. In this case $f_o g$ and $g_o f$ are α-set contractions and*

$$\alpha_{f_o g} \leqq \alpha_f \cdot \alpha_g, \qquad \alpha_{g_o f} \leqq \alpha_f \cdot \alpha_g.$$

Proof. Since the proof is simple we omit it.

LEMMA 5.2.6.   *Let C be a closed, bounded and convex set in a Banach space X. Suppose that*

$$f_i : C \to C$$

is an α-set contraction with α_{f_i}. If (a_i) are positive numbers such that

$$\sum_1^\infty a_i = 1$$

then

$$f = \sum_1^\infty a_i f_i$$

is an α-set contraction and

$$\alpha_f \leqq \sup_i \alpha_{f_i}.$$

Proof. Again we omit the proof since the assertion is an easy consequence of the definition of the number α_f.

From this we obtain a very useful property of retraction mappings.

LEMMA 5.2.7.   *Let (X, d) be a metric space and suppose that $(V_i)_1^m$ is a finite open covering of X. Let $f_i : V_i \to Y$ where Y is a Banach space. Suppose further that f_i are α-set contractions and if (λ_i) is a partition subordinate to the covering (V_i) then*

$$f(x) = \sum_1^m \lambda_i(x) f_i(x)$$

is an α-set contraction and

$$\alpha_f \leqq \sup_i \alpha_{f_i}.$$

Proof. Suppose M is any bounded set in X and since $\sum_1^m \lambda_i(s) = 1$ for all

$s \in X$ we have, obviously,

$$f(x) \leqq \operatorname{conv}\left(\bigcup_1^m f_i(M \cap V_i)\right)$$

and thus

$$\alpha f(M) \leqq \max_i \alpha f_i(M \cap V_i)$$

and the assertion is proved.

Consider now a Banach space X and $B_r = (x, x \in X, \|x\| \leqq r)$ and the retraction associated with B_r,

$$Rx = \begin{cases} x & \text{if} \quad \|x\| \leqq r, \\ rx/\|x\| & \text{if} \quad \|x\| \geqq r. \end{cases}$$

Then we have the following result.

LEMMA 5.2.8.   *The mapping R has $\alpha_R = 1$.*

Proof. Indeed, we can take

$$\lambda_1(x) = \begin{cases} 1/\|x\| & \text{if} \quad \|x\| \geqq r \\ 1 & \text{if} \quad \|x\| b r \end{cases}$$

$$\lambda_2(x) = 1 - \lambda_1(x)$$

and

$$f_1(x) = x, \qquad f_2(x) = 0.$$

The assertion follows now from Lemma 5.2.7.

The following theorem was the extension of Schauder's fixed point theorem given by G. Darbo:

THEOREM 5.2.9.   (G. Darbo, 1955). *Let C be a closed, bounded and convex set in a Banach space X. In this case if $f : C \to C$ is any α-set contraction then f has a fixed point in C.*

Proof. For each integer n we define the following set

$$C_n = \overline{\operatorname{conv}(f(C_{n-1}))}$$

$$C_1 = \overline{\operatorname{conv}(f(C))}.$$

It is clear that $C_n \supseteq C_{n+1}$ for all n and thus we can consider the set $\bigcap_n C_n$. Clearly this a closed and convex set. We show now that it is compact.

Indeed, we have the following estimate for the measure of noncompactness of C_n,

$$\alpha(C_n) \leqq \alpha_f^n \cdot \alpha(C)$$

and since $\alpha_f < 1$, we get

$$\lim \alpha(C_n) = 0.$$

This implies that $\bigcap_n C_n$ is nonempty and compact. Since for each n,

$$f : C_n \to C_{n+1}$$

we obtain easy that $\bigcap_n C_n$ is invariant for f. But in this case we can apply the Schauder fixed point theorem and we get a fixed point for f.

Thus Darbo's theorem is proved.

We present now a generalization of Jones's result given in Theorem 5.1.8.

THEOREM 5.2.10. (V. Istrăţescu (1978) *Let C be a closed, convex and bounded set in a Banach space X and let $f : C \to C$ be a continuous mapping. Suppose that for some integer $n > 1$, the mapping $(fh)^n$ is an α-set contraction. Then f has at least one fixed point in C.*

Here h means the mapping defined on the parts of X as follows:

$$A \to \text{conv } A.$$

Proof. We define on all parts of C a new measure of non-compactness, suggested by a method of Krasnoselskii to define a new norm on Banach spaces, as follows:

$$\alpha^*(M) = \alpha(M) + 1/k\alpha(fh(M)) + \cdots + \alpha(fh)^{n-1}(M)/k^{n-1}.$$

Since the proofs that $\alpha^*(,)$ is a measure of noncompactness are not difficult we omit these. Now we show that with respect to this measure of noncompactness the mapping f is an α-set contraction.

Indeed,

$$\alpha^*(fM) \leqq \alpha(fh)(M) + 1/k\alpha((fh)^2(M) + \cdots + 1/k^{n-1}\alpha((fh)^n(M))$$
$$\leqq k(1/k\alpha(fh)(M) + \cdots + 1/k^{n-1}\alpha((fh)^{n-1}(M) + \alpha(M))$$
$$= k\alpha^*(M)$$

where we have supposed that

$$\alpha(fh)^n M \leqq k^n \alpha(M).$$

From this it is clear that f is an α-set contraction and thus f has a fixed point in C.

Remark 5.2.11.   It seems that a direct proof without using the measure of noncompactness α^* is more complicated.

We give now the extension of Schauder's fixed point theorem obtained by V. N. Sadovskii (1967).

THEOREM 5.2.12.   (V. N. Sadovskii, 1967). *Let C be a closed, convex and bounded set in a Banach space X. If f is any condensing mapping $f : C \to C$ then f has a fixed point in C.*

Proof. Let x_0 be an arbitrary but fixed point in C. We define for any $s \in (0, 1)$ the mapping

$$f_s(x) = sf(x) + (1 - s)x_0$$

and clearly f_s is an α-set contraction. Then by Darbo's theorem there exists at least one fixed point of f_s. Let x_s be a fixed point for f_s.

Since

$$(1 - f)(x_s) = sf(x_s) + (1 - s)x_0 - f(x_s) = (1 - s)(x_0 - f(x_s))$$

we get that for $s \to 1$,

$$\lim(1 - f)(x_s) = 0$$

since C is supposedly bounded.

It is easy to see that since f is condensing, $(1 - f)$ is a closed mapping and, since we can choose $s = 1 - 1/n$, we obtain a sequence of mappings converging uniformly to f, each having a fixed point, x_n. Then it is not difficult to see that f has a fixed point.

Thus Sadovskii's theorem is proved.

Remark 5.2.13.   Similar results hold for other measures of non-compactness. The same remark applies for the case of Darbo's theorem as well as for the result in Theorem 5.2.10.

5.3. KRASNOSELSKII'S, ROTHE'S AND ALTMAN'S THEOREMS

In some applications it appears easy to prove that the mapping has a property concerning certain class of points and, from this property, to conclude that there exists at least one fixed point. Such conditions were given by Krasnoselskii (1953), Rothe (1937), Altman (1957a, 1957b) and others. We remark that a stronger condition than that of Altman was proposed by Petryshyn (1967).

We list now these conditions.

In what follows $B_r = (x, \|x\| \leq r)$ and $\partial B_r = (x, \|x\| = r)$ and $f: B_r \to X$ where X is a real Banach space.

(K)   (Krasnoselskii) for $x \in \partial B_r$

$$\langle f(x), x \rangle \leq \|x\|^2.$$

(P)   (Petryshyn) for $x \in \partial B_r$

$$\|x - f(x)\| \geq \|f(x)\|.$$

(R)   (Rothe) for $x \in \partial B_r$

$$\|f(x)\| \leq \|x\|.$$

(A)   (Altman) for $x \in \partial B_r$

$$\|x - f(x)\|^2 \geq \|f(x)\|^2 - \|x\|^2.$$

First we prove the following result about the connections between these conditions.

THEOREM 5.3.1.   *The following assertions hold*:

1. $(K) \to (A)$,

2. $(R) \to (A)$,

3. $(P) \to (A)$,

4. $(R) \to (K)$,

5.   *if X is a Hilbert space then* $(A) \to (K)$.

Proof. First we remark that 4. follows from the Cauchy–Schwarz inequality and the assertion in 3. is obvious.

Suppose now that (R) holds. Then we have, for $x, \|x\| = r$,

$$\|x - f(x)\|^2 \geq (\|x\| - \|f(x)\|)^2$$
$$= \|x\|^2 - 2\|x\|\,\|f(x)\| + \|f(x)\|^2$$
$$\geq \|x\|^2 - 2\|x\|^2 + \|f(x)\|^2$$
$$= \|f(x)\|^2 - \|x\|^2,$$

and thus (A) is satisfied.

Suppose now that (K) is satisfied. Then for x, $\|x\| = r$, we have

$$\|x - f(x)\|^2 = \|x\|^2 - 2\langle x, f(x)\rangle + \|f(x)\|^2$$
$$\geq \|x\|^2 - 2\|x\|^2 + \|f(x)\|^2$$
$$= \|f(x)\|^2 - \|x\|^2$$

and thus we have (A).

Suppose now that (P) holds. Then for each x, $\|x\| = r$ we have obviously the condition (A).

Suppose now that X is a Hilbert space. Then Altman's condition (A) implies

$$\|x - f(x)\|^2 - \|f(x)\|^2 + \|x\|^2 \geq 0$$

or

$$2\langle f(x), x\rangle - 2\|x\|^2 \leq 0$$

which is clearly equivalent to (K). Thus 5 is proved.

Remark 5.3.2.   If X is a Banach space and $[\,,\,]$ denotes a semi-inner product in the sense of Lumer (1961), then a condition of Krasnoselskii type has sense, i.e. for each x, $\|x\| = r$ and $f : B_r \to X$, we say that f satisfies the Krasnoselskii condition if

$$[f(x), x] \leq \|x\|.$$

The problem is: if f satisfies this Krasnoselskii condition then does f satisfy the Altman condition?

The following theorem was proved in A. Istrățescu and V. Istrățescu (1970b) under a slightly stronger condition, the authors being unaware of the fact that the retraction mapping is a 1-set contraction mapping.

THEOREM 5.3.3.   *Let $f : B_r \to X$ be a condensing mapping and suppose that f satisfies the Altman condition* (A). *Then f has a fixed point in B_r.*

Proof. We consider the retraction mapping on B_r defined as follows:

$$Rx = \begin{cases} x & \|x\| \leq r, \\ rx/\|x\| & \|x\| \geq r. \end{cases}$$

and we define now a mapping on B_r with values in B_r by the relation

$$g(x) = Rf(x)$$

and clearly this is a condensing mapping. Then there exists at least one fixed point in B_r. Let this fixed point be x_0.

In this case we show that x_0 is a fixed point of f.
Indeed, we have,

Case 1. $\|x_0\| \leqq r$,
Case 2. $\|x_0\| = r$.

and thus

$$x_0 = g(x_0) = Rf(x_0) = \begin{cases} f(x_0) & \text{if} \quad \|f(x_0)\| \leqq r \\ rf(x_0)/\|f(x_0)\| & \text{if} \quad \|f(x_0)\| \geqq r, \end{cases}$$

and we thus have to discuss only the case $\|f(x_0)\| \geqq r$.

Now the first case is not possible since we obviously have a contradiction. Let us write

$$f(x_0) = ax_0$$

and it is clear that $a \geqq 1$. Now we show that we have in fact $a = 1$. Indeed, since f satisfies Altman's condition we have,

$$\|x_0 - f(x_0)\|^2 = (1 - a)^2 r^2$$

and

$$\|f(x_0)\|^2 - \|x_0\|^2 = (a^2 - 1)r^2$$

and Altman's condition is not possible unless $a = 1$.

The theorem is proved.

COROLLARY 5.3.4.   (Krasnoselskii (1953), Rothe (1937), Altman (1957a, b), Petryshyn (1967)). *If* $f : B_r \to X$ *is compact and satisfies one of the conditions* (K), (A), (R), (P) *then* f *has a fixed point in* B_r.

5.4. BROWDER'S S AND FAN'S GENERALIZATIONS OF SCHAUDER'S AND TYCHONOFF'S FIXED POINT THEOREM

Tychonoff (1935) proved a generalization of Schauder's fixed point theorem for the case of compact operators on locally convex linear spaces. We do not prove this very important extension because we give some theorems which contain Tychonoff's result as a particular case. We present here some extensions of Schauder's fixed point theorem as they were given by F. Browder and Ky Fan. We give both because the methods of proof, although they have some common points are, we think, very instructive and, perhaps, will be useful for new extensions.

DEFINITION 5.4.1. Let C be a closed and convex set in a locally convex linear topological space X. Then a point $x \in C$ is said to be in $\delta(C)$ if there is a finite-dimensional subspace F of X such that x lies in the boundary of $F \cap C$, $\partial F \cap C$.

We present now a theorem which contains as a particular case the Schauder and Tychonoff theorems.

THEOREM 5.4.2. (Browder, 1967). *Let X be a locally convex topological linear space, C be a compact convex subset of X and $f : C \rightarrow X$ be a continuous mapping. Suppose that for each $x \in \delta(C)$ there is an element $v \in C$ and $s > 0$ (both v and s depending on x) such that*

$$f(u) = u + s(v - u).$$

Then f has a fixed point in C.

We note now that this theorem contains the Schauder and Tychonov results.

COROLLARY 5.4.3. *If X, C are as above and $f : C \rightarrow C$, continuous, is compact then f has a fixed point in C.*
Proof. The assertion follows from Theorem 5.4.2. for $s = 1$ and $v = f(u)$.

THEOREM 5.4.4. (Browder, 1967). *Let X, C be as above and $f : C \rightarrow X$ continuous. Suppose that for each $x \in \delta(C)$ there exists an element $v \in C$ and $s < 0$ such that*

$$f(u) = u + s(v - u)$$

then f has a fixed point in C.

For the proof of these results we need some information about mappings defined on X with values in X^*. These are given in the following

THEOREM 5.4.5. *Let X, C be as above and let $g : C \rightarrow X^*$ be a continuous mapping. Then there exists $u_0 \in C$ such that*

$$(g(u_0), u - u_0) \geqq 0$$

holds for all $u \in C$.

We note for later use the following consequence of the Theorem 5.4.5.

THEOREM 5.4.6. *Let X, C be as above, and let $f : C \rightarrow X^*$ be a continuous map* *and $R : (I - f)C \to X^*$ continuous. Then there exists $u_0 \in C$ such that*

$$(R(I - f)(u_0), u - u_0) \geqq 0$$

holds for all $u \in C$.

Proof of Theorem 5.4.6.   We take, as in Theorem 5.4.5, the function g as $R(I - f)$ and clearly the assertion follows.

For the proof of Theorem 5.4.5 we need the following useful

LEMMA 5.4.7.   *Let X, C be as above and let $f : C \to X^*$ be continuous. Let $y \in X$ and define the function*

$$g_y : C \to R$$

by the relation

$$g_y(x) = (f(x), y - x).$$

Then g_y is continuous.

Proof.   Let $\varepsilon > 0$ and x_1 arbitrary in C. Since f is supposedly continuous we find a neighbourhood $V \ni x_1$ such that, for all $x \in V \cap C$ and y, z in C, the following relation holds

$$|(f(x) - f(x_1), y - z)| < \varepsilon/2.$$

Now we remark that we can choose the neighbourhood V so that, for all $x \in V$,

$$|\langle f(x_1), x - x_1 \rangle| \leqq \varepsilon/2$$

and thus, for all $x \in V \subset C$, we get

$$|\langle f(x), y - x \rangle - (f(x_1), y - x_1 \rangle|$$
$$\leqq |\langle f(x_1), x_1 - x \rangle| + |\langle f(x) - f(x_1), y - x \rangle| \leqq \varepsilon$$

and the Lemma is proved.

Proof of Theorem 5.4.5.   Suppose that the assertion of the theorem is false. Then for each $u_0 \in C$ there exists at least one $y \in C$ such that

$$\langle g(u_0), y - u_0 \rangle \geqq 0.$$

Now for each $y \in C$ we consider the following subset V_y,

$$V_y = (x, x \in C, \langle f(x), y - x \rangle > 0)$$

which, according to the Lemma, is an open set.

Since C is compact there exist a finite number of V_y which cover C, say, $V_{y_1}, \ldots, V_{y_m}$. Thus

$$C = \bigcup_{i=1}^{m} V_{y_i}$$

and we consider now a partition of C subordinated to this finite covering, i.e. the continuous positive functions $(\alpha_i)_{i=1}^{m}$ with the following properties:

1. $\alpha_i(x) = 0$ outside of V_{y_i},

2. $\sum_1^m \alpha_i(x) = 1$ for all x.

Now we define the function

$$h(x) = \sum_{i=1}^{m} \alpha_i(x) y_i$$

on C and it is clear that the values are in C. Then by Brouwer's fixed point theorem h has a fixed point in $C \cap \mathrm{conv}\,(y_i)$.

Let p_0 be this fixed point and consider the function

$$k(x) = \sum \alpha_i(x) \langle f(x), x - y_i \rangle$$

which is strictly positive, since

$$k(x) = \langle f(x), x - h(x) \rangle.$$

But,

$$k(p_0) = 0$$

which is a contradiction and the theorem is proved.

To prove Theorem 5.4.2 we need the following

LEMMA 5.4.8. *Let C_0 be a compact set in a locally convex linear space and let $0 \notin C_0$. Then there exists a continuous mapping $R : C_0 \to X^*$ such that $\langle R(u), u \rangle > 0$.*
    Proof. Since $0 \notin C_0$, for each $u \in C_0$ we find $u^* \in X^*$ such that $u^*(u) > 0$ and for fixed $u^* \in X^*$ we consider the set

$$V_{u^*} = (u, u \in C_0, u^*(u) > 0)$$

which is clearly an open set in X. The sets (V_{u^*}) clearly cover C_0 and, since this is supposed to be a compact set, we find a (u_i^*) such that

$$C_0 = \bigcup_{i=1}^{m} V_{u_i^*}$$

Then, as above, we consider a continuous partition associated with this finite covering and we define the mapping

$$R(x) = \sum_{1}^{m} \alpha_i(x) u_i^*$$

which clearly satisfies the assertion of the lemma.

Now we are in a position to prove Theorem 5.4.2. Suppose that f has no fixed points in C. In this case the set

$$C_0 = (I - f)C$$

is a compact set which contains no 0. We can apply Lemma 5.4.8 and we find that there exists $R : C_0 \to X^*$ with the property that $\langle R(u), u \rangle > 0$. This implies that $R(x) \neq 0$ for all x.

In this case according to Theorem 5.4.6. we find $x_0 \in C$ such that for all $x \in C$,

$$R(x_0 - f(x_0)), (x - x_0) \geqq 0.$$

We have two cases to discuss:

Case 1. x_0 does not lie in $\delta(C)$,
Case 2. x_0 lies in $\delta(C)$.

Now, in the Case 1, for each v in X, we find a positive constant s such that $x = x_0 + sv \in C$. Then we have,

$$s \langle R(x_0 - f(x_0)), v \rangle \geqq 0$$

and since s is positive, we get

$$\langle R(x - f(x)), v \rangle \geqq 0.$$

But v is arbitrary and thus

$$R(x_0 - f(x_0)) = 0$$

which is a contradiction.

Now suppose that x is as in Case 2. Then there exists $x_1 \in C$ such that for some $s > 0$,

$$f(x_0) = x_0 + s(x_1 - x_0)$$

or

$$-(x_1 - x_0) = 1/s(x_0 - f(x_0))$$

which implies that

$$0 \leq \langle R(x_0 - f(x_0)), x_1 - x_0 \rangle$$
$$= -1/s \langle R(x_0 - f(x_0)), x_0 - f(x_0)) \rangle$$
$$< 0$$

which is again a contradiction and the theorem is proved.

Proof of Theorem 5.4.4. We take $R^* = -R$ and the proof is similar to the above proof and the details are ommited.

An interesting generalization of both Schauder's and Tychonoff's theorems was obtained by Ky Fan (1961). Ky Fan (1969) proved other extensions suggested in part by the above extensions of Browder. For other extensions see Ky Fan (1961), Ky Fan (1966), Ky Fan (1972) where some very interesting applications are also given.

The extension given by Ky Fan depends upon a lemma which is essentially the infinite-dimensional version of the Knaster-Kuratowski-Mazurkiewicz theorem (1929).

We give now the Knaster-Kuratowski-Mazurkiewicz theorem in the finite-dimensional case. For a proof we refer to the original paper or that of C. Berge (1963).

THEOREM 5.4.9. *Let* $S_n = [a_1, \ldots, a_{n+1}]$ *be an* n-*simplex and* $F_1, \ldots, F_{n+1}$ *be* $n + 1$ *closed subsets. If for each set*

$$(i, j, \ldots, l) \subset (1, 2, \ldots, n + 1)$$

we have

$$[a_i, a_j, \ldots, a_l] \subset F_i \cup F_j \cup \cdots \cup F_l$$

then

$$\bigcap_{i=1}^{n+1} F_i \quad is \ nonempty.$$

Here $[a_1, \ldots, a_N]$ means the convex closure of $a_1, \ldots, a_N$.
The infinite-dimensional version of this result is as follows:

THEOREM 5.4.10. (Ky Fan, 1961). *Let* X *be a topological linear space and let*

$$F : X \to P(X)$$

be a set-valued function with the following properties:

1. $F(x)$ *is closed for each* $x \in X$,

2.   *The convex hull of any finite set* $(a_1, \ldots, a_{n+1})$ *in* X *is contained in* $\bigcup_1^{n+1} F(a_i)$,

3. $F(x)$ *is compact for at least one* $x \in X$.

Then $F(x)$ *is nonempty.*

Proof. First we remark that it suffices to prove that there is an $F(x_i)$ for any finite set $(x_1, \ldots, x_{n+1})$ in X. Now we consider the n simplex in i and the $n + 1$-dimensional spaces with vertices

$$v_i = (0, 0, 0, \ldots, 0, \underset{i}{1}, 0, \ldots)$$

and define the mapping $f(x) = \sum a_i x_i$ for $x = \sum a_i v_i$, x in the n-simplex.

Clearly this is a continuous mapping, and we consider the sets

$$F_i = f^{-1} F(x_i), \qquad i = 1, 2, \ldots, n+1.$$

Then clearly these sets satisfy the Knaster–Kuratowski–Mazurkiewicz theorem. Obviously $\bigcup_1^{n+1} F_i \neq \varnothing$ implies that $\bigcup_1^{n+1} F(x_i) \neq \varnothing$.

THEOREM 5.4.11.   *Let* X *be a topological linear space and* C *be a compact convex subset in* X. *Suppose that* A *is a subset of* $X \times X$ *with the following properties*:

1.   *for any* $x \in X$, $(x, x) \in A$,

2.   *for any fixed* $y \in X$, $(x, x \in X, (x, y) \notin A)$ *is convex or empty.*

Then there exists a point $y_0 \in X$ *such that* $X \times (y_0) \subset A$.

Proof. For each $x \in X$ we set

$$F(x) = (y \in X, (x, y) \in A)$$

which is a convex set and also compact. From the definition and the hypotheses 1 and 2 it follows that

$$x \to F(x)$$

satisfies all the conditions of Theorem 5.4.12. Then we have

$$\bigcap F(x_i) \neq \varnothing$$

and the theorem is proved.

From this theorem we can obtain Tychonoff's fixed point theorem.

THEOREM 5.4.12 (Tychonoff, 1935). *Let X be a locally convex topological space with the topology given by the family $(p_i)_{i \in I}$ of continuous seminorms. If C is any compact and convex subset of X and $f : C \to C$ be any continuous mapping then f has a fixed point in C.*

Proof. For each $i \in I$ we consider the set

$$F_i = (x \in C, \, p_i(x - f(x)) = 0)$$

and clearly a point $x \in C$ is a fixed point of f iff $x \in \bigcap_i F_i$.

Since C is supposedly compact, we show that the family of sets (F_i) has the finite intersection property. Let $i_1, \ldots, i_{n+1}$ in I, a finite subset, and consider the set

$$A = ((x, y) \in C \times C, \, \sum_1^{n+1} p_{i_j}(x - f(y)) \geqq \sum_1^{n+1} p_{i_j}(y - f(y))).$$

and we apply the above theorem and thus find that there exists a y_0 such that

$$C \times (y_0) \subset A.$$

Since $(f(y_0), y_0) \in A$ we obtain

$$p_{i_j}(y_0 - f(y_0)) = 0, \qquad 1 \leqq j \leqq n$$

i.e. $y_0 \in \bigcup_{j=1}^{n+1} F_{ij}$. The theorem is proved.

The following theorem contains as a special case the Schauder fixed point theorem.

THEOREM 5.4.13. *Let X be a Banach space and let C be a compact convex subset of X. Let $f : C \to X$ be a continuous mapping. Then there exists $y_0 \in C$ such that*

$$\| y_0 - f(y_0) \| = \min_{x \in C} \| x - f(y_0) \|$$

Proof. We consider the following set

$$A = ((x, y) \in C \times C, \quad \| x - f(y) \| \geqq \| y - f(y) \|)$$

and it is easy to see that this satisfies the conditions of Theorem 5.4.11 and this gives the assertion of the theorem.

The following result refers to the case of locally convex spaces.

THEOREM 5.4.14. *Let C be a compact convex and nonempty set in a locally convex Hausdorff space X and let $f : C \to X$ be any continuous mapping. Then either f has a fixed point in C or there exists a point y_0 in C and a continuous seminorm p on X such that*

$$0 < p(y_0 - f(y_0)) = \min_{x \in C} (p(x - f(y_0))).$$

Proof. Suppose that the second assertion is not true and thus for every point $y \in C$ and for any continuous seminorm p on X there exists a point $x \in C$ such that

$$p(x - f(y)) < p(y - f(y))$$

and of course y satisfies the inequality

$$p(y - f(y)) > 0.$$

Let $(p_i)_{i \in I}$ be the family of continuous seminorms on X which defines the topology of X. For each $i \in I$ we consider the set

$$F_i = (y, y \in C, p_i(y - f(y)) = 0)$$

and clearly this is closed and thus compact. We prove now that the family of compact sets $(F_i)_{i \in I}$ has the intersection property. Indeed, for this it suffices to show that for $i_1, \ldots, i_m$, a finite number of elements in I, $F_{ij} \neq \varnothing$. Let $p_k = p_{i_1} + \cdots + p_{i_m}$ and let $k \in I$ and set

$$A = ((x, y) \in C \times C, \qquad p_k(x - f(y)) \geqq p_k(y - f(y))$$

which is closed and thus compact. It is easy to see that we can apply Theorem 5.4.11 and we obtain the existence of a $y_0 \in C$ such that

$$C \times (y_0) \subset A.$$

Further we obtain,

$$p_k(x - f(y_0)) \geqq p_k(y_0 - f(y_0))$$

for all $x \in C$. Then we get $p_k(y_0 - f(y_0)) = 0$. This implies that $y_0 \in \bigcap_j F_{i_j}$. The theorem is proved.

Remark 5.4.15. If C is an invariant set for f then, from Theorems 5.4.13 and 5.4.14 the theorems of Schauder and Tychonoff, respectively, follow.

THEOREM 5.4.16. *Let X be a locally convex topological space and let C be a compact convex nonempty subset of X. Suppose that $f : C \to X$ is a continuous mapping satisfying the following property:*

for each $y_0 \in C$ there exists a (real or complex) number z such that

$$|z| < 1 \quad \text{and} \quad zy_0 + (1 - z)f(y_0) \in C.$$

Then f has a fixed point in C.

Proof. Suppose that f has no fixed points in C. Then according to Theorem 5.4.14 there exists a $y_0 \in C$ and a seminorm p such that

$$0 < p(y_0 - f(y_0)) = \min_{x \in C}(p(x - f(y_0))).$$

Now since f has the above property, we find a z such that $zy_0 + (1 - z)$ $f(y_0) \in C$. But

$$x^* - f(y_0) = z(y_0 - f(y_0)), \quad \text{where} \quad x^* = zy_0 + (1 - z)f(y_0)$$

and thus

$$0 < p(y_0 - f(y_0)) \leq p(x^* - f(y_0)) = |z|p(y_0 - f(y_0)).$$

This clearly implies $|z| \geq 1$ and thus we have a contradiction and the theorem is proved.

For other interesting fixed points of this type we refer the reader to the paper of Ky Fan (1969).

5.5. SOME APPLICATIONS

The Schauder fixed point theorem as well as the generalizations obtained have numerous applications of which we note here only a few. The first refers to the existence theorem of Peano for differential equations. The second refers to a problem concerning the geometry of Banach spaces, namely to a sufficient condition for a Chebyshev set to be a sun, and the last application refers to a very important problem in operator theory: the invariant subspace problem.

First we give Peano's theorem.

THEOREM 5.5.1. *Let $f(x, y)$ be a continuous real-valued function defined on*

$$D = ((x, y), |x - x_0| \leq a, |y - y_0| \leq b)$$

and consider the differential equation

$$dy/dx = f(x, y) \qquad y(x_0) = y_0.$$

Then there exists a $y = y(x)$ satisfying the equation.

Proof. We consider the Banach space of all continuous real-valued functions defined on $[x_0 - h, x_0 + h]$ where

$$h = \min(a, b/\|f(x, y)\|).$$

with the sup norm. We remark that the above differential equation is equivalent to the following integral equation

$$y(x) = y_0 + \int_{x_0}^{x} f(s, y(s)) \, ds.$$

This integral equation suggests the consideration of the following operator

$$(Ty)(s) = y_0 + \int_{x_0}^{x} f(t, y(t)) \, dt.$$

We consider the following set in the above Banach space

$$C = (y(,), \|y - y_0\| \leq b)$$

which is obviously closed, bounded and convex. We show now that C is invariant for T. Indeed, we have,

$$(Ty) - y_0 = \int_{x_0}^{x} f(t, y(t)) \, dt$$

and thus

$$\|(Ty) - y_0\| \leq \int_{x_0}^{x} |f(x, y)| \, dx \leq b.$$

Now if y is a function we have,

$$|(Ty)(s) - (Ty)(s')| \leq 2\|f(x, y)\| \, |s - s'|$$

and, by the Arzela–Ascoli theorem T is compact. Thus we can apply the Schauder fixed point theorem to T and thus the existence of a solution of the integral equation is proved.

For the next application we need some preliminary notions.

If X is a Banach space a subset C is called a Chebyshev set if each x has a unique closest point x' in C. The Chebyshev set M is called a sun if for every x, x' is the closest point in M for every point on the ray xx'.

DEFINITION 5.5.2.   A set C in a Banach space is boundedly compact if it intersects each closed ball in a compact set.

THEOREM 5.5.3   (L. P. Vlasov, 1961). *Every boundedly compact Chebyshev set is a sun.*

Proof. Suppose that the set C is boundedly compact and Chebyshev and is not a sun. Then there exists a point z such that no point beyond z on the ray zz' has z' for its closest point. We define

$$x \to f(x) = \tfrac{4}{3}z - \tfrac{1}{3}x'.$$

Let us consider the convex closure of $(f(x), \|x - z\| \leqq \|z - z'\|)$, K. Clearly K is a compact set and it is easy to see that K is an invariant set for f. Then by the Schauder fixed point theorem there exists an x^* such that

$$f(x^*) = x^*$$

and from the form of f this implies immediately that $x^* = z$. This contradicts the definition of z and the theorem is proved.

For other results about such questions we refer the reader to Cheney (1976), and Vlasov (1961).

The next application refers to one of the problems in operator theory which has attracted the attention of many mathematicians.

The problem refers to some subspaces and to operators acting on them.

Let X be a Banach space and $L(X)$ be the set of all bounded linear operators on X. A closed linear subspace X_1 of X is said to be invariant for $T \in L(X)$ if, whenever $x \in X_1$, then $Tx \in X_1$. The invariant subspace is called nontrivial if it is different from X and (0). We say that an operator is not a scalar if there exists no complex number z such that $T = zI$.

An invariant subspace for a nonscalar operator T is said to be hyperinvariant if it is invariant for all operators commuting with T.

The following result, proved by Victor Lomonosov (1973), represents the most spectacular and the most general result concerning the invariant subspace problem.

THEOREM 5.5.4.   *Let $T \in L(X)$ commute with a nonzero compact operator $K \in L(X)$. Then T has a nontrivial hyperinvariant subspace.*

The proof of this theorem is short compared to other existence proofs of invariant subspaces. It uses the Schauder fixed point theorem in an essential way.

Proof. We suppose that the assertion of the theorem is not true. We can suppose, without loss of generality, that $\|K\| = 1$ and we consider now a point x_0 in X such that $\|Kx_0\| > 1$. This obviously implies that $\|x_0\| > 1$. We consider now the set $S_{x_0}(1) = (x, \|x - x_0\| \leq 1)$ and let $\overline{K} = \operatorname{conv} KS_{x_0}(1)$. which is a closed, convex and bounded set in X. For each R in $L(X)$ we consider the set

$$V_R = (y, y \in X, \|Ry - x_0\| < 1)$$

which is obviously an open set. We show now that $\bigcup_R V_R$ is $X \backslash 0$. Indeed, for each y,

$$X_y = (Ry, R \text{ a polynomial in } T)$$

is obviously a linear space invariant for T. From our hypothesis it follows that $\bar{X}_y = X$. This proves our assertion. Now, the set K is compact and thus we find a finite set of V_R's which cover K. Let

$$K \subset V_{R_1} \cup \cdots \cup V_{R_m}$$

and we define now the functions

$$a_j(y) = \max(0, 1 - \|R_j y - x_0\|)$$

for $j = 1, \ldots, m$. Clearly these are continuous functions and for each y there exists a j such that $a_j(y) \neq 0$. Thus we can define the following functions:

$$b_j(y) = a_j(y) \Big/ \sum_1^m a_i(y).$$

Now we consider the function f on K where f is defined as follows:

$$f(y) = \sum_1^m b_j(y) R_j y,$$

which has the values in the space generated by $R_j y$.

But we have also

$$\|f(Kx) - x_0\| < 1$$

and thus S is an invariant set for $f(K)$. Then, by the Schauder fixed point theorem we find a fixed point, say x^*, i.e.

$$f(Kx^*) = x^*.$$

Consider now the linear operator

$$S = \sum_{1}^{m} b_j(Kx^*)R_j$$

which, by the result just obtained, has an eigenvalue 1.

This contradicts the fact that T has an invariant eigenspace. The theorem is thus proved.

For interesting applications to differential equations see Lazer (1968), J. B. Keller and S. Antman (1969), C. Miranda (1970).

Chapter 6

Fixed Point Theorems for Nonexpansive Mappings and

Related Classes of Mappings

6.0. INTRODUCTION

A natural generalization of the class of contractive mappings is the following: if (X, d) is a complete metric space and $f : X \to X$ then we consider those f such that,

$$d(f(x), f(y)) \leq d(x, y)$$

for all $x, y \in X$, and we ask about the existence of fixed points of f in X. Simple examples show that, in general, for such mappings the existence of fixed points is not assured on the one hand, and on the other hand there exist mappings satisfying the above property in which the fixed points abound.

Example 6.0.1. Let $X = R$ and

$$f : R \to R, \qquad f(x) = x + 1$$

which clearly has the property

$$d(f(x), f(y)) = |x - y|$$

and obviously is without fixed points.

Example 6.0.2. Let $X = R \times R$ and define $f((x, y))$ by

$$f((x, y)) = (x, 0).$$

Then clearly f has the above property and every point of the form $(x, 0)$ is a fixed point for this mapping.

The first important result in the theory of fixed points for nonexpansive mappings was obtained by R. de Marr (1963) who has proved an interesting extension of the famous result of Kakutani–Markov.

For nonexpansive mappings define on convex sets in Banach spaces interesting results have also been obtained by Milman and Brodski (1948), Browder (1965, 1967), Petryshin (1967), Kirk (1965), etc.

In what follows we give these result together with other results, and some applications.

6.1. NONEXPANSIVE MAPPINGS

Let X be a Banach space and let C be a set in X.

DEFINITION 6.1.1.   A mapping $f : C \rightarrow X$ is said to be nonexpansive if, for any $x, y \in C$, $\|f(x) - f(y)\| \leqq \|x - y\|$.

It is clear, from the very definition, that a nonexpansive mapping is continuous. A more general class of mappings is considered in the following

DEFINITION 6.1.2.   A mapping $f : C \rightarrow X$ is said to be of Lipschitz class (Lip) if there exists a constant $K > 0$ such that, for all $x, y \in C$, $\|f(x) - f(y)\| \leqq K\|x - y\|$.

Remark 6.1.3.   It is clear that if $K < 1$, we have the class of mappings studied in Chapter 1; also if $K = 1$ we have the class of nonexpansive mappings.

In what follows we give a useful and important example of mappings of the above type.

Let X be a Banach space and consider the set

$$C_r = \{x, \|x\| \leqq r\}.$$

The mapping R_{C_r} defined in what follows, is called the 'Retraction Mapping':

$$R_{C_r}x = \begin{cases} x & \text{if} \quad x \in C_r \\ x/\|x\| \cdot r & \text{if} \quad \|x\| \geqq r. \end{cases}$$

We can prove the following results about the mapping R_{C_r}:

THEOREM 6.1.4.   *If X is a Hilbert space then R_{C_r} is nonexpansive.*
   Proof. Let $x, y \in H$ and denote by $x^\sim = R_{C_r} x$, $y^\sim = R_{C_r} y$. In this case, for any $z \in C_r$ and any $t > 0$

$$\|x + (tz + (1 - t)x^\sim)\|^2 \geqq \|x - x^\sim\|^2.$$

Expanding we find

$$\langle x - x^\sim, x^\sim - z \rangle \geqq - t/2\|z - x^\sim\|^2$$

and thus for $t \rightarrow 0^+$ we get

$$\langle x - x\tilde{} , x\tilde{} - z \rangle \geq 0.$$

Now for $z = y\tilde{}$ we obtain

$$\langle x - x\tilde{} , x\tilde{} - y\tilde{} \rangle \geq 0$$

and this gives further,

(*) $\qquad \langle x\tilde{} , x\tilde{} - y\tilde{} \rangle \leq \langle x, x\tilde{} - y\tilde{} \rangle.$

Now from

$$\| y + (tz + (1 - t)y\tilde{}) \|^2 \geq \| y - y\tilde{} \|^2$$

we obtain, as above, that

$$\langle y - y\tilde{} , y\tilde{} - z \rangle \geq 0$$

and for $z = x\tilde{}$ we get

(**) $\qquad - \langle y\tilde{} , x\tilde{} - y\tilde{} \rangle \leq \langle - y\tilde{} , x\tilde{} - y\tilde{} \rangle$

and adding the inequalities (*) and (**) we get

$$\| x\tilde{} - y\tilde{} \|^2 \leq \langle x - y, x\tilde{} - y\tilde{} \rangle \langle \| x - y \| \cdot \| x\tilde{} - y\tilde{} \| \rangle$$

which implies

$$\| x\tilde{} - y\tilde{} \| \leq \| x - y \|.$$

For the case of Banach spaces we can prove the following

THEOREM 6.1.5. *Let X be a Banach space. Then the mapping R_{C_r} is Lip. More precisely, for any $x, y \in X$ we have,*

$$\| R_{C_r} x - R_{C_r} y \| \leq 2 \| x - y \|.$$

Proof. The case when x, y are in C_r is clear. Suppose now that $x \in C_r$ and $y \notin C_r$. In this case we have,

$$R_{C_r} x - R_{C_r} y = x - y / \| y \| \cdot r = (\| y \|(x - y) + (\| y \| - r)y) / \| y \|$$

and this implies,

$$\| R_{C_r} x - R_{C_r} y \| \leq \| x - y \| + (\| y \| - r) \leq 2 \| x - y \|$$

since

$$\| y \| - r \leq \| y \| - \| x \| \leq \| x - y \|.$$

Now, to complete the proof, we consider the case when x, y are not in C_r. In this case we have,

$$R_{C_r}x - R_{C_r}y = (x/\|x\| - y/\|y\|)r.$$

Now

$$\|R_{C_r}x - R_{C_r}y\| \leq r\|x/\|x\| - y/\|y\| \|$$
$$\leq r(\|x - y\| \cdot 1/\|x\| + \|y\| \cdot$$
$$\cdot (\|x - y\|/\|x\| \cdot \|y\|))$$
$$\leq 2\|x - y\|.$$

6.2. THE EXTENSION OF NONEXPANSIVE MAPPINGS

Let X, Y be real normed spaces and let C be an arbitrary subset of X.

DEFINITION 6.2.1. A mapping $T : C \to Y$ is called nonexpansive if for all $x, y \in C$

$$\|Tx - Ty\| \leq \|x - y\|$$

and a map $T^\sim : X \to Y$ is called an extension of T if

1. $Tx = T^\sim x.$ $x \in C,$

2. $\|T^\sim a - T^\sim b\| \leq \|a - b\|,$ $a, b \in X.$

The problem of the existence of an extension for any nonexpansive mapping is an interesting problem and the conditions for its existence are connected with the structure of Banach spaces. In the case of $X = Y = \mathbf{R}^n$ the problem was first considered by M. D. Kirszbraun (1934).

S. Schönbeck (1966, 1967) calls a pair of normed spaces as having the 'contraction-extension property' if, whenever C is an arbitrary subset of X and $T : C \to Y$ is a nonexpansive mapping, then T has an extension. This is abbreviated $\langle$ property $E \rangle$.

Another property of pairs of normed spaces is introduced in connection with the famous result of Kirszbraun. We say that a pair of normed spaces (X, Y) has the Kirszbraun property if, whenever $\{x_i\}_{i \in I} \subset X$, $\{y_i\}_{i \in I} \subset Y$, where I is an arbitrary index set such that $\|y_i - y_j\| \leq \|x_i - x_j\|$, $i \neq j$, and the intersection of the balls $S(x_i, r_i), r_i > 0$ is nonempty in X, then the intersection of balls $S(y_i, r_i)$ in Y is also nonempty. This is abbreviated as $\langle$ property $K \rangle$.

We have the following result of Valentine (1945).

THEOREM 6.2.2.   *The pair (X, Y) has $\langle$property $E\rangle$ if and only if it has $\langle$property $K\rangle$.*

Proof. First we suppose that the pair (X, Y) has $\langle$property $K\rangle$ and we show that it has $\langle$property $E\rangle$:

Let C be an arbitrary set in X and let $T : C \to Y$ be an arbitrary non-expansive mapping. Let x_0 be an arbitrary point of X which is not C.

Consider the balls $S(x, r_x)$, where $r_x = \|x - x_0\| > 0$ and, in this case, we take as index set the set C itself. Since the pair (X, Y) has $\langle$property $K\rangle$ it follows that, since the intersection of the balls defined above is nonvoid, the intersection of the balls $S(Tx, r_x)$ is also nonvoid. Consider an arbitrary point y_0 of this intersection and define a map on $C \cup \{x_0\}$ by:

1. $T^\sim x = Tx$ if $x \in C$,

2. $T^\sim x_0 = y_0$.

In this case it is clear that this is again nonexpansive. Let us consider the family F of all extensions of T. We define a partial order in this family as follows: two extensions T_1 and T_2 of T are in the relation $T_1 < T_2$ if the sets on which they are defined, say C_1 and C_2 respectively, are as follows $C_1 \subset C_2$ and $T_1 x = T_2 x$ for all $x \in C_1$.

It is not difficult to see that we can apply Zorn's Lemma and we denote by $T^\sim$ one maximal element. We show that the map $T^\sim$ satisfies the theorem. Indeed, if the domain on which $T^\sim$ is defined is not X then we can obtain, using the construction described above, a proper extension of $T^\sim$ which is thus not maximal. This contradiction proves the assertion.

We suppose now that the pair (X, Y) has $\langle$property $E\rangle$ and we show that it has $\langle$property $K\rangle$. Let $S(x_i, r_i)$ be a family of balls in X, and let $S(y_i, r_i)$ be a family of balls in Y such that $\|y_i - y_j\| \leq \|x_i - x_j\|$ for $i \neq j$ and the intersection of the balls $S(x_i, r_i)$ is nonempty.

We show now that the intersection of the balls $S(y_i, r_i)$ is also nonempty. Indeed, let T be defined as follows:

$$Tx_i = y_i$$

on the set $\{x_i\}$. Since the pair (X, Y) has the extension property we find an extension $T^\sim$ of T. Let x_0 be an arbitrary point in the intersection of the balls $S(x_i, r_i)$. In this case it is clear that the point $y^\sim = T^\sim x_0$ is in the intersection of the balls $S(y_i, r_i)$.

The assertion of the theorem is proved.

We show now that some pairs have the $\langle$property $E\rangle$ (or equivalently, the $\langle$property $K\rangle$).

THEOREM 6.2.3.   *The pair* $(\mathbf{R}^n, \mathbf{R}^n)$ *has the* $\langle$ *property* $K \rangle$.

This is the famous result of Kirszbraun (1934).

In what follows we give the proof of this result given by I. J. Schoenberg (1953). First we remark that Theorem 6.2.3. is equivalent to the following one.

THEOREM 6.2.4.   *Let* X *be a finite dimensional Hilbert space,* $\dim X = n$ *and* $\{x_i\}_1^m, \{y_i\}_1^m$ *be arbitrary sets of elements in* X *such that*

$$\|y_i - y_j\| \leqq \|x_i - x_j\|, \qquad i \neq j$$

and let p *be a given element in* X. *Then there exists a* $q \in X$ *such that*

$$\|q - y_i\| \leqq \|p - x_i\|, \qquad i = 1, 2, \ldots, m.$$

Proof. We may suppose without loss of generality that $p \neq x_i$, $i = 1, 2, 3, \ldots, m$, since in the contrary case we can take $q = y_{i_0}$, for some i_0 in $[1, m]$, and the assertion of the theorem is satisfied.

For the case when $p \neq x_i$, $i \in [1, m]$ we consider the function

$$\varphi(x) = \max_i \{\|x - y_i\|/\|p - x_i\|\}$$

and clearly this is a continuous function. Let λ be the absolute minimum of this function, which is attained since X is of finite dimension. Let q be a point in X such that $\varphi(q) = \lambda$. We show now that q satisfies the assertion of the theorem.

We have two cases to discuss.

Case 1. $\lambda = 0$. In this case $q = y_1 = y_2 = \cdots = y_m$ and the theorem is proved.

Case 2. $\lambda > 0$ and then

$$\|q - y\|/\|p - x\| = \begin{cases} \lambda, & i = 1, 2, \ldots, k. \\ < \lambda & i = k+1, \ldots, m. \end{cases}$$

for some index k.

First we prove that $q \in \text{conv} \{y_1, y_2, \ldots, y_k\}$. Indeed, in the contrary case we can take the numbers $\|q - y_i\|$, $i = 1, 2, \ldots, k$ such that

$$\|q - y_i\|/\|p - x_i\| < \lambda$$

for $i = 1, 2, \ldots, k$ and this represents a contradiction. We remark also that this implies that $k \geqq 2$.

To prove the theorem it is sufficient to show that $\lambda < 1$. Suppose that this is not so. We consider the vectors,

$$a_i = x_i - p, \qquad b_i = y_i - q, \qquad i = 1, 2, 3, \ldots, m$$

and thus

$$\|b_i\|^2 \geqq \|a_i\|^2$$

and since

$$\|y_i - y_j\| \leqq \|x_i - x_j\|$$

we get

$$\|b_i - b_j\|^2 \leqq \|a_i - a_j\|^2$$

and thus

$$\langle b_i, b_j \rangle \geqq \langle a_i, a_j \rangle.$$

Now since $q \in \text{conv}\{y_1, y_2, \ldots, y_k\}$ we find the positive numbers $\alpha_1, \alpha_2, \ldots, \alpha_k, \sum_1^k \alpha_i = 1$ such that

$$q = \sum_1^k \alpha_i y_i.$$

From these we obtain,

$$0 = -q + \sum_1^k \alpha_i y_i = -\sum_1^k \alpha_i (q - y_i) = \sum_1^k \alpha_i b_i.$$

Multiplying the relation

$$\langle b_i, b_j \rangle > \langle a_i, a_j \rangle$$

by $\alpha_i \alpha_j$ and adding, we get

$$0 = \|\sum_1^k \alpha_i b_i\|^2 > \|\sum_1^k \alpha_i a_i\|^2$$

and this contradiction proves that $\lambda < 1$.

In the following theorem of Valentine we show that the pair (X, Y), where X, Y are Hilbert spaces, has $\langle$ property $K \rangle$.

THEOREM 6.2.5. *Let X be a Hilbert space. Then the pair (X, X) has* $\langle$ property $K \rangle$.

Proof. Let $\{x_i\}_{i \in I}$ and $\{y_i\}_{i \in I}$ be families of elements in X such that the balls $S(x_i, r_i)_{i \in I}, r_i > 0$ have nonempty intersection, and also the following inequality holds:

$$\|y_i - y_j\| \leq \|x_i - x_j\|, \qquad i \neq j,$$

and we consider also the balls $S(y_{i_j}, r_{i_j})$. We show that the intersection of these balls is nonempty. Indeed, let $i_1, i_2, \ldots, i_n$ be the finite part of I and consider the finite-dimensional Hilbert space generated by the elements $x_{i_1}, \ldots, x_{i_n}$ and $y_{i_1}, \ldots, y_{i_n}$, respectively.

According to Theorem 6.2.4. the balls $S(y_{i_j}, r_{i_j})$, $j = 1, \ldots, n$ have a nonempty intersection. Now we remark that our balls are weakly closed and, since they are bounded, they are in fact weak compact sets.

This implies, by the finite intersection property, that the balls $S(y_i, r_i)$ have nonempty intersection.

The assertion of the theorem is proved.

Theorems 6.2.4 and 6.2.5 gave some examples of pairs of Banach spaces which have $\langle$ property $E \rangle$ (or, equivalently, $\langle$ property $K \rangle$).

The following example of Schönbeck (1966) shows that there exist pairs of Banach spaces (even finite-dimensional) without the $\langle$ property $E \rangle$.

Example 6.2.6.   An example of a pair of finite-dimensional normed spaces without $\langle$ property $E \rangle$. Let us consider $X = \mathbf{R}^2$ and $Y = \mathbf{R}^2$ and we consider X with the norm induced by the unit ball which is a regular hexagon, and Y with the usual norm (i.e. the Euclidean norm).

Let the norms so defined be denoted by $\|, \|_X, \|, \|_Y$ and

$$H = \{x, \|x\|_X \leq 1\}$$
$$C = \{y, \|y\|_Y \leq 1\}.$$

Consider now x_1 and x_2, being two consecutive points of contact between H and C. In this case,

$$\|x_1\|_X = \|x_1\|_Y = \|x_2\|_X = \|x_2\|_Y$$
$$= \|x_1 - x_2\|_X = \|x_1 - x_2\|_Y = 1$$

and on the set

$$\{0, x_1, x_2\}$$

we define a nonexpansive mapping with values in Y by the relations

$$T0 = 0, \qquad Tx_1 = x_1, \qquad Tx_2 = x_2.$$

From the definition of T it is clear that we have a nonexpansive mapping and we suppose that the pair (X, Y) has the property E. In this case we consider the point $x = \frac{1}{2}(x_1 + x_2)$. If $T\tilde{}$ is the extension of T, any $y = T\tilde{}x$, then we have,

$$\|T\tilde{}x\|_Y = \|y\|_Y \leqq \tfrac{1}{2},$$

$$\|y - x_1\|_Y \leqq \tfrac{1}{2},$$

$$\|y - x_2\|_Y \leqq \tfrac{1}{2}$$

which are impossible since,

$$\|x\|_X = \|x - x_1\|_X = \|x - x_2\|_Y \leqq \tfrac{1}{2}.$$

Thus the following problem appears naturally:

Problem 6.2.7. Characterize the Banach spaces X, Y such that the pair (X, Y) has $\langle$property $E\rangle$.

The following result of Schönbeck gives a partial answer to this problem.

THEOREM 6.2.8. *If the pair of Banach spaces (X, Y) has $\langle$property $E\rangle$ and Y is strictly convex then X and Y are Hilbert spaces.*

Before the proof, we recall that a Banach space Z is said to be strictly convex (rotund) if, whenever $\|x\| = \|y\|$, then $\|x + y\| < \|y\| + \|x\|$.

For the proof, we first show that, if x_1, x_2 are in X and y_1 and y_2 are in Y, and if

$$\|x_1\| = \|y_1\|, \qquad \|x_2\| = \|y_2\|$$

$$\|x_1 - x_2\| = \|y_1 - y_2\|$$

then, for all a and b,

$$\|ax_1 + bx_2\| \geqq \|ay_1 + by_2\|.$$

First we suppose that a and b are positive numbers and $a + b = 1$. We consider the sets,

$$S_X^1 = \{x, \|x - x_1\| \leqq bd\}$$

$$S_X^2 = \{x, \|x - x_2\| \leqq ad\}$$

$$S_X^3 = \{x, \|x\| \leqq \|ax_1 + bx_2\|\}$$

where

$$d = \|x_1 - x_2\|.$$

We consider also the following sets in Y,

$$S_Y^1 = \{y, \|y - y_1\| \leq bd\}$$

$$S_Y^2 = \{y, \|y - y_1\| \leq ad\}$$

$$S_Y^3 = \{y, \|y\| \leq \|ax_1 + bx_2\|\}.$$

We set

$$F = \{S_X^1, S_X^2, S_X^3\}, \qquad G = \{S_Y^1, S_Y^2, S_Y^3\}.$$

We use the following order relation in the families of sets of the form $\{x, \|x - x^\sim\| \leq r\}$ (i.e. balls): let R and S be two families of sets and we say that $R > S$ if there exists a one-to-one function f from the centres of the balls in R with values in the set of centres of S such that for all $i, j, i \neq j$

$$\|f(x_i) - f(x_j)\| \leq \|x_i - x_j\|$$

and the corresponding balls have the same radius.

In our case we clearly have $F > G$ and obviously the intersection of balls in F is nonempty since the point $ax_1 + bx_2$ is in all balls.

Now, since the pair has the extension property, the intersection of balls in G is also nonempty.

We remark that

$$S_Y^1 \cap S_Y^2 = \{ay_1 + by_2\}.$$

Indeed, we have

$$\|y_1 - ay_1 - by_2\| = bd$$

$$\|y_2 - ay_1 - by_2\| = ad,$$

and thus $ay + by$ is in $S_Y^1 \cap S_Y^2$. Now, if z is another point in the intersection, then

$$\|z - y_1\| = bd, \qquad \|z - y_2\| = ad$$

and since Y is strictly convex, $z = ay_1 + by_2$.

Furthermore, we have

$$\|ay_1 + by_2\| \leq \|ax_1 + bx_2\|$$

and clearly from this follows the relation for all positive numbers a and b. To obtain the relation for all a and b we apply the result just obtained for the pairs

$$x_1 - x_2, -x_2; \qquad y_1 - y_2, -y_2$$

and we get

$$\|a(x_1 - x_2) - bx_2\| \geq \|a(y_1 - y_2) - by_2\|$$

for $a \geq 0$, $b \geq 0$ which is equivalent to

$$\|ax_1 + bx_2\| \geq \|ay_1 + by_2\|$$

for $a \geq 0$, $b \leq 0$ and $a + b \leq 0$. Repeating the application of the above result for the set of elements

$$(x_2 - x_1, \ -x_1, \ y_2 - y_1, \ -y_1)$$

we obtain

$$\|ax_1 + bx_2\| \geq \|ay_1 + by_2\|$$

which holds for $a \geq 0$, $b \leq 0$ and $a + b \geq 0$. Thus the inequality

$$\|ax_1 + bx_2\| \geq \|ay_1 + by_2\|$$

holds for all a, b.

Following Birkhoff (1935) we define, in a Banach space, the concept of orthogonality as follows: Let X be a Banach space and let $x \in X$. We say that $x \perp y$, if, for all z, $\|x + zy\| \geq \|x\|$. It is not difficult to see that in the case of Hilbert spaces this reduces to the well-known notion of orthogonality.

We recall the following important characterization of Hilbert spaces obtained by M. M. Day (1947a, 1947b): *A Banach space X is a Hilbert space iff '$\perp$' is symmetric.*

Thus to prove the theorem it suffices to show that '$\perp$' is symmetric.

First we prove the following assertion: if X, Y are as in the theorem and $x_1, x_2 \in X$, $y_1, y_2 \in Y$ have the property

$$\|x_1\| = \|y_1\|, \qquad \|x_2\| = \|y_2\|, \qquad x_1 \perp x_2, \qquad y_1 \perp y_2$$

then for all a, b the following relation holds

(*) $\|ax_1 + bx_2\| \geq \|ay_1 + by_2\|$.

First we remark that we can suppose without loss of generality that

$$\|x_1\| = \|x_2\| = \|y_1\| = \|y_2\| = 1.$$

Then we find two sequences (x_n^1), (x_n^2) in the linear span of (x_1, x_2) such that:

1. $\|x_n^1\| = 1 = \|x_n^2\|$,

2. $\qquad x_n^2 - x_n^1 = \|x_n^2 - x_n^1\| x_2 \neq 0,$

3. $\qquad \lim_n x_n^1 = \lim_n x_n^2 = x_1$

and similarly for y_1, i.e. we find two sequences $(y_n^1), (y_n^2)$ such that these lie in the linear span of y_1 and y_2 and have the following properties:

1. $\qquad \|y_n^1\| = \|y_n^2\| = 1,$

2. $\qquad y_n^2 - y_n^1 = \|y_n^2 - y_n^1\| y_2 \neq 0$

3. $\qquad \lim_n y_n^1 = \lim_n y_n^2 = y_1.$

Now from the inequality (*) we get

$$\|ax_n^1 + bx_n^2\| \geq \|ay_n^1 + by_n^2\|$$

for all a and b and thus,

$$\|ax_n^1 + b(x_n^1 - x_n^2)/\|x_n^2 - x_n^1\|\| \geq \|ay_n^1 + b(y_n^2 - y_n^1)/\|x_n^1 - x_n^2\|\|$$

or

$$\|ax_n^1 + bx_2\| \geq \|ay_n^1 + by_2\|$$

for all a and b. Now for $n \to \infty$ we obtain the inequality

$$\|ax_1 + bx_2\| \geq \|ay_1 + by_2\|$$

for all a and b and thus the assertion is proved.

Now using this relation we show that the relation '$\perp$' is symmetric. First we prove this for X. Let x_1, x_2 in X and $x_1 \perp x_2$. In Day (1947b) it is proved that there exist y_1, y_2 in Y such that

$$\|x_1\| = \|y_1\| \qquad \|x_2\| = \|y_2\| \quad \text{and} \quad y_1 \perp y_2$$

and thus by the above assertion

$$\|ax_1 + x_2\| \geq \|ay_1 + y_2\| \geq \|y_2\| = \|x_2\|$$

i.e. $x_2 \perp x_1$. Thus '$\perp$' is symmetric in X.

We prove now that '$\perp$' is symmetric in Y. Let y_1, y_2 in Y and $y_1 \perp y_2$. We find x_1 and x_2 in X such that

$$\|x_1\| = \|y_1\|, \qquad \|x_2\| = \|y_2\|, \qquad x_1 \perp x_2.$$

Clearly we have, for all $s \neq 0$,

(**) $\qquad \|x_1 + sx_2\| > \|x_1\|.$

Let u be an arbitrary point in the linear span of y_1 and y_2 and thus if

$$u = Ay_1 + By_2, \qquad v = Ax_1 + Bx_2$$

we get

$$\|x_2 + rv\| = \|rAx_1 + (1 + rB)x_2\| \geq \|rAy_1 + (1 + rB)y_2\|$$
$$= \|y_2 + ru\| \geq \|y_2\| = \|x_2\|$$

and thus $x_2 \perp z$. Since '$\perp$' is symmetric in X we obtain that $z \perp x_2$. From (**) we get that $B = 0$ and thus $u = Ay_1$. Then we clearly have $y_2 \perp y_1$ and thus '$\perp$' is symmetric in Y. Then by Day's result (see also James, 1947) we obtain that X and Y are Hilbert spaces.

For further information concerning these problems we refer to Minty (1967, 1970, 1974), D. G. de Figueiredo and coworkers. (1967, 1970, 1972) S. Karamardian (1969).

6.3. SOME GENERAL PROPERTIES OF NONEXPANSIVE MAPPINGS

We present now several general properties of nonexpansive mappings. The fact that these are general refers to the fact that these properties are valid in arbitrary Banach spaces.

THEOREM 6.3.1. *Let X be a Banach space, and let C be a closed, convex and bounded subset of X. If $f : C \to C$ is nonexpansive and $(I - f)(C)$ is a closed subset of X, then f has a fixed point in C.*

Proof. We can suppose, without loss of generality, that 0 is in C. Let (r_n) be a sequence of positive numbers, $r_n \to 1$ and $r_n < 1$. We consider the mapping

$$f_n(x) = r_n f(x)$$

which, it is easy to see, is a contraction mapping. Therefore there exists a unique point $x_n \in C$ such that

$$f_n(x_n) = x_n$$

for all n. We show now that

$$(I - f)(x_n) \to 0.$$

Indeed, we have,

$$x_n - f(x_n) = x_n - r_n f(x_n) + r_n f(x_n) - f(x_n) = (r_n - 1)f(x_n)$$

and since $r_n < 1$, we obtain the assertion. Thus $0 \in \text{Cl}(I - f)(C)$. But the last set is closed and thus there exists a point $x_0 \in C$ such that $f(x_0) = x_0$ and the theorem is proved.

THEOREM 6.3.2. *If* $f : C \to C$ *is a nonexpansive mapping then any power* $f^n = f_0 f_0 \cdots {}_0 f$ *is again a nonexpansive map.*
    $\underset{n\text{-times}}{}$
Proof. Obvious.

We show now by an example that on general Banach spaces nonexpansive mappings may fail to have fixed points.

Example 6.3.3. (Sadovski, 1971). Let c_0 be the Banach space of all sequences converging to zero with the norm,

$$x = (x_i) \to \|x\| = \sup |x_i|$$

and let C be the set of all $x \in c_0$ such that $\|x\| \leq 1$. Then C is a closed, convex and bounded set in c_0. We define now the following map on C,

$$f(x) = (1, x_1, x_2, x_3, \ldots)$$

where $x = (x_i)$. It is obvious that f is a nonexpansive mapping on C with values in C and, moreover, that f is an isometry, i.e., for any x, y in C,

$$\|f(x) - f(y)\| = \|x - y\|.$$

If f has a fixed point in C then this is necessarily of the form $(1, 1, 1, \ldots)$ which is not in c_0.

This example suggests that to obtain positive results in the problem of existence of fixed points for nonexpansive mappings it is necessary to impose some restrictions either on f or the Banach space X.

6.4. NONEXPANSIVE MAPPINGS ON SOME CLASSES OF BANACH SPACES

In what follows we give the results obtained concerning the existence of fixed points for nonexpansive mappings for the case when the Banach space satisfies some additional properties such as uniform convexity, reflexivity, possession of normal structure, etc.

The following lemma gives an interesting and useful property of nonexpansive mappings. This property, together with some additional properties of the space, readily gives some fixed point theorems.

LEMMA 6.4.1. *Let X be a Banach space and let C be a weakly compact subset of X. Let $f : C \rightarrow C$ be a nonexpansive mapping and we suppose that C is minimal in the sense that it contains no proper closed subsets which are convex and invariant for f.*

If (x_n) is a sequence of approximate fixed points of f, i.e. $\lim (x_n - f(x_n)) = 0$, then for all x in C we have,

$$\lim \|x_n - x\| = \operatorname{diam} C.$$

Proof. Let x in C be arbitrary and set $s = \limsup \|x - x_m\|$. Consider the set

$$D = (x', x' \in C, \limsup \|x' - x_n\| \leq s)$$

which is a nonempty, closed and convex subset of C.

We prove that $f(D) \subset D$. Indeed, if x is in D then we have

$$\limsup \|f(x) - x_n\| = \limsup \|f(x) - f(x_n) + f(x_n) - x_n\|$$
$$\leq \limsup \|f(x) - f(x_n)\| +$$
$$+ \limsup \|f(x_n) - x_n\|$$
$$\leq \limsup \|f(x) - f(x_n)\|$$
$$\leq \limsup \|x - x_n\|.$$

and the assertion is proved. Since C is minimal, in the sense described above, $D = C$. Let (x_n) be a subsequence of (x_n) such that

$$s' = \lim \|x - x_{n'}\|.$$

Now, if there exists a z in C such that, for some subsequence $(x_{n''})$ of (x_n),

$$\lim \|x - x_{n''}\| = t \neq s$$

we consider the set

$$D' = (x, x \in C, \limsup \|x - x_{n''}\| \leq \min (t, s))$$

which is obviously closed and convex. It is also nonempty and invariant for f. Since C is minimal we have that $D' = C$. In this case, for each x in C, $\lim \|x - x_{n'}\|$ exists and it is equal to s'. We show now that $s = \operatorname{diam} C$. For this we consider the set F defined as follows:

$$F = (u, u \in C, \|u - x\| \leq s' \quad \text{for each } x \text{ in } C)$$

and we remark that this set is nonempty. Indeed, let $(x'_{n'})$ be a weakly convergent subsequence of $(x_{n'})$, $\lim x'_n = z$. Then we have,

1. $\qquad \lim \|x - x'_{n'}\| = s'$,

2. $\qquad \|x - z\| \leq \limsup \|x - x'_{n\cdot}\| \leq s$.

This gives us that z is in F. Suppose now that $r = \operatorname{diam} C$ is greater than s', i.e. $r > s'$. Since C is supposed to be minimal in the above sense, we conclude from this that $\overline{\operatorname{conv hull} f(C)} = C$.

Let u be an arbitrary element in C and from the above relation we have that, given there exists an element, say v, $v = \sum a_i f(x_i)$, $a_i \geq 0$, $\sum a_i = 1$, and $\|u - v\| \leq \varepsilon$. Consider now an arbitrary element in C and thus we have,

$$\|f(w) - u\| \leq \|f(w) - v + u - v\|$$
$$\leq \sum a_i \|f(w) - f(x_i)\| + \|u - v\|$$
$$\leq \sum a_i \|w - x_i\| + \|u - v\|$$
$$\leq s' + \varepsilon.$$

Since ε is arbitrary as well as u in C we have that $f(w) \in F$.
Thus by the minimality of C we have $F = C$. Thus $r = s'$.
In this case we have

$$\lim \|x - x_{n'}\| = r$$

whenever

$$(\|x - x_{n'}\|)$$

is convergent. Then $(\|x - x_n\|)$ converges to r and the theorem is proved. From this result we obtain the following result, given by Kirk (1965):

THEOREM 6.4.2. *Let X be a reflexive Banach space and suppose that X has normal structure. Let C be a closed convex and bounded set in X and $f : C \to C$ be a nonexpansive mapping. Then f has a fixed point in C.*

Proof. Using Zorn's Lemma we find a set K in C which is closed, convex and invariant for f and minimal with respect to the order defined by the inclusion. In this case it is clear that f and K satisfy the conditions of the above Lemma. If K contains more than one point we clearly have a contradiction since K has normal structure. Thus $K = (x_0)$ and this is a fixed point of f.

Remark 6.4.3. As we know, every uniformly convex Banach space is reflexive and has normal structure. In this case Theorem 6.4.3. was proved by F. Browder (1965). Since the proof of this result in the case of Hilbert spaces, as is given in the paper of Browder–Petryshyin (1967) presents some interesting features we give the proof below.

THEOREM 6.4.4. *Let H be a Hilbert space and C be a closed convex and bounded set in H. If $f: C \to C$ is a nonexpansive mapping on C then f has a fixed point in C.*

Proof. Let x_0 be a fixed point in C and consider a sequence of positive numbers (r_n) converging to 1 and $r_n < 1$.

For each n we consider the mappings $f_n: C \to C$ defined as follows:

$$f_n(x) = r_n f(x) + (1 - r_n)\bar{x}_0$$

and it is easy to see that these are contraction mappings. Then we find a unique point x_n in C such that

$$f_n(x_n) = x_n.$$

Since C is closed and convex as well as a bounded set in a Hilbert space we find a weakly convergent subsequence of (x_n).

We can suppose, without loss of generality, that (x_n) has just this property, i.e. (x_n) is weakly convergent. Let

$$x_n \rightharpoonup x_0$$

and then x_0 is in C since C is also weakly closed. We shall prove that x_0 is a fixed point of f.

Let x be an arbitrary point of H and thus we have,

$$\|x_n - x\|^2 = \|(x_n - x_0) + (x_0 - x)\|^2$$
$$= \|(x_n - x_0)\|^2 + \|(x_0 - x)\|^2 + 2\langle (x_n - x_0), (x_0 - x) \rangle$$

and we remark that

$$2\langle (x_n - x_0), (x_0 - x) \rangle \to 0$$

since (x_n) converges weakly to x_0. Now the sequence (r_n) converges to 1 and this implies that,

$$f(x_n) - x_n = (r_n f(x_n) + (1 - r_n)\bar{x}_0) - x_n + (1 - r_n)(f(x_n) - \bar{x}_0)$$
$$= (f_n(x_n) - x_n + (1 - r_n)(f(x_n) - \bar{x}_0)$$
$$= (1 - r_n)(f(x_n) - \bar{x}_0)$$

tends to zero

We now take $x = f(x_0)$ and we obtain

$$\lim (\|(f(x_0) - x_n)\|^2 - \|(x_n - x_0)\|^2) = \|x_0 - f(x_0)\|^2.$$

Since f is nonexpansive we have

$$\|f(x_n) - f(x_0)\| \leq \|x_n - x_0\|$$

and thus

$$\|x_n - f(x_0)\| \leq \|x_n - f(x_n)\| + \|f(x_n) - f(x_0)\|$$
$$\leq \|x_n - f(x_n)\| + \|x_n - x_0\|$$

which implies the following relation:

$$\overline{\lim} (\|x_n - f(x_0)\| - \|x_n - x_0\|) \leq 0$$

and, therefore,

$$\overline{\lim} (\|x_n - f(x_0)\|^2 - \|x_n - x_0\|^2) \leq 0.$$

This clearly gives us

$$\|x_0 - f(x_0)\|^2 = 0$$

i.e., x_0 is a fixed point of f.

Following Browder and Petryshyn [1967] we now present some results concerning the iterative methods for computation of fixed points for nonexpansive mappings.

First we give some definitions.

DEFINITION 6.4.5. Let H be a Hilbert space and let C be a closed and convex subset of H. A mapping $f : C \to C$ is called asymptotically regular at x if and only if

$$\lim_{n \to \infty} \|f^n(x) - f^{n+1}(x)\| = 0.$$

DEFINITION 6.4.6. A mapping $f : C \to C$ is called a reasonable wanderer in C if, starting at any $x_0 \in C$, the sequence (x_n) has the property that

$$\sum \|x_{n+1} - x_n\|^2 < \infty$$

where

$$x_n = f^n(x_0).$$

The following class of mappings was introduced by W. Petryshyn and plays an important role in the problem of extension of various fixed point theorems.

DEFINITION 6.4.7.   Let X be a Banach space, let C be a closed, convex subset of X and let $f : C \to X$. Then f is called demicompact if it has the property that, whenever (x_n) is a bounded sequence in C and $(f(x_n) - x_n)$ is strongly convergent, then there exists a subsequence (x_{n_i}) of (x_n) which is strongly convergent.

THEOREM 6.4.8.   *Let H be a Hilbert space and C be a closed and convex set in H. Suppose that the set of Fixed points of f, $F(f)$ is nonempty and choose s in (0, 1), where $f : C \to C$.*
   In this case the mapping

$$f_s(x) = sx + (1 - s)f(x)$$

maps C into C and has the same fixed points as f. For each s f_s is asymptotically regular.

   Proof. Let x be a fixed point of f. In this case we have,

$$f_s(x) = sx + (1 - s)f(x) = x + (1 - s)x = x,$$

i.e., x is a fixed point of f_s. Conversely, if x is a fixed point of f_s then we have,

$$f_s(x) = sx + (1 - s)f(x) = x$$

which gives us

$$(1 - s)f(s) = (1 - s)x$$

or

$$x = f(x).$$

   The assertion that f_s maps C into C is obvious. We show now that f_s is asymptotically regular. Indeed, let x be an arbitrary point of C and consider the sequence (x_n) defined by the relation

$$x_{n+1} = f(x_n), \qquad n = 0, 1, 2, 3, \ldots, x_0 = x.$$

and we show that

$$\sum \|x_{n+1} - x_n\|^2 < \infty.$$

   If y is a fixed point of f we have,

$$x_{n+1} - y = sx_n + (1 - s)f(x_n) = s(x_n - y) + (1 - s)(f(x_n) - y)$$

and for any constant a we have,

$$a(x_n - f(x_n)) = a(x_n - y) - a(f(x_n) - y).$$

But

$$\|x_{n+1} - y\|^2 = s^2 \|x_n - y\|^2 + (1-s)^2 \|(f(x_n) - y)\|^2 +$$
$$+ 2(1-s)\langle (f(x_n) - y), x_n - y \rangle$$

and

$$a^2 \|x_n - f(x_n)\|^2 = a^2 \|x_n - y\|^2 + a^2 \|f(x_n) - y\|^2 +$$
$$+ (-2s(1-s)\langle (f(x_n) - y), x_n - y \rangle$$

hence we obtain, by adding the corresponding sides and noting that $f(y) = y$.

$$\|x_{n+1} - y\|^2 + a^2 \|x_n - f(x_n)\|^2$$
$$\leq (2a^2 + s^2 + (1-s)^2 \|x_n - y\|^2 +$$
$$+ 2(s(1-s) - a^2))\langle f(x_n) - y, x - y \rangle.$$

Since the constant a is arbitrary we can assume, without loss of generality, that $a^2 = s(1-s)$ and from the last inequality we get,

$$\|x_{n+1} - y\|^2 + a^2 \|x_n - f(x_n)\|^2$$
$$\leq (2a^2 + s^2 + (1-s)^2 + 2s(1-s) - 2a^2) \cdot$$
$$\cdot \|x_n - y\|^2 = \|x_n - y\|^2.$$

$$\|x_{n+1} - y\|^2 + a^2 \|x_n - f(x_n)\|^2$$
$$\leq (2a^2 + s^2 + (1-s)^2 + 2s(1-s) - 2a^2) \cdot$$
$$\cdot \|x_n - y\|^2 = \|x_n - y\|^2.$$

Now summing up from $n = 0$ to $n = N$ we get

$$s(1-s) \sum_{n=0}^{N} \|x_n - y\|^2 \leq \sum_{n=0}^{N} (\|x_n - y\|^2 - \|x_{n+1} - y\|^2)$$
$$= \|x_0 - y\|^2 - \|x_{N+1} - y\|^2 \leq \|x_0 - y\|^2$$

which clearly implies that the series $\sum \|x_n - f(x_n)\|^2$ converges. Since $x_{n+1} - x_n = (1-s)(f(x_n) - x_n)$ we obtain that

$$\sum_{n=0}^{\infty} \|x_{n+1} - x_n\|^2 \leq (1-s)\|x_0 - y\|^2 / s$$

and thus f_s is a reasonable wanderer in C.

COROLLARY 6.4.9. *Let* $f : C \to C$ *be a nonexpansive mapping of a closed, convex and bounded set in a Hilbert space H. In this case for each s, f_s is asymptotically regular.*

Proof. It is easy to see that any mapping which is a reasonable wanderer is asymptotically regular and thus the assertion follows from the above theorem.

THEOREM 6.4.10. *Let* $f : C \to C$ *be nonexpansive and demicompact on the closed, convex and bounded set C in a Hilbert space or a strictly convex space. In this case the set of fixed points of f, F(f) is a nonempty, closed and convex set. Also, for each $s \in (0, 1)$ the sequence* (x_n), *where*

$$x_n = sf(x_{n-1}) + (1 - s)x_{n-1}, \qquad n = 1, 2, 3, \dots,$$

converges strongly to a fixed point of f.

Proof. From the above results it is clear that $F(f)$ is a nonempty set and we show now that it is a convex set. Indeed, for any x_1 and x_2 in $F(f)$ and any $t \in (0, 1)$ we have,

$$\| f(tx_1 + (1 - t)x_2) - x_1 \| \leq \| tx_1 + (1 - t)x_2 - x_1 \|$$

and similarly

$$\| f(tx_1 + (1 - t)x_2) - x_2 \| \leq \| tx_1 + (1 - t)x_2 - x_2 \|.$$

From these we obtain that

$$\| x_1 - x_2 \| \leq x_1 - f(tx_1 + (1 - t)x_2) \| +$$
$$+ \| f(tx_1 + (1 - t)x_2) - x_2 \|$$
$$\leq \| x_1 - x_2 \|$$

and since H is a strictly convex space, for some a, b in $(0, 1)$ we have

$$x_1 - f(tx_1 + (1 - t)x_2) = a(x_1 - tx_1 - (1 - t)x_2)$$

and

$$x_2 - f(x_1 t + (1 - t)x_2)) = b(x_2 - tx_1 - (1 - t)x_2$$

which implies the convexity of $F(f)$. Since the set C is bounded the sequence (x_n) is bounded and, by the Corollary 6.4.10, the sequence $(x_n - f(x_n)) = s^{-1}(x_n - x_{n+1})$ converges strongly. But f is supposed to be demicompact and thus there exists a convergent subsequence say (x_{n_i}). Let $x' = \lim x_{n_i}$ and thus

$$f(x_{n_i}) \to f(x'), \qquad f(x') = x'.$$

Now, since for each integer n we find $n_i \geq n$, and

$$\|x_n - x'\| \leq \|x_{n_i} - x'\|$$

we have that the entire sequence (x_n) converges to x' and the theorem is proved.

The following theorem of Browder and Petryshyn gives a result for nonexpansive mappings for which the set of fixed points reduces to a single point.

THEOREM 6.4.11. *Let $f : C \to C$ be a nonexpansive mapping of a closed, convex and bounded set C in a Hilbert space H. Suppose that $F(f) = (x')$. Then for any x_0 in C, the sequence $(f^n(x_0))$ converges weakly to x'.*

Proof. For the proof it suffices to prove that if for some (n_j), $(f^{n_j}(x_0))$ converges weakly to x_0', then x_0' is a fixed point of f. The condition about $F(f)$ implies that $x_0' = x'$.

Let

$$x_{n_j} = f^{n_j}(x_0) \to x_0'$$

and thus, f being nonexpansive,

$$\|x_{n_j} - f_s(x_0')\| = \|f_s(x_{n_j}) - f_s(x_0')\| + \|x_{n_j} - f_s(x_{n_j})\|$$
$$\leq \|x_{n_j} - x_0'\| + \|x_{n_j} - f_s(x_{n_j})\|.$$

According to Corollary 6.4.9,

$$\lim_{j \to \infty} (x_{n_j} - f_s(x_{n_j})) = 0$$

and from the last inequality we get,

$$\limsup(\|x_{n_j} - f_s(x_{n_j})\| - \|x_{n_j} - x_0'\|) = 0.$$

However,

$$\|x_{n_j} - f_s(x_0')\|^2 = \|(x_{n_j} - x_0' + (x_0' - f_s(x_0'))\|^2$$
$$= \|x_{n_j} - x_0'\|^2 + \|x_0' - f_s(x_0')\|^2 +$$
$$+ 2\langle (x_{n_j} - x_0'), x_0' - f_s(x_0')\rangle$$

and from the convergence of (x_{n_j}) to x_0' (weak convergence) we obtain that,

$$\lim_n (\|x_{n_j} - f_s(x_0')\|^2 - \|(x_{n_j} - x_0')\|^2) = \|x_0' - f_s(x_0')\|^2.$$

Since

$$\|x_{n_j} - f_s(x_0')\|^2 - \|x_{n_j} - x_0'\|^2 = (\|x_{n_j} - f_s(x_0')\| - (\|x_{n_j} - x_0'\|) \cdot$$
$$\cdot (\|x_{n_j} - f_s(x_0')\| + \|x_{n_j} - x_0'\|)$$

and

$$(\|x_{n_j} - f_s(x_0')\| + \|(x_{n_j} - x_0')\|) \le 2 \operatorname{diam} C$$

the above relations imply

$$\|x_0' - f_s(x_0')\| = 0$$

i.e., x_0' is a fixed point of f_s and thus a fixed point of f.

In the following theorem we show that the assumption on the fixed points of f in the above theorem can be dropped.

THEOREM 6.4.12.   *Let $f : C \to C$ be a nonexpansive mapping of a closed, convex and bounded set in H. Then for any s in $(0, 1)$ and any x_0 in C, the sequence $(f^n(x_0))$ converges weakly to a fixed point of f.*

Proof. First we remark that, according to the above results about nonexpansive mappings, the fixed point set $F(f)$ of f is nonempty. Also we know that it is convex, too. Now, for each y in $F(f)$ we have, for $x_n = f^n(x_0)$,

$$\|x_n - y\| \le \|x_{n-1} - y\|$$

by the nonexpansivity of f. We set

$$g(y) = \lim \|x_n - y\|$$

which clearly exists. For $d_0 = \inf(g(y), y \in F(f))$ we set

$$F_r = (y, g(y) \le d_0 + r)$$

for $r > 0$ and this set is obviously closed, convex, bounded and nonempty. Then it is also weakly compact and thus the set

$$F_0 = \bigcap_{r > 0}^{F} r$$

is nonempty.

We show now that F_0 contains only a single point. Indeed, F_0 is clearly convex and closed, and we suppose that there exist two distinct points in F_0, say y_0 and y_1. In this case we have, for s in $(0, 1)$ and $y_s = sy_0 + (1 - s)y_1$,

$$g^2(y_s) = \lim_n \|y_s - x_n\|^2$$

$$= \lim_n (\|s(y_0 - x_n) + (1 - s)(y_1 - x_n)\|^2)$$

$$= \lim_n (s^2 \|y_0 - x_n\|^2 + (1 - s)^2 \|y_1 - x_n\|^2 +$$

$$+ 2s(1 - s)\langle (y_0 - x_n), (y_1 - x_n)\rangle)$$

$$= \lim_n (s^2 \|y_0 - x_n\|^2 + (1 - s)^2 \|y_1 - x_n\|^2 +$$

$$+ 2s(1 - s)\|y_0 - x_n\| \, \|y_1 - x_n\|) +$$

$$+ \lim_n (2s(1 - s)\langle y_0 - x_n, y_1 - x_n\rangle -$$

$$- (\|y_0 - x_n\| \, \|y_1 - x_n\|))$$

$$= g^2(y_s) + \lim_n (2s(1 - s)\langle y_0 - x_n, y_1 - x_n\rangle -$$

$$- (\|y_0 - x_n\| \, \|y_1 - x_n\|))$$

and thus

$$\lim_n (2s(1 - s)(\langle y_0 - x_n, y_1 - x_n\rangle - \|y_0 - x_n\| \, \|y_1 - x_n\|)) = 0.$$

But

$$\lim\|y_0 - x_n\| = \lim\|y_1 - x_n\| = d_0$$

and thus

$$\|y_0 - y_1\|^2 = \lim \|(y_0 - x_n) + (x_n - y_1)\|^2 = 0$$

and this is contradiction.

As in Theorem 6.4.11, we can prove that, if $x_{n_j} \rightharpoonup y_0$, then y_0 is a fixed point of f and, in fact, the sequence (x_n) converges weakly to y_0. From the definition of the function g we have

$$\lim \|x_{n_j} - y_0\|^2 = \lim \|x_{n_j} - y + y - y_0\|^2 = g^2(y) + \|y - y_0\|^2$$

and, since

$$g(y) \geq d_0,$$

because $F_0 = (y)$, we get $y = y_0$ and the theorem is proved.

We give now some results for the case of nonexpansive mappings on uniformly convex Banach spaces.

First we give the following definition.

DEFINITION 6.4.13. Let $f : C \to C$ be a continuous mapping. We say that f satisfies the condition (α) if $(I - f)$ maps every bounded closed set of C into a closed set.

THEOREM 6.4.14. *Let $f : C \to C$ be a nonexpansive mapping on a closed, convex and bounded subset C of a uniformly convex Banach space X and s in $(0, 1)$. In this case the mapping*

$$f_s(x) = sx + (1 - s)f(x)$$

is asymptotically regular.

Proof. Let x_0 be an arbitrary point in C and we consider the sequence (x_n), $x_n = f_s^n(x_0)$. It is easy to see that the set of fixed points of f and f_s are the same. Let x' be a fixed point of f and we have:

$$\| f_s^{n+1}(x_0) - x' \| = \| f_s^{n+1}(x_0) - f(x') \| \leq \| f_s^n(x_0) - x' \|,$$

and thus the sequence $(\| f_s^n(x_0) - x' \|)$ is monotone. Let $d_0 = \lim \| f_s^n(x_0) - x' \|$. If $d_0 = 0$ then we have

$$\lim (f_s^{n+1}(x_0) - f_s^n(x_0)) = 0$$

and the assertion of the theorem is proved. Now, if $d_0 > 0$ since x' is a fixed point of f we get,

$$f_s^{n+1}(x_0) - x' = sf_s^n(x_0) + (1 - s)f_0f_s^n(x_0) - x'$$
$$= s(f_s^n(x_0) - x') + (1 - s)(f_0f_s^n(x_0) - x')$$

and because $\lim \| f^{n+1}(x_0) - x' \| = d_0 = \lim \| f_s^n(x_0) - x' \|$, from the uniform convexity of X it follows that

$$\lim (f_s^{n+1}(x_0) - f_s^n(x_0)) = 0,$$

since

$$f_s - 1 = (s - 1)(I - f).$$

From this result we obtain, in the presence of an additional property of f, the convergence of the sequence (x_n) to a fixed point of f.

THEOREM 6.4.15. *Let $f : C \to C$ be a nonexpansive mapping of a closed, convex and bounded set in a uniformly convex Banach space X. Suppose that f*

satisfies the condition (α) *of Definition* 6.4.13. *In this case the set of fixed points* $F(f)$ *of f is a closed, convex set and, for each s in* (0, 1) *and any x_0 in C, the sequence* (x_n) *converges strongly to a point of* $F(f)$.

Proof. We know that $F(f)$ is nonempty. The fact that $F(f)$ is a convex set can be proved as in Theorem 6.4.10 by remarking that, in fact, the proof given there is valid for any rotund space.

Let x' be a point in $F(f)$ and thus for each point x_0 of C the sequence $(\|f_s^n(x_0) - x'\|)$ is decreasing. To prove the assertion of the theorem it suffices to show that there exists a strongly convergent subsequence of $(f_s^n(x_0))$. Let M be the strong closure of $(f_s^n(x_0))$, and from the above theorem we know that

$$(1 - f_s)(f_s^n(x_0)) \to 0.$$

But f satisfies the condition (α) and this implies that f_s satisfies the condition (α). Then we get that 0 lies in the strong closure of $(1 - f_s)C$ which is actually closed. This proves the existence of the desired subsequence and thus the theorem is proved.

Remark 6.4.16   Every demicompact mapping satisfies the condition (α).

Stronger results of this type will be given in Chapter 7 using duality functions.

Remark 6.4.17.   The first result about the convergence of sequence (x_n) for the case $s = \frac{1}{2}$ was obtained by M. A. Krasnoselskii (1955). The result of Krasnoselskii was extended by H. Schaefer (1957) by proving the convergence for any fixed s in (0, 1) and then weakening the assumptions about the mapping f. Next, the second result of Schaefer was extended by M. Edelstein (1966) to the case of rotund spaces. The results which were given above are given by F. Browder and W. Petryshyn (1967). We also mention that interesting and important iteration methods have been proposed by Z. Opial (1967), S. Kaniel (1971), A. Pazy (1971,1977), etc.

As we know, the original result of M. Krasnoselskii refers to the case when the nonexpansive mapping is compact and the sequence is for $s = \frac{1}{2}$; i.e.,

$$x_{n+1} = (x_n + f(x_n))/2, \qquad n = 0, 1, 2, 3, \ldots.$$

The results about the convergence of the sequence (x_n) naturally suggest the following problem: is it possible that for any nonexpansive mapping $f : C \to C$, the sequence (x_n) defined above converges strongly to a fixed point of f?

J. Lindenstraus has constructed an example of a nonexpansive mapping defined on a closed, convex and bounded set of a Banach space such that the sequence (x_n) does not converge. We now give this interesting example, following the presentation given by A. Genel and J. Lindenstraus (1975).

Example 6.4.18 (Genel and Lindenstraus). There exists a closed, convex and bounded set in l^2 and a nonexpansive mapping defined on the set such that, for some x_0, the sequence (x_n) does not converge strongly to a fixed point of f.

The construction of this example uses the Kirzbraun–Valentine theorem.

Let $(e_k)_1^\infty$ be an orthonormal basis of l^2. We define a sequence $(x_n)_1^\infty$ in the following manner: first we set $x_1 = e_1$ and, for a suitable integer n, the points $x_2, \ldots, x_{n_1}$ are chosen in the plane determined by the points $(0, e_1, e_2)$ according to the following rules:

1. $\|x_i\| = \|Tx_i\|, \qquad i = 1, 2, \ldots, n_1 - 1,$

2. $\langle x_i, Tx_i \rangle / \|x_i\|^2 = \cos 2\varphi_1, \qquad i = 1, 2, \ldots, n_1 - 1.$

Here the angle φ_1 is independent of i. From these it is clear that the triangle determined by $(0, x_i, x_j)$ is congruent to the triangle determined by $(0, Tx_i, Tx_j)$ for $i, j = 1, 2, \ldots, n_1 - 1$.

Now, to choose the integer n_1 and the angle φ_1, we set φ_1 such that

$$\varphi_1 = \pi/3 \cdot (n_1 - 1), \quad n_1 > 10, \quad (\cos \varphi_1)^{n_1} \geq \tfrac{3}{4}$$

and we remark that this is possible, since $\lim_n (\cos \pi/3n)^n = 1$.

We now define the point Tx_{n_1} as follows: first we consider a point y_1 in the plane determined by $(0, e_1, e_2)$ with the following properties:

$y_1 = Tx_{n_1}$, the angle between x_{n_1} and y_1 is $2\varphi_1$.

We consider the point $z_1 = \tfrac{1}{2}(y_1 + x_{n_1})$ and we remark that

$$\|z_1 - Tx_i\| < \|x_{n_1} - x_i\|, \qquad i = 1, 2, \ldots, n_1 - 1$$

as is easy to see from a diagram in which the points are represented.

Since we have a finite number of inequalities, we find a small positive number λ_1 such that if we set $Tx_{n_1} = z_1 + \lambda_1 e_3$ the following inequalities are valid:

$$\|Tx_{n_1} - Tx_1\| \leq \|x_{n_1} - x_i\|, \qquad i = 1, 2, \ldots, n_1 - 1,$$

and, since $\|z_1\| < \|x_{n_1}\|$ we can choose λ_1 such that $\|Tx_{n_1}\| < \|x_{n_1}\|$,

Also we have the following inequality:

$$\|x_{n_1 + 1}\| = \|(x_{n_1} + Tx_{n_1})/2\| \geq \|z_1\| \geq \tfrac{3}{4},$$

which follows from the construction of the points and the property of λ_1 and n_1.

Now we see that the angle between e_1 and the unit vector w_1 in the direction of $(x_{n_1} + z_1)/2$ is between $\pi/3$ and $2\pi/5$ and, for $a > 0$,

$$\|aw_1 - Tx_i\| < \|aw_1 - x_i\|, \qquad i = 1, 2, \ldots, n_1 - 1.$$

We now consider the plane determined by $(0, w_1, e_3)$ and thus the point x_{n_1+1} belongs to this plane. Also in this plane will be the points $(x_i)_{i=n_1+2}^{n_2}$ and we choose these points as follows:

Let $\bar{x}_{n_1}$ and $\bar{T}x_{n_1}$ be the orthogonal projection of x_{n_1} and Tx_{n_1} on the plane $(0, w_1, e_3)$ respectively. In this case we have,

$$\|\bar{x}_{n_1} - x_{n_1}\| < \|Tx_{n_1} - \bar{T}x_{n_1}\|.$$

From the property of a, and from the fact that every point in the set $\mathbf{R}_2$ is of the form $aw_1 + be_3$ with $a \geq \frac{1}{2}(\cos(2\pi/5))/2$, we get

$$\varepsilon_1 = \min_{\substack{u \in R_2, \\ 1 \leq i \leq n_1 - 1}} (\|u - x_i\| - \|u - Tx_i\|) > 0.$$

Now we repeat the procedure used for constructing $(x_i)_1^{n_1}$ by starting with x_{n_1+1} and rotating always in the plane $(0, w_1, e_3)$ by a fixed angle $2\varphi_2$. Thus we take, for $n_1 < i < n_2$,

$$\|Tx_i\| = \|x_i\|, \qquad \langle x_i, Tx_i \rangle = \|Tx_i\|^2 \cos 2\varphi_2$$

where φ_2 and n_2 satisfy the following conditions:

$$\varphi_2 = (\pi/3)(n_2 - n_1 - 1), \qquad n_2 > n_1 + 10, \qquad \tfrac{3}{4}(\cos(\varphi_2))^{n_2 - n_1}$$
$$> \tfrac{5}{8},$$
$$4\sin\varphi_2 < \varepsilon_1, \qquad \|Tx_i - \bar{T}x_{n_1}\| < \|x_i - \bar{x}_{n_1}\| \quad \text{for } i \in (n_1, n_2).$$

We remark that the relations are possible since

$$\lim_k (\cos \pi/3)^k = 1$$

and the fact that all points (x_i), i in (n_1, n_2) are in R_2 and for each u in R_2 there exists $\delta(u) > 0$ such that if $\|u\| = \|v\|$ and the angle between u and v is positive and not larger than $\delta(u)$ then $\|v - \bar{T}x_{n_1}\| < \|u - \bar{x}_{n_1}\|$. We show now that on the set (x_i), i in $(1, n_2)$ the mapping T is nonexpansive. From the above considerations it remains only to consider the difference

$$Tx_i - Tx_j$$

with $1 \leq i < n_1$ and $n_1 < j < n_2$. Since for j in the above set we have by construction

$$\|Tx_j\| = \|x_j\|, \qquad \langle x_j, Tx_j \rangle = \|Tx_j\|^2 \cos 2\varphi_2$$

which implies that

$$\|x_j - Tx_j\| < 2 \sin \varphi_2.$$

Since $\varepsilon_1 > 0$ and from the considerations about φ_2 and n_2 we get

$$\|Tx_i - Tx_j\| \leq \|Tx_i - x_j\| + \|x_j - Tx_j\|$$
$$= \|Tx_i - x_j\| + \varepsilon_1/2 < \|x_i - x_j\|$$

i.e. T is nonexpansive. We continue using this procedure to obtain a sequence (x_i) and the mapping T nonexpansive.

Let us consider the point y_2 in the plane determined by $(0, w_1, e_3)$ such that $\|y_2\| = \|x_{n_2}\|$ and setting $z_2 = \frac{1}{2}(y_2 + x_{n_2})$ and $w_2 = z_2/\|z_2\|$. We define now Tx_{n_2} as the point $z_2 + \lambda_2 e_4$ with λ_2 chosen appropriately. We note that

$$\|x_{n_2 + 1}\| = \tfrac{5}{8}.$$

The points (x_i), $n_2 + 1 \leq i \leq n_3$ will be chosen in the plane determined by the points $(0, w_2, e_4)$ in such a way that T remains nonexpansive. The construction of these points is similar to the above constructions. First we note that in $(0, w_2, e_4)$ there is a domain analogous to the domain R_2 and we denote this domain by R_3. In this case, as above, for any u in R_3 we have $u = aw_1 + be_3 + ce_4$ where $a \geq \frac{1}{2}(\cos(2\pi/5))^2$ and thus

$$\min_{\substack{u \in R \\ 1 \leq i \leq n_1 - 1}} (\|u - x_i\| - \|u - Tx_i\|) > 0$$

and

$$\|aw_2 - Tx_i\| < \|aw_2 - x_i\| \quad \text{if } a > 0 \quad \text{and } n_1 < i < n_2$$
$$\|aw_2 - \bar{T}x_{n_1}\| < \|aw_2 - \bar{x}_{n_1}\| \quad \text{if } a > 0.$$

From these we get

$$\lambda_2 = \min_{\substack{u \in R \\ 1 \leq i \leq n_2 - 1}} (\|u - x_i\| - \|u - Tx_i\|) > 0.$$

As in the case of n_2 we continue to define the entire sequence (x_i) and on this sequence T is nonexpansive. Now we can use Kirzbraun's extension theorem to obtain a nonexpansive mapping on the set $\mathrm{Cl\,conv}\,(x_i) \cup (Tx_i))$. In this case the sequence (x_i) is bounded and also bounded from below in norm. Since (x_i) converges weakly to zero and is not norm convergent to zero this shows the property of the example.

For other fixed points for nonexpansive mappings, especially for spaces which do not have normal structure we refer to Browder (1967), (1976), Browder and Petryshyn (1967), Petryshyn and Williamson (1972, 1973), Kirk (1971 a, 1971 b, 1975) and the references quoted there.

The following problem has remained unsolved for several years:

Problem 6.4.19.   If K is a closed and convex subset of a real Hilbert space and $T : K \to K$ is nonexpansive then does T necessarily have fixed points in K?

This problem has been solved by William O. Ray (1979) with the following theorem:

THEOREM 6.4.20.   *Let K be a closed and convex set in a real Hilbert space H. Then K has the fixed point property for nonexpansive mappings iff K is bounded.*

This actually means that if K is not bounded then there exists at least one nonexpansive mapping on K without fixed points. We note that Problem 6.4.19 has sense also for subsets in other classes of Banach spaces, i.e., the problem whenever boundedness is necessary for the existence of fixed points for nonexpansive mappings: does it apply, for example, in sets K situated in Banach spaces with normal structure?

We begin with an example which is very instructive for the proof of Theorem 6.4.20.

First we give the definition of an interesting class of closed and convex sets in Banach spaces.

DEFINITION 6.4.21.   Let (e_i) be an orthonormal set in $H = l^2$. Then a set of the form

$$(x, |\langle x, e_i \rangle| \leq M_i)$$

is called a block set.

Example 6.4.22.   The block set

$$K = (x, |\langle x, e_i \rangle| \leq 1)$$

does not have the fixed property with respect to nonexpansive mappings.

Proof. For any $x \in K$ we set $Tx = e_1 + \sum_1^\infty \langle x, e_i \rangle e_{i+1}$ which is obviously nonexpansive and without fixed points.

For the proof of Theorem 6.4.21 we need the following result stated as

THEOREM 6.4.23. *If K is a closed, convex and linearly bounded and unbounded subset of a real Hilbert space H and if $0 \in K$, then there exists an orthonormal set $(e_i) \subseteq H$ such that for each n, $\sum_1^n e_i \in K$.*

For the proof of this theorem we need some results which are given as lemmas. First we recall the notion of linearly bounded sets in Banach spaces or in topological vector spaces.

DEFINITION 6.4.24. A set K is said to be linearly bounded if the intersection with each line in the space is a bounded set.

DEFINITION 6.4.25. Any nonempty subset M of X is called admissible if it is unbounded, closed and convex and linearly bounded.

Let M be a subset of a Hilbert space H, $y \in H$ and $\mu \in (0, 1)$. we set

$$C(y, \mu, M) = (z, z \in M, \mu \|z\|^2 \leq \langle z, y \rangle \leq \|y\|^2)$$

and

$$E(y, \mu, M) = (z, z \in M, \mu \|y\|^2 = \langle z, y \rangle).$$

Then we have the following useful result:

THEOREM 6.4.26. *Let M be an admissable set in H and $y \in H$. If y is not the unique point of minimal norm of M then $C(y, \mu, M)$ is admissable for each $\mu \in (0, 1)$.*

Proof. From the definition of $C(y, \mu, M)$ it is obvious that this is nonempty and closed as well as convex and linearly bounded. Thus it remains to prove the unboundedness. Since M is supposed to be unbounded, then there exists a sequence (x_n) in M such that $\|x_n\| \to \infty$. Take $R > 0$, arbitrary but fixed, and let (a_n) be a sequence of positive numbers such that

$$\|a_n x_n + (1 - a_n)y\| = R$$

(which is possible from the continuity of the norm). Since $\|x_n - y\| \to \infty$ we get

$$a_n = \|a_n(x_n - y)/(\|x_n - y\|) + y/(\|x_n - y\|)\|$$
$$\leq R/(\|x_n - y\|) + \|y\|/(\|x_n - y\|)$$

and thus $a_m \to 0$. This implies that, for n large, $a_n x_n + (1 - a_n)y \in M$.

But M is supposed to be linearly bounded and thus $\{a_n x_n + (1 - a_n)y\}$ is a bounded sequence in H. Then we can suppose, without loss of generality, that it converges (weakly) to an element of H, say w_0. We now show that in fact we have $y = w_0$. Fix b in $\mathbf{R}^+$ and consider $bw_0 + (1-b)y$. If N is sufficiently large, from $\lim a_n = 0$ we get that $ba_n \leq 1$. But $a_n x_n + (1-a)y \rightharpoonup w_0$ and thus

$$w_0 \in \overline{\text{conv}}\,(a_n x_n + (1-a_n)y : n \geq N).$$

Further, since $ba_n \leq 1$,

$$bw_0 + (1-b)y \in \overline{\text{conv}}\,(b(a_n(x_n - y) + y) + (1-b)y : n \geq N)$$

$$= \overline{\text{conv}}\,(ba_n x_n + (1-a_n)y : n \geq N) \subseteq M$$

and thus, if $y \neq w_0$, the ray from y through w_0 is contained in M, which is a contradiction with the linearly bounded property of M.
Thus

$$a_n x_n + (1-a_n)y > y$$

and this implies that for each $R > 0$ and each 0 there exists a $w \in M$ such that:

1. $\quad \|w\| = R$,

2. $\quad \|y\|^2 - \varepsilon \leq \langle w, y \rangle \leq \|y\|^2 + \varepsilon.$

Now the assertion will be proved if we show that, for each $R \geq 0$ and each $\varepsilon \geq 0$, there exists a $w \in M$ such that:

1. $\quad \|w\| = R$,

2. $\quad \|y\|^2 - \varepsilon \leq \langle w, y \rangle \leq \|y\|^2.$

Since M is a Hilbert space then there exists a unique element of minimal norm in M, say u. Let (ε_n) be a sequence in $(0, \frac{1}{2}(\|y^2\| - \|w\|\,\|y\|))$ and $\lim \varepsilon_n = 0$. We consider now the following sequence of numbers (b_n) defined by

$$b_n = \varepsilon_n/(\|y\|^2 + \varepsilon_n - \|u\|\,\|y\|)$$

and it is clear that b_n is in $(0, 1)$ as well as $\lim b_n = 0$. Let $(w_n) \subset M$ such that

$$\|y\|^2 - \varepsilon_n \leq \langle w_n, y \rangle \leq \|y\|^2 + \varepsilon_n$$

and define further the sequence (z_n) in M by

$$z_n = (1 - b_n)w_n + b_n u.$$

It is clear that $\|z_n\| \to R' \geq R$ and further we have,

$$\langle z_n, y \rangle = \langle (1 - b_n)w_n, y \rangle + b_n \langle u, y \rangle \leq (1 - b_n)(\|y\|^2 + \varepsilon_n) +$$
$$+ b_n \|u\| \|y\|$$
$$\leq \|y\|^2 - b_n(\|y\| - \|u\| \|y\| + \varepsilon_n) + \varepsilon_n = \|y\|^2.$$

Thus we have

$$\langle z_n, y \rangle \leq \|y\|^2.$$

Obviously $\langle z_n, y \rangle \to \|y\|^2$ and thus, for n sufficiently large,

1. $\qquad \|z_n\| \geq R,$

2. $\qquad \|y\|^2 - \varepsilon \leq [z_n, y\rangle \leq \|y\|^2$

and the theorem is proved.

LEMMA 6.4.27.   *If M is an admissable set in a Hilbert space, then for each $b \in (0, 1)$ and $y \in M$ such that $\|y\| \geq \|u\|/b$, u being the unique element of M of minimal norm, the set $E(y, b, M)$ is admissable.*

Proof. Let us consider $a = u, y$ and then $b\|y\|^2 > a$. Let $T \geq 0$ fixed and then according to Theorem 6.4.26 there exists a $w \in M$ such that

$$\|w\| \geq ((\|y\|^2 - a)/(b\|y\|^2 - a))(T + \|u\|)$$

and

$$\|y\|^2 \leq \langle w, y \rangle \leq r\|y\|^2 \leq b\|y\|^2.$$

Then we define

$$\lambda = (b\|y\|^2 - a)(r\|y\|^2 - a)^{-1}$$

and since $r \geq b$ we have that $\lambda \leq 1$, Further we get that

$$\lambda = ((b\|y\|^2 - a)/(\|y\|^2 - a) > 0.$$

We now consider the element

$$v = \lambda w + (1 - \lambda)u \in M$$

and we prove that $v \in E(y, b, M)$ and $\|v\| \geq T$.

First we compute the number $\langle v, y \rangle$. We have

$$\langle v, y \rangle = \langle \lambda w + (1 - \lambda)u, y \rangle$$
$$= \lambda \langle w, y \rangle + (1 - \lambda) \langle u, v \rangle$$
$$= \lambda r \|y\|^2 + (1 - \lambda)a$$

$$= \lambda r \|y\|^2 + (r-b)\|y\|^2 a/(r\|y\|^2 - a)$$
$$= (rb\|y\|^2 - ra + (r-b)/(r\|y\|^2 - a)\|y\|^2 = b\|y\|^2$$

and, since w satisfies the above inequality, we then have,

$$\|v\| = \|\lambda w + (1-\lambda)u\|$$
$$\geq \lambda \|w\| - (1-\lambda)\|u\|$$
$$\geq (b\|y\|^2 - a)/(\|y\|^2 - a)\|w\| - \|u\|$$
$$\geq T + \|u\| - \|u\| = T.$$

This implies the assertion of the Lemma.

LEMMA 6.4.28. *Let M be a closed and convex set with $0 \in M$. Suppose that (y_n) is a sequence in H with the property that $\sum_1^n y_j \in M$ for each n. If (a_n) is a nonincreasing sequence in $[0, 1]$ then*

$$\sum_{j=1}^n a_j y_j \in M$$

for each n.

Proof. The assertion is true for $n = 1$ since $0 \in M$; and then we prove its further validity by induction: suppose that the result is true for $i \leq m-1$, then prove that it is true for $i = m$. Let $a_1, \ldots, a_m$ be nonincreasing and bounded by 1; and we may suppose, without loss of generality, that $a_1 \geq a_m$. Then we have

$$\sum_{i=1}^m a_i y_i = a_1 \sum_{i=1}^m a_i/a_i y_i$$
$$= a_1((a_1 - a_m)) \Big/ a_1 \sum_{i=1}^{m-1} ((a_i - a_m))/(a_1 - a_m)y_i) +$$
$$+ a_m \Big/ a_1 \sum_{i=1}^m y_i$$

and, since $a_i \leq a_{i+1}$, we have

$$(a_i - a_m)/(a_1 - a_m) \geq (a_{i+1} - a_m)/(a_1 - a_m).$$

The hypothesis of induction implies that

$$x = \sum_{i=1}^m ((a_i - a_m)/(a_1 - a_m)) y_j \in M,$$

$$x^+ = \sum_{i=1}^m y_i \in M$$

and, by the convexity of M, we get $\sum_{i=1}^m a_i y_i \in M$. The Lemma is proved.

COROLLARY 6.4.29.   *Let M be a closed, convex set with $0 \in M$ and suppose that (y_n) is a sequence which is an orthonormal set in H and has the property that $\sum_1^m y_i \in M$ for all m. Define*

$$M_0 = (w, w \in \overline{sp((y_n))}:(w, y_n) \text{ is a nonincreasing sequence of positive numbers bounded above by } 1).$$

Then M_0 is a closed convex subset of M.

Now we are in a position to prove Theorem 6.4.23.

Let us consider b in $(0, 1)$ and choose $y_1 \in K$ such that $\|y_1\| > 1/b$ and set $K_1 = E(y_1, b, K)$. If u_1 is the unique element of K_1 of minimal norm, then since K_1 is admissable according to Lemma 6.4.27, there exists w_1 in K_1 with the property that $\|w_1\| \geq 1/b\|u_1\|$ and $\|w_1\| \geq (b+1)\|u_1\|$. We define y_2 by

$$y_2 = w_1 - \langle w_1, y_1 \rangle / \|y_1\|^2 y_1.$$

Then we obviously have the relations

$$\langle y_2, y_1 \rangle = 0, \qquad \|y_2\| \geq \|y_1\|, \qquad w_1 = by_1 + y_2.$$

Now if λ_2 is defined by

$$(b\|y_2\|^2 + b^2\|y_1\|^2)/(\|y_2\|^2 + b^2\|y_1\|^2)$$

then $K_2 = E(w, \lambda_2, K_1)$ is admissable and for $w \in K_2$, $\langle w, y_1 \rangle = b\|y_1\|^2$ and $\langle w, y_2 \rangle = b\|y_2\|^2$. Indeed, from Lemma 6.4.27, K_2 is admissable if $\|w_1\| > 1/\lambda_2\|u_1\|$ since

$$1/\lambda_2\|u_1\| = \|u_1\|/b(\|y_2\|^2 + b^2\|y_1\|^2)/$$
$$(\|y_2\|^2/(\|y_2\|^2 + b\|y_1\|^2)) < \|u_1\|/b < \|w_1\|$$

which proves the admissability. Further, since $w_1 \in K_1$ and $\langle w, y_1 \rangle = b\|y_1\|^2$ we obtain

$$\langle w, y_2 \rangle + b^2\|y_1\|^2 = \langle w, y_2 \rangle + \langle w, by_1 \rangle$$
$$= \langle w, w_1 \rangle = \lambda_2\|w_1\|^2$$
$$= \lambda_2(\|y_2\|^2 + b_2\|y_1\|^2)$$
$$= b(\|y_2\|^2 + b\|y_1\|^2)$$

which implies that

$$\langle w, y_2 \rangle = b\|y_2\|^2.$$

Now we proceed by induction to construct the sequence (y_n). Suppose for this purpose that the elements $y_1, \ldots, y_m$ are constructed. Then we set

$$\lambda_j = \left(b\|y_j\|^2 + b^2 \sum_{k=1}^{m-1} \|y_k\|^2 \right) \bigg/ \left(\|y_j\|^2 + b^2 \sum_{k=1}^{m-1} \|y_k\|^2 \right)$$

for $j = 1, 2, 3, \ldots, m$ and

$$w_j = Y_{j+1} + b \sum_{k=1}^{j} y_k$$

for $j = 1, 2, 3, \ldots, m-1$.

Suppose further that the following assertions are true:

(a) $\|y_1\| \le \|y_2\| \le \|y_3\| \le \cdots \le \|y_m\|$,

(b) $\langle y_i, y_j \rangle = 0$ for $i \ne j$,

(c) $K_j = E(w_{j-1}, \lambda_j, K_{j-1})$ is admissable and has the unique element of minimal norm u_j,

(d) $w_j \in K_j$, $\|w_j\| > 1/b \|u_j\|$,

(e)   if $w \in K_i$, $i = 1, 2, 3, \ldots, m$ and if $j = 1, 2, 3, \ldots i$ then $\langle w, y_j \rangle = b \|y_j\|^2$.

We note that the above results show that if $m = 2$ these assertions are true and thus permit us to apply the method of induction.

Now the set K_m is unbounded and thus there exist $w_m \in K_m$ with the norm satisfying the inequality $\|w_m\| > 1/b \|u_m\|$ and if

$$y_{m+1} = w_n - \sum_{k=1}^{m} (w_m, y_k)/\|y_k\|^2 y_k$$

then $\|y_{m+1}\| \ge \|y_m\|$ and thus we have (a) valid for $m+1$.

Now, since (e) is true from our choice of w_m and y_{m+1} we get that

$$w_m = y_{m+1} + b \sum_{k=1}^{m} y_k$$

and this gives us that (b) holds for $m+1$.

Now by Lemma 6.4.27 $K_{m+1} = E(w_m, \lambda_{m+1}, K_m)$ is admissable if

$$1/\lambda_{m+1} \|u_m\| = \|u_m\|/b (\|y_{m+1}\|^2 + b^2 \sum_{k=1}^{m} \|y_k\|^2) \div$$

$$\div (\|y_{m+1}\|^2 + b \sum_{k=1}^{m} \|y_k\|^2)$$

$$< \|u_m\|/b < \|w_m\|$$

which is true just by construction. Thus (c) is verified and clearly (d) holds as well.

Now if $w \in K_{m+1} \subseteq K_m$ then (e) implies that $\langle w, y_j \rangle = b \|y_j\|^2$ for $j = 1, \ldots, m$ and thus if we prove that $\langle w, y_{m+1} \rangle = b \|y_{m+1}\|^2$ (e) is fulfilled for the sequence $(y_1, \ldots, y_m, y_{m+1})$. Now we have

$$\langle w, y_{m+1} \rangle + b \sum_{k=1}^{m} \|y_k\|^2 = \langle w, y_{m+1} \rangle + b \sum_{k=1}^{m} \langle w, y_k \rangle$$

$$= \langle w, y_{m+1} + b \sum_{k=1}^{m} y_k \rangle$$

$$= \langle w, w_m \rangle = \lambda_{m+1} \|w_m\|^2$$

$$= b \|y_{m+1}\|^2 + b \sum_{k=1}^{m} \|y_k\|^2$$

and this proves (e) for $m + 1$.

Thus all considerations for applying the induction method are fulfilled and we obtain an orthonormal sequence (y_m) with the property that, for each m,

$$y_m + b \sum_{k=1}^{m-1} y_k$$

is in K. The sequence also has the property that $\|y_i\| \leq \|y_{i+1}\|$ for all i. Now we can apply Lemma 6.4.29 to the sequence $(by_1, \ldots, by_{m-1}, y_m)$ and an induction argument gives that $\sum_{k=1}^{m} by_k$ is in K for all m.

Once again applying Lemma 6.4.29 we get $e_i = by_i / \|by_i\|$ which satisfies the Theorem 6.4.23.

Proof of Theorem 6.4.20. The fact that boundedness is a sufficient condition is a result proved in Theorem 6.4.4. We prove now that the condition is also necessary. If H_e denotes the closed Hilbert subspace generated by e_i's then we set

$$K_0 = (w, \; w \in H_e, \; (|\langle w, e_i \rangle|) \; \text{is a nonincreasing sequence of nonnegative numbers bounded above by 1})$$

and by the Corollary 6.4.29, K_0 is a closed convex subset of K. For $w \in K_0$,

$$w = \sum_{1}^{\infty} \langle w, e_i \rangle e_i$$

we set

$$f(w) = e_1 + \sum_1^\infty \langle w, e_i \rangle e_{i+1}$$

and clearly K_0 is an invariant set for f which is obviously and isometry.

It is obvious that f has not fixed points in K_0 and if, for x in K, $p(x)$ denotes the nearest point in K_0 to x then we define

$$F(x) = f(p(x))$$

which is nonexpansive on K. Since f is fixed point free, F is fixed point free and the theorem is proved.

6.5. CONVERGENCE OF ITERATIONS OF NONEXPANSIVE MAPPINGS

An interesting class of nonexpansive mappings for which the Cauchy–Picard sequence of iterations converges was discovered by J. J. Moreau (1978). His result refers to the Hilbert space nonlinear mappings. The extension of Moreau's result to the case of uniformly convex Banach spaces was obtained by B. Beauzamy (1978).

The result obtained by Moreau is as follows:

THEOREM 6.5.1.   *Let H be a real Hilbert space and C be a set in H. Suppose that $f : C \to C$ is nonexpansive and that the set of fixed points $F(f)$ of f contains an open set. In this case the Cauchy–Picard sequence $\{f^n(x)\}$ for $x \in C$ arbitrary, converges to a point in C.*

The extension given by B. Beauzamy is the following:

THEOREM 6.5.2.   *Let X be a uniformly convex real Banach space and take f, C as in the Theorem 6.3.1. In this case the Cauchy–Picard sequence $\{f^n(x)\}$, x arbitrary in C, converges to a point in $F(f)$.*

In what follows we remark that the above result is in fact valid for a more general class of mappings, namely, the class of mappings studied first by F. Tricomi (1916) and named T-mappings.

We recall that a continuous mapping f defined on a set of a Banach space is called a T-mapping if the following conditions are satisfied:

1. $F(f)$, the set of fixed points of f, is a nonempty set;

2.   for any x in the set on which f is defined and any $p \in F(f)$,

$$\|f(x) - p\| \leq \|x - p\|.$$

From the definition it is clear that this class of mappings is larger than the class of nonexpansive mappings. First we prove the following result.

THEOREM 6.5.3.   *Let H be a real Hilbert space and let C be a set in H and $f : C \to C$ be a T-mapping such that $F(f)$ contains an open set. In this case the Cauchy–Picard sequence $\{f^n(x)\}$ converges to a point of $F(f)$.*

Proof. Since $F(f)$ contains an open set, then there exist an $r > 0$ and a point $p \in F(f)$ such that $S_r(p) = \{x, \|x - p\| \leq r\}$ is in C.

If we consider an arbitrary point $x \in C$, then the point

$$p + r(x - f(x))/(\|x - f(x)\|)$$

is in $S_r(p)$. Of course we can suppose, without loss of generality, that $f(x) \neq x$. Since f is a T-mapping we obtain,

$$\|p + r(x - f(x))/(\|x - f(x)\|) - f(x)\|$$
$$\leq \|x - p - r(x - f(x))/(\|x - f(x)\|)$$

and thus, since H is a Hilbert space,

$$\|x - p\|^2 + 2\langle (x - p), r(f(x) - x)\rangle/(\|x - f(x)\|) + r^2$$
$$\geq \|f(x) - p\|^2 + r^2 + 2\langle (f(x) - p), r(f(x) - p) \div$$
$$\div (\|x - f(x)\|)\rangle$$

or

$$2r\|x - f(x)\| \leq \|x - p\|^2 - \|f(x) - p\|^2.$$

This implies that for any $x \in C$,

$$\|x - f(x)\| \leq 1/2r\{\|x - p\|^2 - \|f(x) - p\|^2.$$

Now we remark that for any fixed point p and any $x \in C$, the sequence $\{\|f^n(x) - p\|^2\}$ is monotonic and thus convergent. From the above inequality we obtain that, for any n and p,

$$\|f^n(x) - f^{n+p}(x)\| \leq \sum_{i=0}^{p-1} \|f^{n+i}(x) - f^{n+1+i}(x)\|$$
$$\leq \sum_{i=0}^{p-1} 1/2r(\|p - f^{n+i}(x)\|^2 -$$
$$- \|p - f^{n+1+i}(x)\|^2)$$
$$\leq 1/2r\{\|p - f^n(x)\|^2 - \|p - f^{n+p}(x)\|^2\}$$

and, since the sequence $\{\|p - f^n(x)\|\}$ is convergent, we get that $\{f^n(x)\}$ is a Cauchy sequence and this obviously implies that it converges to a fixed point of f. The theorem is thus proved.

Remark 6.5.4. The above proof is adapted from Moreau's proof, which proved the assertion for the case of nonexpansive mappings.

In what follows we give the extension of the above result to the case of uniformly convex Banach spaces.

THEOREM 6.5.5. *Let X be a uniformly convex Banach space and $f : C \to C$ be a T-mapping. Suppose that $F(f)$ contains a nonempty open set. Then for any $x \in C$, $\{f^n(x)\}$ is a convergent sequence to a fixed point of f.*

For the proof we need several results which are given as lemmas.

First we recall that a Banach space is called uniformly convex if there exists a function

$$\varepsilon \to \delta(\varepsilon)$$

on the set of strictly positive numbers such that for all $x, y \in X$,

$$\|(x + y)/2\|^2 \leq \tfrac{1}{2}\{1 - \delta((\|x - y\|)/((\max\{\|x\|, \|y\|\})) \cdot$$
$$\cdot (\|x\|^2 + \|y\|^2).$$

For any $x \in C$ we define

$$m = \|p - x\|$$

where p is the centre of $S_r(p, r)$ which is contained in $F(f)$.

We consider now another number defined as follows:

$$\|p - a\| = \inf\{\|p - (tx + (1 - t)f(x))\|\}$$

i.e. a is the projection of the best approximation of p on the line through x and $f(x)$.

LEMMA 6.5.6. *Choose z such that $\|z - p\| < 2m$. Then*

$$\|p - z\|^2 \geq \|p - a\|^2 + \tfrac{1}{2}\|a - z\|^2 \delta((\|a - z\|)/2m).$$

Proof. Since X is supposed to be uniformly convex we have,

$$\|p - \tfrac{1}{2}(z + a)\|^2 \leq \tfrac{1}{2}(1 - \delta((\|z - a\|/2m))(\|p - z\|^2 + \|p - a\|^2)$$

and from the definition of a we get

$$\| - \tfrac{1}{2}(z + a)\|^2 \geq \|p - a\|^2$$

and

$$\tfrac{1}{2}(\|p-z\|^2 + \|p-a\|^2) \geq \|(a-z)/2\|^2$$

which implies the assertion of the lemma.

For $x \in C$ fixed we consider the function of a real variable,

$$t \to \|p - (tx + (1-t)f(x)\|^2$$

which is easy to see is a convex function.

Let t_0 be the real number such that

$$a = t_0 x + (1-t_0)f(x).$$

LEMMA 6.5.7.   *Let t_1 and t_2 be two points such that $t_i \leq t_0$ for $i = 1, 2$ or $t_i \geq t_0$ for $i = 1, 2$. Suppose that*

$$\varphi(t) = \|p - tx - (1-t)f(x)\|$$

has the properties that $\varphi(t_1) \leq 4m^2$ and $\varphi(t_2) \leq 4m^2$.

In this case

$$|\varphi(t_1) - \varphi(t_2)| \geq \tfrac{1}{2}|t_1 - t_2|^2 \|x - f(x)\|^2 \frac{\delta((t-t)\|x - f(x)\|)}{2m}$$

Proof. Suppose that $t_0 \leq t_1 \leq t_2$ and since φ is a convex function we get,

$$\varphi(t_2) - \varphi(t_1) \geq \varphi(t_0 + t_2 - t_1) - \varphi(t_0)$$

and

$$\varphi(t_0 + (t_2 - t_1)) \leq 4m^2.$$

Using Lemma 6.3.6 we obtain,

$$\varphi(t_2) - \varphi(t_1) \geq \varphi(t_0 + t_2 - t_1) - \varphi(t_0)$$
$$\geq \tfrac{1}{2}|t_2 - t_1|^2 \|x - f(x)\|^2 \delta(h)$$

where

$$h = |t_2 - t_1| \|x - f(x)\|/2m.$$

This proves the assertion of the lemma if $t_i \geq t_0$, and we can similarly prove the assertion if $t_i \leq t_0$.

The following lemma is fundamental for the proof of the theorem.

LEMMA 6.5.8.   *For any $f : C \to C$, where f is a T-mapping we have the* *following inequality*

$$\|p - x\|^2 - \|p - f(x)\|^2 \geq r^r/8m^2 \|x - f(x)\|^2 \delta(h^\sim)$$

where

$$h^\sim = r/4m^2 \cdot \|x - f(x)\|.$$

Proof. We have two cases to consider:

Case 1. $t_0 \leq 0$. In this case Lemma 6.3.7 for $t_1 = 0$, $t_2 = 1$ gives

$$\varphi(1) - \varphi(0) \geq \tfrac{1}{2}\|x - f(x)\|^2 \delta((\|x - f(x)\|)/2m).$$

Case 2. $t_0 \in (0, 1)$. In this case let $a^\sim = (x - f(x))/2m$; then clearly we have $\|a^\sim\| \leq 1$. This gives further:

$$\|p + ra^\sim - f(x)\| \leq \|p + ra^\sim - x\|$$

and thus

$$\varphi(-r/2m) \leq \varphi(1 - r/2m).$$

Since the minimum of φ is attained in t_0, $t_0 \in (0, 1)$ we get

$$\varphi(-r/2m) \geq \varphi(0)$$

and

$$\varphi(1) \geq 1 - r/2m.$$

But

$$1 - r/2m \geq 0$$

and thus

$$t_0 \leq 1 - r/2m$$

and we can apply Lemma 6.3.7. for $t_1 = 1 - r/2m$, $t_2 = 1$ to get $\varphi(1) - \varphi(0) \geq \varphi(1) - \varphi(1 - r/2m) \geq (r^2/8m^2)\|x - f(x)\|^2 \delta(r/4m^2\|x - f(x)\|)$ and, since the estimate in Case 1 is optimal, the inequality of the lemma is proved.

Now we are in a position to prove Theorem 6.3.5. First we remark that for any $x \in C$ the sequence $\{\|p - f^n(x)\|\}$ is a monotonic sequence and this implies that it is in fact a Cauchy sequence.

From Lemma 6.5.8. we obtain that $\{f^n(x)\}$ is a Cauchy sequence.

Remark 6.5.9. It appears to be of interest to know if a similar result holds when f is a T-mapping and the set of fixed points reduces to a single point.

The following example shows that this is not the case. The example was used by Petryshin and Williamson in connection with the asymptotic regularity of the nonexpansive mappings.

Example 6.5.10. Let $\bar{S}(0, 1)$ be the unit ball in $\mathbf{R}^2$ with the usual norm. We define the mapping f as follows:

$$f(x, y) = (-x/2, -y)$$

and it is easy to see that the unique fixed point of f is $(0, 0)$.

Also we remark that f is nonexpansive. Indeed, for any (x, y) and $(x\tilde{}, y\tilde{})$ in $S(0, 1)$ we have,

$$\begin{aligned} \|f(x, y) - f(x\tilde{}, y\tilde{})\|^2 &= \|(-x/2, -y) - (-x\tilde{}/2, y\tilde{})\|^2 \\ &= \tfrac{1}{4}(x - x\tilde{})^2 + (y - y\tilde{})^2 \\ &\leq (x - x\tilde{})^2 + (y - y\tilde{})^2 \\ &= \|(x, y) - (x\tilde{}, y\tilde{})\|^2 \end{aligned}$$

and the assertion is proved.

For any (x, y) in $\bar{S}(0, 1)$ we have,

$$\|f^n(x, y) - f^{n+1}(x, y)\|^2 = (3x/2^{n+1})^2 + (2y)^2$$

which for $y \neq 0$ does not converge to zero.

Remark 6.5.11. Since $\bar{S}(0, 1)$ is a compact set then f is also a compact mapping.

6.6. CLASSES OF MAPPINGS RELATED TO NONEXPANSIVE MAPPINGS

The class of nonexpansive mappings has been further generalized by considering some new classes of mappings satisfying weak conditions. We now give the definitions and some properties of related classes of mappings. We mention that there exists a vast literature which is concerned with various extensions of the nonexpansive mappings.

DEFINITION 6.6.1.   Let X be a Banach space and C be a convex closed and bounded set in X. The mapping $f: C \to C$ is called strictly pseudocontractive if there exists a constant k in $(0, 1)$ such that

$$\|f(x) - f(y)\|^2 \leq \|x - y\|^2 + k\|(1 - f)(x) - (1 - f)(y)\|^2$$

for all x, y in C.

DEFINITION 6.6.2.   The mapping $f : C \to C$ is called pseudocontractive if

$$\| f(x) - f(y) \|^2 \leq \| x - y \|^2 + \| (1 - f)(x) - (1 - f)(y) \|^2$$

for all x, y in C.

DEFINITION 6.6.3.   (V. Istrățescu). The mapping $f : C \to C$ is called convex-nonexpansive of order r if there exist the positive numbers a_i, $i \leq r - 1$, $\sum_0^{r-1} a_i = 1$ such that

$$\| f^r(x) - f^r(y) \| \leq \sum_0^{r-1} a_i (\| f^i(x) - f^i(y) \|).$$

DEFINITION 6.6.4.   The mapping $f : C \to C$ is called a T-quasi-nonexpansive map if the set of fixed points $F(f)$ is nonempty and, for any x in C and p in $F(f)$,

$$\| f(x) - p \| \leq \| p - x \|.$$

DEFINITION 6.6.5.   (V. Istrățescu). The mapping $f : C \to C$ is called convex T-quasi-nonexpansive of order r if there exist positive numbers a_i as in Definition 6.6.3. such that

$$\| f^r(x) - p \| \leq \sum_0^{r-1} a_i (\| f^i(x) - p \|) \quad p \in F(f)$$

The class of T-quasi-nonexpansive mappings was first introduced by Francesco Tricomi for the case of real functions.

It is clear from the above definitions that these classes of mappings contain the class of nonexpansive mappings.

We give now some fixed point theorems for the mappings in the above classes.

THEOREM 6.6.6.   *Let C be a bounded closed and convex set of a Hilbert space H and $f : C \to C$ be strictly pseudocontractive mapping. Then for any x_0 in C and any r in $(k, 1)$ the sequence $(f_{s'}^n(x_0))$ where*

$$s' = 1 - (1 - k)s \qquad s \text{ fixed in } (0, 1)$$

and

$$f_{s'} = s'I + (1 - s')f_s, \qquad f_s = sf + (1 - s)I$$

converges weakly to a fixed point of f and if f is demicompact then the convergence is strong.

Proof. First we remark that for each s in $(0, 1)$ the mapping f_s has the same fixed point set as f and, more importantly, it is nonexpansive. In this case the assertions of the theorem follow from the results in Section 6.4.

For the case of mappings considered in Definition 6.6.2, i.e., pseudocontractive mappings, F. Browder and W. Petryshyn (1967) give methods to compute the fixed points. For the results and the details we refer to the paper of Browder–Petryshyn (1967).

We give now a result which characterizes the convergence of iterations in the case of T-quasi-nonexpansive mappings.

THEOREM 6.6.7   (Petryshyn and Williamson, 1972, 1973). *Let* $f : C \to X$ *be a T-quasi-nonexpansive mapping and suppose that there exists a point x_0 in C such that*

$$x_n = f^n(x_0) \in C$$

(*we suppose that C is only a closed set of a Banach X space and that f is continuous). In this case (x_n) converges to a fixed point of f in C if and only if*

$$\lim d(x_n, F(f)) = 0.$$

Proof. It is obvious that the condition is necessary. Suppose now that

$$\lim d(x_n, F(f)) = 0$$

and we show that (x_n) is a Cauchy sequence. Indeed, let $\varepsilon > 0$ and N_ε such that for all $n \geq N_\varepsilon$

$$d(x_n, F(f)) \leq \varepsilon/2.$$

In this case for $n, m \geq N_\varepsilon$ we have

$$\|x_n - x_m\| \leq \|x_n - p\| + \|x_m - p\|$$

where $p \in F(f)$ and, since f is T-quasi-nonexpansive, we get

$$\|x_n - p\| = \|f^n(x_0) - p\| \leq \|f^N(x_0) - p\|$$

and similarly

$$\|x_m - p\| = \|f^m(x_0) - p\| \leq \|f^N(x_0) - p\|$$

or

$$\|x_n - x_m\| \leq 2\|x_N - p\|$$

We take infimum over p in $F(f)$ and we get

$$\|x_n - x_m\| \leq 2d(x_N, F(f))$$

which gives us that (x_n) is a Cauchy sequence. Let $x' = \lim x_n$, which is a point of C and, from the continuity of f, $F(f)$ is a closed set, and since

$$\lim d(x_n, F(f)) = 0$$

we obtain that x' is in $F(f)$.

The following example shows that this class is strictly larger than the class of nonexpansive mappings for which the fixed point set $F(f)$ is nonempty.

Example 6.6.8.   Let $X = \mathbf{R}$ and let

$$f : X \to X$$

be defined as follows:

$$f(x) = \begin{cases} 0 & \text{if } x = 0 \\ (x/2)\sin(1/x) & \text{if } x \neq 0; \end{cases}$$

then clearly the only fixed point of f is $x = 0$. Now f is T-quasi-nonexpansive since

$$|f(y) - f(0)| = |f(y)| = |y/2| \, |\sin(1/y)| \leq |y/2|$$
$$\leq |y - p|/2 < |y - p|.$$

We show now that f is not expansive. Indeed, we take the points $p_1 = 2/\pi$ and $p_2 = 2/3\pi$ and then

$$|f(p_1) - f(p_2)| = |(2/\pi)\sin(\pi/2) - (2/3\pi)\sin(3\pi/2)| = 8/3\pi$$

and since

$$|p_1 - p_2| = 4/3\pi$$

the nonexpansivity of f is proved.

THEOREM 6.6.9.   *Let C be a closed subset of a Banach space and $f : C \to X$ be a continuous T-quasi-nonexpansive mapping of C satisfying the following conditions:*

1.   *for every $x \in C - F(f)$ there exists p_x in $F(f)$ such that*

   $$\|f(x) - p_x\| < \|x - p_x\|,$$

2.   *there exists x_0 in C such that $f^n(x_0)$ are in C for all integers $n \geq 1$ and the sequence $(x_n = f^n(x_0))$ contains a convergent subsequence converging to a point x' in C.*

Then x' is in F(f) and $\lim x_n = x'$.

Proof. Since f is T-quasi-nonexpansive we get the existence of

$$\lim d(x_n, F(f)) = d.$$

Clearly it suffices to show that $d = 0$. If x' is in $F(f)$ then $d = 0$. Now, if x' is not in $F(f)$, then by condition 1 we find p'_x in F such that $\| f(x') - p_{x\cdot} \| < \| x' - p_{x\cdot} \|$.

Now, since f is continuous and satisfies condition 2 we obtain,

$$\| f(x') - p_{x\cdot} \| = \| f \lim_j (x_{n_j}) - p_{x'} \|$$

$$= \lim_j \| f^{n_j + 1}(x_0) - p_{x'} \|$$

$$= \lim_n \| f^n(x_0) - p_{x'} \|$$

$$= \lim_j \| f^{n_j}(x_0) - p \|$$

$$= \lim_j \| x_{n_j} - p \|$$

$$= \lim_j \| x_{n_j} - p \| = \| x' - p_{x\cdot} \|.$$

This is a contradiction and thus x' is in $F(f)$.

The following result follows immediately from this theorem of Petryshyn and Williamson and was obtained by Diaz and Metcalf (1967).

THEOREM 6.6.10.   *Let C be a closed subset in a Banach space X and let f be a continuous mapping of C into X satisfying the following properties:*

1. $F(f) \neq \emptyset$,

2.   *for every x in* $C \setminus F(f)$ *and every* $p \in F(f)$

$$\| f(x) - p \| < \| x - p \|.$$

Let x_0 *be an arbitrary point in C and set* $x_n = f^n(x_0)$. *Suppose that the sequence* (x_n) *contains a convergent subsequence. Then* (x_n) *converges to a fixed point of f.*

Theorem 6.6.9. has an interesting application to condensing non-expansive mappings.

First we give the following definition.

DEFINITION 6.6.11. A mapping $f: C \to C$ is said to be strictly non-expansive if for all x, y in C,

$$\|f(x) - f(y)\| < \|x - y\|.$$

THEOREM 6.6.12. *Let X be a Banach space and let C be a closed, convex and bounded set in X and $f: C \to C$. We suppose that f is condensing and nonexpansive and that either X is rotund or f is strictly nonexpansive. In this case, for any s in $(0, 1)$ and any x_0 in C, the sequence $(x_n = f_s^n(x_0))$, $f_s = sI + (1 - s)f$, converges strongly to a fixed point of f.*

Proof. Since f is condensing, the set $F(f)$ is nonempty. We show now that f satisfies the conditions of Theorem 6.5.9.

Indeed, if f is strictly nonexpansive the first condition is satisfied. Now suppose that X is a rotund space and thus, for any x in $C - F(f)$ and $p \in F(f)$ we have:

$$\|s(x - p) + (1 - s)(f(x) - p)\| = \|f_s - p\| < \|x - p\|,$$

and in inequality must be strict, since X is rotund. Now we show that condition 2 is satisfied, too. Indeed, we remark that the mappings f_s are also condensing and we consider the set $(f_s^n(x_0))$ which, it is not difficult to see, is relatively compact. This obviously implies condition 2 of Theorem 6.6.9. Then the assertion of the theorem follows from Theorem 6.6.9.

Remark 6.6.13. If we suppose, instead of f being condensing, that the following conditions hold:

1. f is nonexpansive;

2. $F(f)$ is nonempty and, for each s, f_s is locally power condensing;

3. X is rotund or f is strictly nonexpansive; then the conclusion of Theorem 6.6.12. remains valid for such f. The proof is essentially the same.

For further information about classes of mappings related to non-expansive mappings we refer to the papers of Browder–Petryshyn (1967), Browder (1967a, 1967b, 1976) as well as to the literature quoted there.

6.7. COMPUTATION OF FIXED POINTS FOR CLASSES OF NONEXPANSIVE MAPPINGS

Let X be a Banach space and let C be a bounded, closed and convex set in X. Suppose that $f : C \to C$ is a nonexpansive mapping with $F(f) \neq \emptyset$.

An important problem is to give methods which permit the computation of the fixed points of f with a given accuracy.

In this respect we have already mentioned the paper of Browder and Petryshyn (1967). Now we give other results in this area: results obtained by D. de Figueiredo (1967) and W. Petryshyn (1967).

THEOREM 6.7.1.   (D. de Figueiredo, 1967). *Let H be a Hilbert space and C be a closed, bounded and convex set in H containing 0.*

If $T : C \to C$ is any nonexpansive mapping, then for any x_0 in C the sequence (x_n), with

$$x_n = T_n^{n^2} x_{n-1}, \qquad n = 1, 2, 3, \ldots$$

and

$$T_n x = n/(n+1) \, T x$$

converges strongly to a fixed point of T.

Proof. It is easy to see that each T_n is a contraction mapping and thus there exists a unique fixed point x'_n of T_n.

Thus we have, for any $x \in C$,

$$\| T_n^k x - x'_n \| \leq ((n/(n+1)^k/(1 - n/(n+1)) \cdot \| T_n x - x \|$$

and thus

$$\| T_n^k x - x'_n \| \leq d \cdot n^k/(n+1)^{k-1}$$

where $d = \operatorname{diam} C$.

Now the sequence (x'_n) converges strongly to a fixed point x^* of T. We now show that (x_n) converges strongly to x^*. Indeed, since we have,

$$\| x_n - x^* \| \leq \| T_n^{n^2} x_{n-1} - x'_n \| + \| x'_n - x^* \|$$

we get

$$\| x_n - x^* \| \leq d \cdot n^{n^2}/(n+1)^{n^2-1} + \| x'_n - x^* \|$$

and the right member converges to zero.

This result was extended to the class of Banach spaces having a weakly continuous duality mapping[1] by Petryshyn (1967).

[1] See Chapter 8

THEOREM 6.7.2.   *Let C be a closed convex and bounded set in a Banach space X having a weakly continuous duality mapping $J : X \to X^*$ and suppose that $0 \in C$. If $T : C \to C$ is a nonexpansive mapping such that there exists $x_0^* \in F(T)$ with the property $(x_0^*, J(x_0^* - p)) \leqq 0$ for all $p \in F(T)$ then the sequence (x_n) defined by*

$$x_n = T_n^{n^2} x_{n-1}, \qquad x_0 \in C$$

and

$$T_n = n/(n+1)T$$

converges strongly to a fixed point of T.

Proof. Let x_n' be the fixed point of T_n and, since T_n is a contraction mapping, we have

$$\| T_n^k x - x_n' \| \leqq ((n/(n+1)^k/(1 - n/(n+1))) \| T_n x - x_0 \|$$
$$\leqq d \cdot n^k/(n+1)^{k-1}$$

where $d = \text{diam } C$.

Now we know that $x_n' \to x_n^*$ and as above we show that (x_n) converges to x_0^*.

For other results concerning the convergence of some sequences defined using the iterations of a nonexpansive mapping we refer to the papers of A. Pazy (1971, 1977), S. Kaniel (1971), F. Browder (1967a, 1967b, 1976) and Browder and Petryshyn (1966).

6.8. A SIMPLE EXAMPLE OF A NONEXPANSIVE MAPPING ON A ROTUND SPACE WITHOUT FIXED POINTS

As we know, every nonexpansive mapping on a convex, bounded and closed subset of a uniformly convex space has a fixed point in that set. In what follows we show that this assertion is not true for rotund spaces and moreover it remains untrue for locally uniformly convex spaces.

To this end we consider the Banach space c_0 of all sequences converging to zero with Day's norm:

$$x \in c_0, \qquad \| x \| = \sup \left(\sum_{i=1}^{\infty} 1/2^{2i} x^2 (\alpha_i) \right)^{1/2}$$

where the supremum is taken over all permutations α of the integers and it is known (Day, 1958) that this norm is equivalent to the usual norm of c_0.

Reinwater (1969) has shown that this norm has a very useful property, namely, $(c_0, \| , \|)$ is a locally uniformly convex space in the sense of Lovaglia (1955).

It is easy to see, however, that every locally uniformly convex space is rotund. Now we consider a set C in c_0, the set of all $x = (x_i)$ with the property that

$$\sup_i |x_i| = 1.$$

From the equivalence of Day's norm with the usual norm of c_0, we see that C is a closed, bounded set in c_0. It is obvious that C is convex. The mapping $f : C \to C$ is defined as follows:

$$f(x) = (1, x_1, x_2, x_3, \dots)$$

if $x = (x_1, x_2, x_3, \dots)$. If x_0 is a fixed point of f then we easily obtain that it is not in c_0, which is a contradiction.

Remark 6.8.1.   This mapping has been used many times (Sadovskii, (1972)) (Meyer (1964/65)) and perhaps by many others.

Remark 6.8.2.   This example is taken from V. Istrǎţescu (1979b).

Chapter 7

Sequences of Mappings and Fixed Points

7.0. INTRODUCTION

Let (X, d) be a complete metric space and suppose that we have a sequence $\{T_n\}_1^\infty$ of mappings of X into itself with the following properties:

1. T_n is a contraction mapping for each integer n,

2. $\{T_n\}_1^\infty$ converges uniformly to a contraction mapping T_0.

In this case for each $n = 0, 1, 2, \ldots$, we have a unique point x_n such that $T_n x_n = x_n$ and it is natural to ask about the convergence of the sequence $\{x_n\}_1^\infty$ to the point x_0. The first result in this direction seems to be that obtained by F. F. Bonsall (1962) in the case when the contraction T_n satisfies the relation

$$d(T_n x, T_n y) \leq kd(x, y)$$

for some $k \in (0, 1)$ and all $n = 0, 1, 2, \ldots$.

Further interesting results about this problem or related ones were obtained by S. Nadler, (1968) R. Fraser, (1969) H. Covitz, M. Furi and A. Vignoli (1969).

In what follows we give results on the convergence of fixed points. Some of the results are generalizations and extensions of known results.

7.1. CONVERGENCE OF FIXED POINTS FOR
      CONTRACTIONS OR RELATED MAPPINGS

Let (X, d) be a complete metric space and consider the problem of convergence of fixed points for a sequence of contraction mappings.

THEOREM 7.1.1. *If $\{T_n\}_1^\infty$ is a sequence of mappings each mapping with at least one fixed point x_i, $i = 1, 2, 3, \ldots$, and $T_0 : X \to X$ is a contraction mapping with $d(T_0 x, T_0 y) \leq k_0 d(x, y)$ and the sequence $\{T_n\}_1^\infty$ converges uniformly to T_0, then $\{x_n\}_1^\infty$ converges to x_0.*

Proof. Let $\varepsilon > 0$ and choose N_ε such that, for all $n \geq N_\varepsilon$,

$$d(T_n x, T_0 x) \leqq \varepsilon(1 - k_0),$$

which is possible because the sequence $\{T_n\}$ converges uniformly to T_0. Then we have for $n \geq N_\varepsilon$:

$$d(x_n, x_0) = d(T_n x_n, T_0 x_0)$$
$$\leqq d(T_n x_n, T_0 x_n) + d(T_0 x_n, x_0)$$
$$\leqq \varepsilon(1 - k_0) + k_0 d(x_n, x_0);$$

and thus

$$d(x_n, x_0) \leqq \varepsilon,$$

which proves the convergence.

In the following theorem we present a result connected with the iterations of the mappings.

THEOREM 7.1.2.   *Let* $\{T_n\}_1^\infty$ *be a sequence of mappings such that each mapping has at least one fixed point,* $x_n = T_n x_n$. *Let* $T_0 : X \to X$ *such that for some integer* m, T_0^m *is a contraction mapping. If the sequence* $\{T_n\}_1^\infty$ *converges uniformly to* T_0, *then the sequence* $\{x_n\}_1^\infty$ *converges to* $x_0 = T_0 x_0$.

Proof. If $k_0 \in (0, 1)$ has the property that

$$d(T_0^m x, T_0^m y) \leqq k_0^m d(x, y)$$

we define a new metric on X, equivalent to the metric d, by the relation

$$d^{\sim}(x, y) = d(x, y) + 1/kd(T_0 x, T_0 y) + \cdots +$$
$$+ 1/k_0^{m-1} d(T_0 x, T_0 y).$$

The fact that this metric is equivalent to d is obvious and we remark that, with respect to this metric, T_0 is a contraction mapping. Indeed,

$$d^{\sim}(T_0 x, T_0 y) = d(T_0 x, T_0 y) + 1/k_0 d(T_0^2, T_0^2 y) +$$
$$+ \ldots + 1/k_0^{m-1} d(T_0^m x, T_0^m y)$$
$$\leqq k_0 \{d(T_0 x, T_0 y)/k_0 + \cdots$$
$$+ 1/k_0^m d(T_0^m x, T_0^m y))$$
$$\leqq k_0 (d(T_0^m x, T_0 y) + \cdots + d(x, y))$$
$$= k_0 d^{\sim}(x, y)$$

and the assertion is proved.

Now it is clear that the assertion of the theorem follows from Theorem 7.1.1.

The following result refers to a special class of metric spaces.

THEOREM 7.1.3.   *Let (X, d) be a locally compact metric space and let $\{T_n\}_0^\infty$ be a sequence of mappings on X into itself such that*:

1.   *T_n^m is a contraction mapping for some $m = m(n)$,*

2.   *$\{T_n\}_1^\infty$ converges to T_0 pointwise and $\{T_n\}_1^\infty$ is an equi-continuous sequence.*

In this case the sequence $\{x_n\}_1^\infty$ converges to x_0.
Here $x_n = T_n x_n$ for $n = 0, 1, 2, \ldots$.
This result was proved for the case where $m = m(n) = 1$ by S. Nadler.
Proof. Let $\varepsilon > 0$, and we can assume that ε is sufficiently small so that

$$K(x_0, \varepsilon) = \{x, d(x_0, x) \leqq \varepsilon\}$$

is a compact subset of X.

In this case, since the sequence $\{T_n\}_1^\infty$ is equicontinuous and pointwise convergent on a compact set $K(x_0, \varepsilon)$, it converges uniformly on $K(x_0, \varepsilon)$. Choose N_ε such that, for all $n \geqq N_\varepsilon$ and all x in $K(x_0, \varepsilon)$,

$$d(T_0^m 0 x, T_n^m 0 x) \leqq (1 - k_0)\varepsilon$$

where

$$d(T_0^m 0 x, T_0^m 0 y) \leqq k_0 d(x, y).$$

In this case, for $n \geqq N_\varepsilon$ we have,

$$
\begin{aligned}
d(T_n^m 0 x, x_0) &= d(T_n^m 0 x, T_0^m 0 x_0) \\
&\leqq d(T_n^m 0 x, T_0^m 0 x) + d(T_0^m 0 x, T_0^m 0 x_0) \\
&\leqq (1 - k_0)\varepsilon + k_0 \varepsilon = \varepsilon
\end{aligned}
$$

and thus $K(x_0, \varepsilon)$ is an invariant set for $T_n^m 0$ for all $n \geqq N_\varepsilon$.

Since the mappings T_n have some power a contraction mapping it follows that the fixed points of T_n are in $K(x_0, \varepsilon)$ for $n \geqq N_\varepsilon$.

But $K(x_0, \varepsilon)$ is a compact set, and from the definition of $K(x_0, \varepsilon)$ we see that

$$d(x_n, x_0) \leqq \varepsilon$$

for $n \geqq N_\varepsilon$.

Clearly this proves that $\{x_n\}_1^\infty$ converges to x_0.

Remark 7.1.4.   It is of some interest to know if the theorem is valid for sequences without the equicontinuity property.

We now give, following Nadler and Fraser, an example of a compact metric space (X, d_0) and a sequence $\{d_n\}_1^\infty$ of metrics on X, each equivalent to d_0 and converging uniformly to d_0, a sequence $\{f_n\}$ of mappings satisfying the relation

$$d_n(f_n(x), f_n(y)) \leq k d_n(x, y)$$

with $k \in (0, 1)$, $n = 0, 1, 2, \ldots$, and the sequence of fixed points of the sequence $\{f_n\}_1^\infty$, $\{x_n\}_1^\infty$ does not converge to the fixed point of f_0, x_0; f_0 also has the property that $f_n \to f_0$ pointwise.

Example 7.1.5.   Let $X = \{(2^{-i}, 2^{-j}), \ i, j = 0, 1, 2, 3, \ldots\}$ with the convention that $2^{-\infty} = 0$. Let x, y be points in X and we assume that

$$x = (2^{-k}, 2^{-l}), \qquad y = (2^{-m}, 2^{-p}).$$

For each integer n, we define the metric d_n on X as follows:

$$d_n(x, y) = \begin{cases} |2^{-k} - 2^{-m}| & \text{if } l = p = n, \\ 2 - 2^{-k} + 2^{-p} & \text{if } l = n, \ p \neq n \text{ and } m = 0, \\ 4 & \text{if } l = n, \ p \neq n \text{ and } m \neq 0, \\ |2^{-1} - 2^{-p}| & \text{if } l \neq n, \ p \neq n \text{ and } m = k = 0, \\ 4 & \text{if } l \neq n, \ p \neq n, \ m = 0 \text{ and } k \neq 0, \\ |2^{-k} - 2^{-m}| + \\ \quad + |2^{-1} - 2^{-p}| & \text{if } l \neq n, \ p \neq n, \ m = 0 \text{ and } k \neq 0. \end{cases}$$

and

$$d_0(x, y) = \begin{cases} 4 & \text{if } k = 0 \text{ and } m \neq 0, \\ |2^{-k} - 2^{-m}| + |2^{-1} - 2^{-p}| & \text{if } k \neq 0 \text{ and } m \neq 0, \\ |2^{-1} - 2^{-p}| & \text{if } k = 0 \text{ and } m = 0. \end{cases}$$

It is not difficult to see that these are metrics on X and that each d_n is equivalent to d_0, and also that $\{d_n\}_1^\infty$ converges pointwise to d_0.

Now we define the mappings. For each integer $n > 0$, let

$$f_n : X \to X, \quad f_n(2^{-i}, 2^{-j}) = \begin{cases} (2^{-(i+1)}, 2^{-n}) & \text{if } j = n, \\ (1, 2^{-n}) & \text{if } j = n \text{ and } i = 0, \\ (1, 0) & \text{if } j = n \text{ and } i \neq 0. \end{cases}$$

and

$$f_0 : X \to X, \qquad f_0(x) = (1, 0) \quad \text{for all } x \in X.$$

Also it is not difficult to see that the following properties hold:

1. for each $n > 0, f_n$ is a contraction mapping on (X, d_n) and $d_n(f_n(x), f_n(y)) \leq \frac{1}{2} d_n(x, y)$,

2. $\{f_n\}_1^\infty$ converges pointwise to f_0,

3. the fixed point of f_n is $x_n = (0, 2^{-n})$ for $n > 0$ and for $n = 0$, $x_0 = (1, 0)$,

4. the sequence $\{x_n\}_1^\infty$ of fixed points of $\{f_n\}$ converges to $(0, 0)$.

The following theorem of Nadler and Fraser is related to Theorem 7.1.3.

THEOREM 7.1.6.   *Let (X, d) be a locally compact metric space and suppose that the sequence of metrics $\{d_n\}$ and mappings $\{f_n\}$ have the following properties*:

1. $\{d_n\}_1^\infty$ *converges uniformly to* d_0,

2. f_n *is a contraction mapping on* (X, d_n),

3. $\{f_n\}_1^\infty$ *converges pointwise to* f_0.

Then the sequence of fixed points of $\{f_n\}_1^\infty$, $\{x_n\}_1^\infty$ converges to the fixed point of f_0, x_0.

For the proof we need the following

LEMMA 7.1.7.   *Let (X, d_0) be a complete metric space and K be a compact subset of X. If $\{d_n\}$ and $\{f_n\}$ are as in Theorem 7.1.6. then $\{f_n\}_1^\infty$ converges uniformly on K to f_0, with respect to d_0.*
   Proof.   Let $\varepsilon > 0$ and choose $\delta = \varepsilon/3$. Let N_ε be an integer such that for all $n \geq N_\varepsilon$,

$$|d_n(x, y) - d_0(x, y)| < \delta$$

which exists because of uniform convergence. Choose x, y in X such that $d_0(x, y) \leq \delta$ and thus for $n \geq N_\varepsilon$ we get,

$$
\begin{aligned}
d_0(f_n(x), f_n(y)) &\leq |d_0(f_n(x), f_n(y)) - d_n(f_n(x), f_n(y))| + d_n(f_n(x), f_n(y)) \\
&\leq \delta + d_n(x, y) \\
&= \delta + |d_n(x, y) - d_0(x, y)| + d_0(x, y) \\
&\leq 3\delta = \varepsilon.
\end{aligned}
$$

Now the functions $f_1, \ldots, f_{N_\varepsilon}$ are each uniformly continuous on K and thus sequence $\{f_n\}_1^\infty$ is equicontinuous on K with respect to the metric d_0. But K is a compact set and the sequence $\{f_n\}_1^\infty$ converges pointwise to f_0; then $\{f_n\}_1^\infty$ converges uniformly to f_0 on K and the lemma is proved.

We are now in a position to prove the Theorem 7.1.6.
Indeed, let $\varepsilon > 0$ be sufficiently small such that

$$K = K(x_0, \varepsilon) = \{x, d_0(x, x_0) \leq \varepsilon\}$$

is a compact subset in X. Then according to Lemma 7.1.7 the sequence $\{f_n\}_1^\infty$ converges uniformly to f_0 on K. Let N be an integer such that if $n \geq N$ and x is arbitrary in K,

$$d_0(f_n(x), f_0(x)) \leq \varepsilon - \sup_{z \in K} \{d_0(f_0(z), x_0)$$

and thus,

$$d_0(f_n(x), x_0) \leq d_0(f_n(x), f_0(x)) + d_0(f_0(x), x_0)$$

$$\leq \varepsilon\text{-sup} \{d_0(f_0(x), x_0)\} + d_0(f_0(x), x_0) \leq \varepsilon.$$

It follows that for any $n \geq N$, f_n has K as an invariant set and the restriction of f_n to K is again a contraction with respect to d_n. It follows that the fixed point of f_n is in K. From the definition of K it follows that for any $n \geq N$,

$$d_0(x_n, x_0) \leq \varepsilon$$

and thus the convergence of fixed points is proved.

Using Lemma 7.1.7. we can prove the following result:

THEOREM 7.1.8. *Let (X, d_0) be a metric space and $\{d_n\}_1^\infty$ and $\{f_n\}_0^\infty$ as in Theorem 7.1.6. Suppose that for each $n = 1, 2, 3, \ldots, f_n$ has a fixed point x_n and that some subsequence of $\{x_n\}_1^\infty$ converges to x_0. Then $f_0(x_0) = x_0$.*
Proof. Let $\{a_{n_k}\}$ be the subsequence of $\{a_n\}_1^\infty$ which converges to x_0. The closure of the set $\{x_{n_k}\}$ is a compact set and we can apply Lemma 7.1.7. We find, that the sequence $\{f_{n_k}\}$ converges uniformly on K, to f_0. Then clearly $\{f_{n_k}(x_{n_k})\}$ converges to $f_0(x_0)$. This gives us that $f_0(x_0) = x_0$.

We now give an example to show that some conditions in the above results cannot be omitted.

Example 7.1.9. In every separable or reflexive Banach space there exists a sequence of contraction mappings converging pointwise to a contraction mapping (the zero mapping, i.e. $Z(x) = 0$) but the sequence of fixed points of the respective mappings has no convergent subsequences.

In the construction of the example the following lemma is useful:

LEMMA 7.1.10. *Let (X, d) be a metric space and let $\{f_n\}_1^\infty$ be a sequence of contraction mappings with fixed points $\{x_n\}$. Suppose that $f_0 : X \to X$ is a contraction mapping with the fixed point x_0. If the sequence $\{f_n\}_1^\infty$ converges pointwise and if a subsequence $\{x_{n_k}\}$ converges to y_0 then $y_0 = x_0$.*

Proof. Let $\varepsilon > 0$ and choose N_ε such that for all $n > N_\varepsilon$

$$d(x_{n_k}, x_0) \leqq \varepsilon$$

$$d(f_{n_k}(x_0), x_0) \leqq \varepsilon$$

and thus,

$$
\begin{aligned}
d(x_{n_k}, f_0(x_0)) &= d(f_{n_k}(x_{n_k}), f_0(x_0)) \\
&\leqq d(f_{n_k}(x_{n_k}), f_{n_k}(x_0)) + d(f_{n_k}(x_0), f_0(x_0)) \\
&\leqq d(x_{n_k}, x_0) + d(f_{n_k}(x_0), f_0(x_0)) \leqq 2\varepsilon
\end{aligned}
$$

and thus $\{x_{n_k}\}$ converges to x_0. Thus $y_0 = x_0$ and the lemma is proved.

Now the construction of the example is as follows: let X be a Banach space which is separable or reflexive. In this case the set $\{x^*, \|x^*\| \leqq 1\}$ is weak*-sequentially compact and since X is infinite dimensional there exists a sequence $\{x_k^*\}$ of elements in $\{x^*, \|x^*\| \leqq 1\}$ which has no norm convergent sequence. (In the case of l^p, $p \in (1, \infty)$ the existence of a such sequence is obvious.) Now let $\{x_{n_k}^*\}$ be a weak*-convergent subsequence of the sequence $\{x_n^*\}$. Let x^* be the weak* limit of the subsequence.

For each $k = 1, 2, 3, \ldots$ we set

$$y_k^* = (x_{n_k} - x^*)/\|x_{n_k}^* - x^*\|$$

From the definition of this sequence it follows that it converges to zero and $\|y_k^*\| = 1$, $k = 1, 2, 3, \ldots$.

For each $k = 1, 2, 3, \ldots$, we find a_k such that

1. $\|a_k\| = 1$,

2. $|y_k^*(a_k)| \geqq 1 - 1/k^2$.

Now we define the mappings $f_n : X \to X$ by the relations

$$f_n(x) = (1 - 1/n)f_n(x)/f_n(a_n)a_n + 1/n \cdot a_n$$

and we have,

$$\| f_n(x) - f_n(y) \| = \| (1 - 1/n) f_n(x)/f_n(a_n) - (1 - 1/n) f_n(y)/f_n(a_n) a_n \|$$
$$= (1 - 1/n) | f_n(x) - f_n(y) | \cdot \| a_n \| / | f_n(a_n) |$$
$$\leq (1 - 1/n) \| x - y \| / | f_n(a_n) |$$
$$\leq n/(n + 1) \| x - y \|.$$

Thus each f_n is a contraction mapping and, since the sequence $\{y_n^*\}$ is weak* convergent to zero and the sequence $\{f_n(a_n)\}$ is bounded away from zero, the sequence $\{f_n\}$ converges pointwise to zero. From the definition of the mappings f_n it follows that a_n is the fixed point for f_n. Since $\| a_n \| = 1$ it follows from Lemma 7.1.10 that the sequence $\{a_n\}$ has no convergent subsequences.

Using this example we can obtain a characterization of finite dimensional spaces as follows:

THEOREM 7.1.11. *A separable or a reflexive Banach space is finite dimensional if and only if, whenever for a pointwise convergent sequence of contractive mappings, the sequence of fixed points converges to the fixed point of the pointwise limit.*

Proof. The condition is obviously sufficient, since every finite dimensional space is locally compact and thus Theorem 7.1.3 applies.

The converse assertion follows from Example 7.1.9.

We now give some results on the convergence of fixed points for local power contraction mappings.

THEOREM 7.1.12. *Let (X, d) be a complete metric space and $\{f_n\}_0^\infty$ be a sequence of mappings such that, for each $n > 0$, the mapping f_n has at least one fixed point, x_n. Suppose that $f_0 : X \to X$ is a local power contraction mapping and $\{f_n\}_1^\infty$ converges uniformly to f_0. In this case $\{x_n\}_1^\infty$ converges to the fixed point of f_0, x_0.*

Proof. Since f_0 is a local power contraction mapping there exists $k \in (0, 1)$ and for each $x \in X$, an integer $n = n(x)$ such that for all $y \in X$,

$$d(f_0^n(x), f_0^n(x)) \leq (1 - k)\varepsilon$$

Let $\varepsilon > 0$ and choose an integer N_ε such that for all $m > N_\varepsilon$,

$$d(f_m^n(x), f_0^n(x)) \leq \varepsilon k$$

where $n = n(x_0)$ the fixed point of f_0.

We have further,

$$d(x_m, x_0) = d(f_m^n(x_0), f_0^n(x_0))$$
$$\leq d(f_m^n(x_m), f_0^n(x_m)) + d(f_0^n(x_m), f_0^n(x_0))$$
$$\leq (1 - k)\varepsilon + kd(x_m, x_0)$$

which implies that

$$d(x_m, x_0) \leq \varepsilon$$

for all $m \geq N_\varepsilon$ and thus the convergence is proved.

We give now a version of Theorem 7.1.3. for the case of local power contractions.

THEOREM 7.1.13.   *Let (X, d) be a locally compact metric space and let $\{f_n\}_0^\infty$ be local power contractions with the same $n = n(x)$ for all mappings and $\{f_n\}_1^\infty$ equicontinuous.*

Let $\{x_n\}_0^\infty$ be the sequence of corresponding fixed points and suppose that $\{f_n\}_1^\infty$ converges pointwise to f_0. Then $\{x_n\}_1^\infty$ converges to x_0.

Proof. Choose $\varepsilon > 0$ such that the set

$$K = K(x_0, \varepsilon) = \{x, d(x_0, x) \leq \varepsilon\}$$

is a compact subset of X.

From the equicontinuity property it follows that the sequence $\{f_n\}_1^\infty$ is uniformly convergent on K. We choose N_ε such that for $m = n(x_0)$ the following inequalities hold:

$$d(f_n^m(x), f_0^m(x)) \leq (1 - k)\varepsilon$$

where k is the constant for the mapping f_0.

Let $x \in K$, we thus have,

$$d(f_n^m(x), x_0) = d(f_n^m(x), f_0^m(x_0))$$
$$\leq d(f_n^m(x), f_0^m(x)) + d(f_0^m(x), f_0^m(x_0))$$
$$\leq (1 - k)\varepsilon + k\varepsilon = \varepsilon$$

and thus for each $n \geq N_\varepsilon$, f_n has K as an invariant set. Of course the property that f_n is a local power contraction mapping remains valid. This clearly implies that $x_n \in K$ and this, of course, implies the assertion of theorem.

Remark 7.1.14. From the proof it is clear that the assertion of the theorem is valid under the following conditions:

1. f_n has a unique fixed point,

2. f_0 is a local power contraction mapping,

3. $\{f_n\}_1^\infty$ converges pointwise to f_0, X locally compact

or

3'. $\{f_n\}_1^\infty$ converges uniformly to f_0.

Remark 7.1.15. The above theorems can be generalized to the case when we have two sequences of mappings $\{f_n\}_1^\infty$, $\{g_n\}_1^\infty$ and two mappings f_0 and g_0 satisfying conditions of the form

$$d(f_n(x), g_n(x)) \leq a_n d(f_n(x), x)) + b_n d(g_n(y), y))$$

where $a_n, b_n \geq 0$ and $a_n + b_n < 1$.

7.2. SEQUENCES OF MAPPINGS AND MEASURES OF NONCOMPACTNESS

We now give some results concerning the sequences of fixed points using Kuratowski's measure of noncompactness.

Suppose that $\{f_n\}_1^\infty$ is a sequence of single-valued mappings and f_0 is a mapping defined on X with values in X. For each $a > 0$ we consider the sets

$$A_a = \{x, x \in X, d(x, f_0(x)) \leq a\}.$$

It is not difficult to see that when f_0 is uniformly continuous the sets A_a are closed.

The following theorem gives information concerning the sequence of fixed points and the measure of noncompactness:

THEOREM 7.2.1.   *Let $\{f_n\}_1^\infty$ be a sequence of mappings defined on a complete metric space (X, d) and $\{f_n\}$ converges uniformly to a mapping $f_0 : X \to X$ which is supposed to be uniformly continuous.*

Suppose that

$$\lim_{a \to 0} \alpha(A_a) = 0$$

and there exists a sequence of points $\{x_n\}_1^\infty$ such that

$$\lim_{n \to \infty} d(x_n, f_n(x_n)) = 0.$$

In this case the set $\{x_n\}$ is relatively compact and every accumulation point of this set is a fixed point for f_0.

Proof. We consider the set

$$A = \bigcap_{a>0} A_a$$

and from the hypothesis it follows that A is a nonempty and compact set. Let $x_n \in A$ and from the uniform convergence we obtain,

$$d(f_0(x_n), x_n)) \leq d(f_0(x_n), f_n(x_n)) + d(f_n(x_n), x_n) \to 0$$

and since A is a compact set then there exists a y_n such that

$$d(x_n, A) = \inf_{y \in A} d(x_n, y) = d(x_n, y_n).$$

Now the sequence $\{y_n\}$ admits a convergent subsequence and thus $\{x_n\}$ admits a convergent subsequence which obviously converges to a fixed point of f_0 and the proof of the theorem is complete.

From this theorem we obtain the following result:

COROLLARY 7.2.2. *Let $\{f_n\}$ be as in the theorem and let f_0 be a contraction mapping. In this case the sequence of fixed points of f_n's, $\{x_n\}_1^\infty$, converges to the fixed point of f_0, x_0.*

Proof. Obviously f_0 is uniformly continuous and for each $a > 0$ the sets A_a, have the property that

$$\alpha(A_a) \leq 2a/1 - k_0$$

where k_0 is the Lipschitz constant of f_0, because,

$$d(x, y) \leq d(x, f_0 x) + d(f_0(x), f_0(y)) + d(f_0(y), y)$$
$$\leq 2a + kd(x, y)$$

for all $x, y \in X$, and obviously this implies the above inequality.

From the inequality it follows that the conditions in Theorem 7.2.1. are satisfied and the assertion of the corollary follows.

COROLLARY 7.2.3. *Corollary 7.2.2 remains valid if we suppose that f_0 is a local power contraction.*

Proof. Since f_0 has a unique fixed point and for each x in X the sequence of iterations of x, $\{f_0^n(x)\}$ converges to the fixed point of f_0 then there exists an equivalent metric on X with respect to this, and f_0 is a contraction mapping. The corollary then follows from Corollary 7.2.2.

Remark 7.2.4.   It is possible to obtain similar results for the case of pairs of contractions.

The above results were obtained by Furi and Vignoli [1969]. Now we give a result obtained by J. Halle (1974) concerning the continuous dependence of fixed points for densifying mappings.

Suppose that X and Y are Banach spaces, Λ is a subset of Y, Γ is a closed, convex and bounded set in X, and $f : \Gamma \times \Lambda \to X$ is a mapping satisfying the following assumptions:

1. $f(, \lambda)$ is continuous for each $\lambda \in \Lambda$ and there exists λ_0 such that $f(x, \lambda)$ is continuous at (x, λ_0) for each $x \in \Gamma$,

2.   the equation $f(x, \lambda) = x$ has for $\lambda = \lambda_0$ a unique solution $x(\lambda_0) \in \Gamma$,

3.   for any $\Gamma' \subset \Gamma$ noncompact there is a neighbourhood $B = B(\Gamma')$ of $\lambda = \lambda_0$ such that for any precompact set Λ' in $B \cap \Lambda$ we have

$$\alpha(f(\Gamma', \Lambda')) < \alpha(\Gamma').$$

In this case we can prove the following result.

THEOREM 7.2.5.   *If each λ in $f(, \lambda)$ has a fixed point x_λ then $x_\lambda \to x_{\lambda_0}$ for $\lambda \to \lambda_0$*

Proof. Let (λ_k) be a sequence converging to λ_0. If x_λ are the fixed point,

$$f(x_\lambda, \lambda) = x_\lambda$$

we choose k sufficiently large such that $\Lambda' = \{\lambda_k\} \subset B \cap (\Gamma')$ and from precompactness of Λ' and property 3 we have

$$\alpha(\Gamma') = \alpha(f(x_{\lambda_k}, \lambda_k)) = \alpha(f(f(\Gamma', \Lambda'))) < \alpha(\Gamma')$$

if Γ' is not precompact. But this is not possible and thus Γ' is a precompact set. Thus we find a subsequence (v_k) of (λ_k) such that $\lim x(v_k) = z$. Property 1 implies that $z = f(z, \lambda_0)$ and property 2 implies that $z = x(\lambda_0)$. Since every convergent subsequence of the sequence $(x(\lambda_k))$ converges to the same limit, we get that $\lim x(\lambda_k) = z$. From the fact that the sequence (λ_k) is arbitrary the continuity is proved.

Chapter 8

Duality Mappings and Monotone Operators

8.0. INTRODUCTION

The well known theorem of Riesz and Fischer states that : if (c_n) is a sequence of complex numbers such that $\sum |c_n|^2 < \infty$ then there exists a function $f \in L^2_{[0, 2\pi]}(dx)$, where dx is the Lebesgue measure, such that the Fourier co-efficients

$$a_n(f) = \tfrac{1}{2}\pi \int_0^{2\pi} f(s) e^{-ins}\, ds$$

are exactly (c_n).

This theorem was generalized by A. Beurling and A. Livingston to the case of L^p-spaces. We remark that in fact, theorem of Riesz–Fischer gives a mapping between the space L^2 and $(L^2)^* = L^2$. We give now that theorem of Beurling and Livingston (1962):

THEOREM 8.0.1. *Define the set* $(\cdots -2, -1, 0, 1, 2, \ldots) = M \cup N$, *where* M, N *are nonempty sets, and suppose that* $p \in (1, \infty)$. *Let* $(a_m)_{m \in M}$ *and* $(b_n)_{n \in N}$ *be two sets of complex numbers such that there exist the functions* h *and* k *in* L^p *and* $L^{p/(p-1)}$ *respectively, and*

$$a_n(h) = a_n \quad \text{if} \quad n \in M,$$
$$a_n(k) = b_n \quad \text{if} \quad n \in N.$$

Then there exists an f *unique in* L^p *such that*:

1. $\qquad a_n(f) = a_n \quad \text{if} \quad n \in M,$

2. $\qquad a_n(S_{p-1}f) = b_n \quad \text{if} \quad n \in N$

and where

$$S_\alpha : z \to |z|^\alpha z.$$

We remark that the mappings S_p form a semigroup and, for any $f \in L^p$, $S_\alpha f$ is in $L^{p/\alpha}$.

It is also obvious that S_{p-1} maps L^p onto $L^q(1/p + 1/q = 1)$.

The theorem of Beurling and Livingston was extended by F. Browder (1965) to reflexive Banach spaces. E. Asplund (1967) has proved a more general result. The duality mappings are important examples of monotone operators. Following a note in Minty's paper (1969) the method of monotone operators was first used by M. Golomb (1935).

The concept of duality mapping was introduced by A. Beurling and A. E. Livingston (1962) and extensively studied by Asplund (1967), Browder (1965 a), (1965 b) (1965), R. I. Kachurovskii (1960) (1968) in connection with monotone operators, etc.

In what follows we establish some of the most important properties of duality mappings and monotone operators.

8.1. DUALITY MAPPINGS

We give now some essential properties of duality mappings and we prove the Beurling–Livingston Theorem as a consequence of a more general result given by Asplund.

First we introduce the notion of 'duality mapping'.

DEFINITION 8.1.1.   A gauge function μ is a real-valued function defined on $\mathbf{R}^+ = (t, t \geq 0)$ with the following properties:

1. $\mu(0) = 0,$

2. $\lim_{u \to \infty} \mu(t) = +\infty,$

3. μ is strictly increasing.

DEFINITION 8.1.2.   Let X be a Banach space and let X^* be its dual.
A mapping $J : X \to 2^{X*}$ is said to be a 'duality mapping' if the following properties are satisfied:

1. $J(0) = 0,$

2. $Jx = (x^* \in X^*, x^*(x) = \|x\| \cdot \|x^*\|, \|x^*\| = \mu(\|x\|), x \neq 0).$

It is easy to see that for each $x \neq 0$, Jx is nonempty, as a consequence of the extension theorems. There exist Banach spaces (for example l^p, L^p, $p \in (0, \infty)$) such that for each x, Jx contains only one point. Also, for each x, Jx is convex. Indeed, let x_1^* and x_2^* be in Jx. Then for any t in $(0, 1)$ we have,

$$z^* = tx_1^* + (1 - t)x_2^*$$

$$z^*(x) = tx_1^*(x) + (1 - t)x_2^*(x) = \mu(\|x\|)\|x\|$$

and thus

$$\|z^*\| \geq \mu(\|x\|).$$

But

$$\|z^*\| \leq t\|x_1^*\| + (1 - t)\|x_2^*\|$$
$$\leq t\mu(\|x\|) + (1 - t)\mu(\|x\|)$$
$$= \mu(\|x\|)$$

and thus

$$\|z^*\| = \mu(\|x\|)$$

and the assertion is proved.

The following simple result gives a connection between different duality mappings corresponding to different gauges functions.

THEOREM 8.1.3.   *Let X be a Banach space and J_1, J_2 be two duality mappings with the gauge functions μ_1 and μ_2 respectively. Then there exists a nonnegative real-valued function r(t) such that*

$$J_1 x = r(\|x\|)J_2 x.$$

Proof. Let x be an arbitrary point in X and consider the numbers

$$\mu_1(\|x\|), \qquad \mu_2(\|x\|).$$

We define

$$r(\|x\|) = \mu_1(\|x\|)/\mu_2(\|x\|).$$

Then clearly $r(t)$ is a nonnegative real-valued function satisfying the theorem.

Example 8.1.4.   If $\mu(t) = t^{p-1}$ then a duality mapping on $L^p(1 < p < \infty)$ is given by

$$Jx = |x|^{p-1}\mathrm{sgn}\, x.$$

Remark 8.1.5.   A similar assertion holds in the case of Banach spaces in which the Lumer semi-inner product satisfies some additional conditions.

We now give the results of Asplund on duality mappings. For these we need some notions connected with convex functions.

DEFINITION 8.1.6.   Let X be a Banach space and $f : X \to \mathbf{R} \cup \{+\infty\}$.

Then f is said to be convex if

$$f(tx + (1 - t)y) \leqq tf(x) + (1 - t)f(y)$$

holds for all $t \in (0, 1)$ and all x, y in X.

The function f is said to be proper convex if the set

$$(x, f(x) < \infty)$$

is nonempty; this set is called also the 'effective domain' of f.

We now give a very simple example of a convex function.

Let C be an arbitrary nonempty convex set in a Banach space X and define

$$C^{(x)} = \begin{cases} 0 & \text{if } x \in C \\ \infty & \text{if } x \notin C \end{cases}$$

and it is obvious that this is a proper convex function.

DEFINITION 8.1.7.   Let $f : X \to \mathbf{R}$ be a proper convex function defined on a Banach space (or a locally convex or topological linear space) X. The element $x^* \in X^*$ is called a subgradient of f at x if

$$f(y) \geqq f(x) + \langle x^*, y - x \rangle$$

holds. The set of all subgradients of f at x is denoted by $\partial f(x)$.

DEFINITION 8.1.8.   Let $f : X \to \mathbf{R}$ be a proper convex function. The set-valued mapping

$$x \to \partial f(x)$$

of X into $P(X^*)$ is called the subdifferential of f.

Remark 8.1.9.   From the definition of the subdifferential it is clear that for each $x \in X$, $\partial f(x)$ is a weak* closed convex set.

We now give some examples concerning the subdifferential.

Example 8.1.10.   Let K^* be a weak*-closed convex nonempty set in X^* (for example the unit ball of X^*) and define

$$f : X \to \mathbf{R}$$

by the formula

$$f(x) = \mathrm{Sup}\,(\langle x^*, x \rangle, x^* \in K^*)$$

which is the so called 'support function of K'. Clearly this is a lower semicontinuous convex function on X. For each $x \in X$, $\partial f(x)$ consists of all points in K^* where the function

$$x \rightarrow \langle x^*, x \rangle$$

attains its maximum.

Example 8.1.11. Let $f : X \rightarrow \mathbf{R}$ be convex and everywhere differentiable (Fréchet or Gateaux). Then clearly $\partial f(x)$ is single valued.

The following theorem characterizes the duality mappings.

THEOREM 8.1.12 (Asplund, 1967). *A mapping* $T : X \rightarrow P(X^*)$ *is a duality mapping iff each* $x^* \in Tx$ *is a subgradient of the convex function*

$$x \rightarrow \mu(\|x\|),$$

i.e.

$$\mu(\|y\|) \geq \mu(\|x\|) + \langle x^*, y - x \rangle$$

holds for all $y \in X$.

Proof. Suppose that T satisfies the above condition. We take y, $\|y\| = \|x\|$ and thus we get

$$\langle x^*, x \rangle \geq \|x^*\| \, \|x\|$$

and this gives

$$\langle x^*, x \rangle = \|x^*\| \, \|x\|.$$

Now we use this for $y = tz$, $x = sz$, $\|z\| = 1$, $t, s \in \mathbf{R}^+$ and we get,

$$\mu(t) > \mu(s) + \|x^*_{sz}\|(t - s)$$

and thus

$$\|x^*\| = \mu(\|x\|).$$

Thus T is a duality mapping.
Conversely, if T is a duality mapping then we have,

$$\mu(\|x\|) + \langle x^*, y - x \rangle \geq \mu(\|y\|) + \|x^*\|(\|y\| - \|x\|)$$

and

$$\mu(\|x\|) + \|x^*\|(\|y\| - \|x\|) \leq \mu(\|y\|)$$

i.e. x^* is in the subdifferential of μ.

THEOREM 8.1.13. (Browder, 1965). *Suppose that $\mu(t) = \infty$ and Y is a reflexive subspace of X. Let $Y^\perp$ be the $t \to \infty$ annihilator of Y in X, let v be an element of X and $w^* \in X^*$. Then there exists a duality mapping T on X such that $T(x + y)$ is in $Y^\perp + w^*$ for some $x \in Y$.*

Proof. We consider the following function

$$f(x) = \mu(\|x + x\|) - \langle w^*, x \rangle$$

which is obviously convex (restricted to Y). Clearly it is also weakly lower semicontinuous and tends to ∞ as $\|x\| \to \infty$, uniformly in all directions. Then it assumes its minimum on Y at some point, say x. By the Hahn–Banach theorem, f is considered as a function on X and has at x some subgradient in $Y^\perp$. Then Theorem 8.1.12 asserts the existence of a duality mapping satisfying the theorem.

For a detailed exposition of the application of duality mappings to the Beurline-Livingston theorem we refer the reader to D. G. de Figueiredo (1967).

We now introduce the very important notion of monotone set-valued operator. We call this, as for the single valued functions, a monotone operator for short.

DEFINITION 8.1.14. *Let $f : X \to P(X^*)$. Then f is called monotone if for any x_0, x_1 in X and any $x_0^* \in f(x_0)$, $x_1^* \in f(x_1)$,*

$$\langle x_1^* - x_0^*, x_1 - x_0 \rangle \geqq 0.$$

THEOREM 8.1.15. *The subdifferential of any proper convex function is a monotone operator.*

Proof. Indeed, if x_0, x_1 and x_0^*, x_1^* are as above, then,

$$f(x_1) - f(x_0) \geqq x_0^*, x_1 - x_0$$

and

$$f(x_0) - f(x_1) \geqq x_1^*, x_0 - x_1.$$

Adding these inequalities, we get

$$\langle x_1^* - x_0^*, x_1 - x_0 \rangle \geqq 0$$

and the assertion is proved.

DEFINITION 8.1.16. A monotone operator $f : X \to P(X^*)$ is called strictly monotone iff

$$\langle x_1^* - x_0^*, x_1 - x_0 \rangle > 0$$

where x_0, x_1, x_0^* and x_1^* are as in Def. 8.1.14.

The following characterization of strictly convex spaces using the duality mappings was obtained by Petryshyn (1970); the proof we give below is that of Strawther and Gudder (1975).

THEOREM 8.1.17. *A Banach space is strictly convex iff the duality mappings are strictly positive.*

Proof. It is easy to see that X is strictly convex iff every $x^* \in X^*$ attains its maximum at most one point of the unit sphere.

Now if $J : X \to P(X^*)$ is any duality mapping, then we have the following identity due to F. Browder: if $x, y \in X$ and $x^* \in J(x)$, $y^* \in J(y)$,

$$\langle x^* - y^*, x - y \rangle = (\|x\| - \|y\|)^2 + (\|x^*\| \|y\| - x^*(y)) +$$
$$+ (\|y^*\| \|x\| - y^*(x)) \geqq 0$$

From this follows that, since each term in the above identity is positive, J is not strictly monotone iff there exists an $\|x\| = \|y\|$ such that

$$\langle x^* - y^*, x - y \rangle = 0 \qquad x^* \in J(x), \quad y^* \in J(y)$$

and this is equivalent to the existence of an element $x^* \in X^*$ such that

$$x^*(x/\|x\|) = x^*(y/\|y\|) = \|x^*\| = \|x\| = \|y\|$$

and this is equivalent to the existence of an element $x^* \in X^*$ with the property that if attains its maximum at two different points of the unit sphere.

The theorem is proved.

Concerning the duality mappings in the case of strictly convex spaces we have the following result:

THEOREM 8.1.18. *If X^* is a strictly convex space and $J : X \to P(X^*)$ is any duality mapping, then for each x, $J(x)$ contains only one point (i.e. J is single-valued).*

Proof. The statement follows immediately from the strict convexity of X^* and the fact that for each x, $J(x)$ is a convex set.

We now give some results concerning the continuity of duality mappings.

THEOREM 8.1.19. *Let X be a Banach space such that X^* is strictly convex.*

Then any duality mapping $J : X \to X^*$ is continuous when we consider X with the strong topology and X^* with the weak* topology.

Proof. Suppose that (x_n) is a sequence in X converging strongly to x_0. For the weak* convergence of (Jx_n) to Jx_0 it suffices to show that every subsequence of (Jx_n) contains a subsequence which converges weakly* to Jx_0.

Since $x_n \to x_0$ we find a constant K such that

$$\|Jx_n\| = \mu(\|x_n\|) \leq K$$

for all n. Then any subsequence of (Jx_n) contains a subsequence which is convergent. Suppose that this sequence is (Jx_{n_k}) and the weak* limit is x_0^*.

We show now that $Jx_0 = x_0^*$.

Indeed, we have,

$$(*) \qquad \|x_0\|\mu(\|x_0\|) = \lim \|x_{n_k}\|\mu(\|x_{n_k}\|)$$
$$= \lim Jx_{n_k}(x_{n_k}) = x_0^*(x_0)$$

and thus

$$\mu(\|x_0\|) \leq \|x_0^*\|.$$

From the weak* convergence of (Jx_{n_k}) to x_0^* we obtain

$$\|x_0^*\| \leq \liminf \|Jx_{n_k}\| = \lim \mu(\|x_{n_k}\|) = \mu(\|x_0\|)$$

and thus

$$\|x_0^*\| = \mu(\|x\|).$$

This result and (*) imply the assertion of the theorem.

We give now another sufficient condition which implies that the duality mapping J is continuous.

First we recall the following notion introduced by Ky Fan and I. Glicksberg (1958), the so-called property (H) for a Banach space.

DEFINITION 8.1.20.   The Banach space X is said to have the property (H) if the following conditions are satisfied:

1. X is strictly convex,

2.   if (x_n) is a sequence in X converging weakly to x_0 and $\lim \|x_n\| = \|x\|$ then $\lim \|x_n - x\| = 0$.

Remark 8.1.21.   Every uniformly convex space satisfies this property; also the locally uniformly convex spaces of Lovaglia (1955) satisfy this property. Also on every separable Banach space there exists an equivalent norm, say $\| , \|^*$, such that the space with this norm has the property (H). This property is valid also for some classes of nonseparable Banach spaces.

THEOREM 8.1.22. *Let X be a reflexive Banach space with X^* having the property (H). Then any duality mapping is continuous from X with the strong topology to X^* with the strong topology.*

Proof. According to Theorem 8.1.19 if J is any duality mapping then it is continuous from X with the strong topology.

Let $x_n \to x$ and we know that $J x_n \to J_x$. We show now that in fact we have strong convergence. Since the space X^* has the property (H), using the hypothesis, we have only to show that

$$\|J x_n\| \to \|J x\|.$$

But

$$\|J x_n\| = \mu(\|x_n\|) \to \mu(\|x\|) = \|J x\|$$

and the theorem is proved.

We give now a negative result obtained by F. Browder (1967) (for arbitrary p and for $p = 4$ this was proved by F. Browder (1966)):

THEOREM 8.1.23. *Let $p (1, \infty)$, $p \neq 2$. In this case the Banach space $L^p_{[0, 2\pi]}$ does not have a weakly continuous mapping.*

This means that there exists no continuous duality mapping on L^p when we consider the weak topology on L^p.

Proof. Consider the periodic (of period 2π) real-valued function on $[0, 2\pi]$ defined as follows:

$$h(t) = \begin{cases} 1 & \text{if } 0 \leq t \leq \frac{3}{4} \\ -2 & \text{if } \frac{3}{4} < t \leq 2. \end{cases}$$

From the Lebesgue theorem about the Fourier coefficients we obtain that, for any step function f,

$$\lim_{n \to \infty} \int_0^{2\pi} h(nt) f(t) \, dt = 0.$$

Thus, for each p, the sequence $(h(nt))$ converges weakly to zero. Let

$g(t) = c$ be a constant function and thus we have,

$$h_p(c) = \lim_{h \to \infty} \|h(nt) - c\|^p = \lim_{n \to \infty} \int_0^{2\pi} |h(nt) - c|^p \, dt$$

$$= \int_0^{2\pi} |h(t) - c|^p \, dt$$

and thus

$$h_p'(0) = -p \int_0^{2\pi} h(t)^{p-1} \operatorname{sgn} h(t) \, dt.$$

But from the definition of h, $h_p'(0) \neq 0$ (for $p \neq 2$) and thus $h_p(0)$ is not an extremal value of $h_p(c)$. Thus the sequence $(Jh(nt))$ does not converge to zero.

The following theorem gives a duality mapping which is weakly continuous.

THEOREM 8.1.24.   *For every $p \in (1, \infty)$ the Banach spaces l^p have weakly continuous duality mappings.*

   Proof. Let $x \in l^p$, $x = (x_1, x_2, x_3, \ldots)$ and the duality mapping has the form

$$Jx = (x_1^{p-1} \operatorname{sgn} x_1, \; x_2^{p-1} \operatorname{sgn} x_2, \ldots)$$

which is clearly weakly continuous from $X = l^p$ to $X^* = l^q$.

8.2. MONOTONE MAPPINGS AND CLASSES OF NONEXPANSIVE MAPPINGS

The relationship between monotonicity and nonexpansiveness (contractivity) for certain classes of mappings was first noted by F. Browder (1965). In what follows, following Browder and Petryshyn (1967), we give some results about this relationship; for more details we refer the reader to Browder–Petryshyn (1967), Opial (1967), and Browder–de Figueiredo (1967).

   First we give the following

DEFINITION 8.2.1.   Let C be a closed, convex and bounded set in a real Hilbert space. If $S : C \to H$ is any mapping, then we say that a mapping $R : C \to H$ is in Ray (S) if there exists $t > 0$ such that R has the form

$$R = tS + (1 - t)I$$

or

$$R = I + t(I - S).$$

We have the following

THEOREM 8.2.2.   *Let* $T : C \to H$. *Then*

1.   *T is quasi-nonexpansive iff* $I - T$ *is monotone*,

2.   *there exists an* $S \in$ Ray (T), *where, S is a contraction mapping, iff* $I - T$ *is strongly monotone (i.e. there exists a* $k \geq 0$ *such that for all* $x, y \in C$

$$\langle Tx - Ty, x - y \rangle \geq k \|x - y\|^2)$$

and T is a Lipschitzian mapping.

Proof. We recall that a mapping $T : C \to C$ is called quasi-nonexpansive if, for all x, y in C,

$$\|Tx - Ty\|^2 \leq \|x - y\|^2 + \|(I - T)x - (I - T)y\|^2.$$

Now suppose that T is monotone and thus for all x, y in C we get, for $S = I - T$,

$$\begin{aligned}
\|Sx - Sy\|^2 &= \|(I - T)x - (I - T)y\|^2 \\
&= \|(x - y)\|^2 + \|(Tx - Ty)\|^2 - \\
&\quad - 2\langle Tx - Ty, x - y \rangle \\
&\leq \|x - y\|^2 + \|(I - S)x - (I - S)y\|^2
\end{aligned}$$

i.e. S is quasi-nonexpansive.

Suppose now that $I - S$ is quasi-nonexpansive and thus for each x, y in C we obtain:

$$\begin{aligned}
\|Tx - Ty\|^2 &= \|(I - S)x - (I - S)y\|^2 \\
&= \|x - y\|^2 + \|Sx - Sy\|^2 - 2\langle Sx - Sy, x - y \rangle \\
&\leq \|x - y\|^2 + \|Sx - Sy\|^2.
\end{aligned}$$

This implies that

$$\langle Tx - Ty, x - y \rangle \leq 0.$$

For the second assertion, suppose that S is in the ray of T which is also strictly contractive. In this case

$$T = I + t(S - I)$$

is a Lipschitzian mapping. We set

$$S_1 = I - S$$

and thus,

$$\langle S_1 x - S_1 y, x - y \rangle = \|x - y\|^2 - \langle Sx - Sy, x - y \rangle$$
$$\geq \|x - y\|^2 - \|Sx - Sy\| \, \|x - y\|$$
$$\geq (1 - k)\|x - y\|^2$$

and this clearly implies that S_1 is a strict contraction mapping.
But

$$tS_1$$

is then strongly monotone.

For the converse, suppose that $I - T$ is strictly contractive and T is Lipschitzian. Then we have

$$T_t = I + t(T - I) = I - tT'$$

with

$$T' = I - T.$$

Since

$$\|T_t x - T_t y\|^2 = \|x - y\|^2 - 2t\langle Tx - Ty, x - y \rangle +$$
$$+ t^2 \|Tx - Ty\|^2$$
$$\leq \|x - y\|^2 - 2tk\|x - y\|^2 + t^2 L\|x - y\|^2$$
$$= (1 - 2tk + t^2 L)\|x - y\|$$
$$= f(t)\|x - y\|^2$$

which implies the assertion.

COROLLARY 8.2.3. *Let $T_1 : C \to C$ be a nonexpansive mapping. Then, for each $s \geq 0$ the mapping*

$$I - (I + s(I - T_1))^{-1}$$

is strongly monotone.

Proof. First we show that for each s the mapping

$$T_{1,s} = (I + s(I - T_1))^{-1}$$

is nonexpansive.

Indeed, if y is an arbitrary but fixed point in C, the equation

$$x + s(I - T_1)x = y$$

can be written as

$$x = (y + sT_1x)/(1 + s)$$

and the mapping defined on C by

$$x \rightarrow (y + sT_1x)/(1 + s)$$

is obviously a contraction mapping. Then it has a fixed point in C and thus the set

$$(I + s(I - T_1))C$$

contains C. In this case the mapping $T_{1,s}$ is well defined and it remains to show that it is nonexpansive. To this end, let

$$x_1 + s(I - T_1)x_1 = y_1$$
$$x_2 + s(I - T_1)x_2 = y_2$$

and thus,

$$(1 + s)(x_1 - x_2) = -s(T_1x_1 - T_1x_2) + y_1 - y_2$$

which implies that

$$(1 + s)\|x_1 - x_2\| \leqq s\|T_1x_1 - T_1x_2\| + \|y_1 - y_2\|$$
$$\leqq s\|x_1 - x_2\| + \|y_1 - y_2\|$$

or

$$\|x_1 - x_2\| \leqq \|y_1 - y_2\|$$

i.e. the assertion. Now the Corollary follows from Theorem 8.2.2.

8.3. SOME SURJECTIVITY THEOREMS ON REAL BANACH SPACES

The theory of monotone operators has deep applications in the theory of partial differential equations, especially for the case of quasi-linear elliptic equations which, in fact, motivates the introduction of monotone operators. The surjectivity theorems have many applications in proving the existence of solutions as well as in some representation theorems (generalizing the well known Lax–Milgram Lemma) (Lax–Milgram (1954)).

DEFINITION 8.3.1.   Let X be a Banach space, X^* its dual and

$$T : X \to X^*$$

be a mapping. Then T is said to be hemicontinuous if it is continuous from line of X to X^* with the weak topology.

DEFINITION 8.3.2.   The mapping

$$T : X \to X^*$$

is called coercive if

$$\lim_{\|x\| \to \infty} \langle Tx, x \rangle / \|x\| = \infty.$$

THEOREM 8.3.3.   (Browder (1968, 1965/1966, 1966), Minty (1962)). *Let X be a reflexive Banach space and let*

$$T : X \to X^*$$

have the following properties:

1.   *T is monotone,*

2.   *T is hemicontinuous,*

3.   *T is coercive.*

Then $T(X) = X^*$, i.e., T is surjective.

In what follows we give Browder's proof.

First we consider the case of finite-dimensional spaces, and in this case we have the following result:

THEOREM 8.3.4.   *Let X be a finite-dimensional space and let T have the properties 1–3 of Theorem 8.3.3. Then T is continuous.*

Proof. As is well known (and not difficult to prove on any finite-dimensional space) all topologies coincide. Then we consider a sequence (x_n) in X converging to x_0. First we show that (Tx_n) is bounded. Indeed, in the contrary case, we find a subsequence, and we can suppose, without loss of generality, that this is just the sequence (Tx_n) such that $\lim \|Tx_n\| = \infty$. Now, since T is supposed to be monotone, we get

$$\langle Tx_n - Tx, x_n - x \rangle \geqq 0$$

and, if we set

$$y_n = Tx_n/Tx_n, \qquad n = 1, 2, 3, \ldots,$$

then clearly $\|y_n\| = 1$

and

$$\langle y_n - Tx/\|Ty_n\| :, x_n - x \rangle \geq 0.$$

But (y_n) belongs to a finite-dimensional space and, being bounded, it has a convergent subsequence. We can suppose, without loss of generality, that this is just (y_n) and thus we get

$$\langle y - 0, x_0 - x \rangle \geq 0$$

where $y = \lim y_n$. Since x is arbitrary in X we obtain that $y = 0$ which contradicts the fact that

$$1 = \|y\| = \lim \|y_n\|.$$

Thus (Tx_n) is a bounded sequence. Now using this fact we can prove the continuity of T as follows: it is obvious that, to prove this, it suffices to show that any subsequence of (Tx_n) contains a sequence converging to Tx_0. Indeed, since (Tx_n) is a bounded sequence, by the Bolzano–Weierstrass theorem there exists a convergent subsequence, say (Tx_{n_k}). Now let

$$x_0' = \lim Tx_{n_k}.$$

For any $x \in X$, since T is monotone we get

$$\langle Tx - Tx_{n_k}, x - x_{n_k} \rangle \geq 0$$

and thus, for $n_k \to \infty$, we obtain

$$\langle Tx - x_0', x - x_0 \rangle \geq 0.$$

Now we take x of the form

$$x = x_0 + ty, \qquad t > 0$$

with y an arbitrary but fixed element in X, we obtain

$$\langle T(x_0 + ty) - x_0', ty \rangle \geq 0$$

and from the hemicontinuity of T we get

$$\langle Tx_0 - x_0', y \rangle \geq 0.$$

Since y is arbitrary in X we obtain that $Tx_0 = x_0'$ and thus the continuity of T is proved.

Now using the continuity of T we can prove the surjectivity.

First we consider the case when $X = \mathbf{R}^n$ for some integer n and we remark that it suffices to prove that the equation

$$Tx = 0$$

has a solution in X. This is equivalent to the fact that $S = I - T$ has a fixed point in X. But T is coercive and thus we find $r > 0$ such that

$$\langle Tx, x \rangle > 0$$

for all x, $\|x\| = r$. Now we have,

$$\langle Sx, x \rangle = \langle x, x \rangle - \langle Tx, x \rangle < \|x\|^2$$

for all $x \in S_r = (x, \|x\| = r)$. We now define a mapping associated with S_r as follows:

$$Rx = \begin{cases} Sx & \text{if } \|Sx\| \leq r, \\ Sx/\|Sx\| & \text{if } \|Sx\| \geq r \end{cases}$$

and clearly this is a continuous mapping on $\mathbf{R}^n$ with values in $\mathbf{R}^n$.

We remark that it satisfies the conditions in Brouwer's theorem with respect to the convex set

$$B_r = (x, \|x\| \leq r).$$

Then R has a fixed point, say x_0. If $\|x_0\| \leq r$ then clearly it is a fixed point of S. Suppose now that $\|x_0\| \leq r$. Then we have

$$\langle x_0, x_0 \rangle = 1/\|Sx_0\| \langle Sx_0, x_0 \rangle$$

and thus

$$\|x_0\|^2 < 1/\|Sx_0\| \cdot \|x_0\|^2$$

which is a contradiction. Thus x_0 necessarily has the property

$$\|x_0\| \leq r$$

and it is thus a fixed point of S.

This proves the theorem in the case of $X = \mathbf{R}^n$.

Now suppose that X is an arbitrary finite-dimensional space and thus there exists an integer n and a linear mapping

$$i : \mathbf{R}^n \to X$$

which is an isomorphism. This implies that there exist the constants c and C such that for all $x \in \mathbf{R}^n$

$$c\|x\| \leqq \|i(x)\|_X \leqq C\|x\|$$

where $\| , \|_X$ means the norm of X. Suppose that $T : X \to X^*$ is a monotone mapping and then we define the mapping

$$T_1 : \mathbf{R}^n \to \mathbf{R}^n$$

by the relation

$$T_1(x) = i^*(T(i(x)))$$

and since

$$\langle T_1 x - T_1 y, x - y \rangle = \langle T(i(x)) - T(i(y)), i(x) - i(y) \rangle \geqq 0$$

T_1 is a monotone mapping. Similarly we can prove that it is coercive and thus, by the result just proved, T_1 is surjective.

Since $i : \mathbf{R}^n \to X$ is an isomorphism this clearly implies that T is a surjective mapping. Thus Theorem 8.3.4. is proved.

Now we prove Theorem 8.3.3. For this we consider the family F_X of all finite-dimensional subspaces of X with the partial order defined as follows: if V_1 and V_2 are two elements in F_X then we say that $V_1 \leqq V_2$ if $V_1 \subseteq V_2$. On each V in F_X we consider the induced topology and then clearly the mapping

$$J_V : V \to X$$

defined by

$$J_V(x) = x$$

is continuous. For each V we consider the mapping associated with T defined as follows:

$$T_V = J_{V0}^* T_0 J_V,$$

and it is easy to see that T_V

1.   is monotone;

2.   is hemicontinuous; and

3.   is coercive.

In this case for each $V \in F_X$ there exists an x_V such that

$$T_V(x_V) = 0.$$

We now prove now that the set $(x_V)_{V \in F_X}$ is bounded. Indeed, in the contrary case we find a sequence, say (x_{V_n}) with the property that

$$\lim \|x_{V_n}\| = \infty.$$

Since T is supposed to be coercive we get,

$$\lim \langle Tx_{V_n}, x_{V_n} \rangle = \infty$$

and since

$$
\begin{aligned}
\langle Tx_{V_n}, x_{V_n} \rangle &= \langle T(J_{V_n}(x_{V_n}), J_{V_n}(x_{V_n})) \rangle \\
&= \langle (J_{V_n 0} T_0 J_{V_n})(x_{V_n}), x_{V_n} \rangle \\
&= \langle T_{V_n} x_{V_n}, x_{V_n} \rangle = 0
\end{aligned}
$$

and this gives a clear a contradiction with the relation obtained above. Thus $(x_V)_{V \in F_X}$ is a bounded set in X. For each $V \in F_X$ we define the following set

$$M_V = \bigcap_{V_1 \supseteq V} (x_{V_1})$$

and since X is supposed to be reflexive, the weak closure $\bar{M}_V$ of M_V, is weakly compact. Then, as is easy to see, the family of sets $(\bar{M}_V)$ has the finite intersection property and thus, by the weak compactness, M_V is nonempty. Let x' be a point in this intersection. Let us consider x, an arbitrary point of X, and thus we have

$$\langle Tx - Tx_V, x - x_V) \rangle \geqq 0$$

and since

$$
\begin{aligned}
\langle Tx_V, x - x_V \rangle &= \langle Tx_V, J_V(x - x_V) \rangle \\
&= \langle (J_{V0}^* T_0 J_V)(x_V), x - x_V \rangle = 0
\end{aligned}
$$

the above inequality implies that

$$\langle Tx, x - x_V \rangle \geqq 0$$

and thus

$$\langle Tx, x - x' \rangle \geqq 0$$

for all $x \in X$. Now T is supposed hemicontinuous and therefore, for an arbitrary but fixed element of X we have:

$$\langle T(x' + ty), ty \rangle \geqq 0, \qquad t > 0$$

This implies that

$$\langle Tx', y \rangle \geqq 0$$

for all $y \in X$. In this case we have $Tx' = 0$ and the theorem is proved.

We give now some corollaries of this theorem.
First we define the strong monotonicity in the case of Banach spaces.

DEFINITION 8.3.5.   A mapping $T : X \to X^*$ is called strongly monotone if there exists a $k > 0$ such that for all $x, y \in X$,

$$\langle Tx - Ty, x - y \rangle \geqq k \|x - y\|^2.$$

LEMMA 8.3.6.   *Every strongly monotone operator is monotone and coercive.*
   Proof. Obvious.

COROLLARY 8.3.7.   *If $T : X \to X^*$ is a hemicontinuous strongly monotone mapping then $TX = X$.*
   Proof. The assertion follows from the above Lemma and Theorem 8.3.3.

The following theorem gives an important result which contains, among other things, the famous Lax-Milgram Lemma.

THEOREM 8.3.8.   *Let X be a reflexive Banach space and let $T : X \to X^*$ be a closed linear mapping which satisfies the inequality*

$$|\langle Tx, x \rangle| \geqq c \|x\|^2$$

for all $x \in X$ and for some $c > 0$. Then $TX = X$.
   Proof. It is obvious that T is injective, i.e. $x \neq y$ implies that $Tx \neq Ty$. From the inequality,

$$|\langle Tx, x \rangle| \geqq c \|x\|^2$$

it follows that T^{-1} is continuous.
   Thus it remains to show that T is surjective. Since

$$N(T^*) = R(T)^{\perp} = (x^*, x^*(Ty) = 0 \ y \in X)$$

it suffices to prove that $R(T) = X^*$. Now by the reflexivity of X the adjoint T^* of T maps X^* into X. It is clear that T satisfies the same inequality and thus it is injective. But

$$(R(T)^\perp)^\perp = R(T) = X^*$$

and the theorem is proved.

THEOREM 8.3.9.   *Let X be a reflexive Banach space and*

$$B : X \times X \to C$$

supposed to be linear in the first variable and antilinear in the second variable and satifying the following properties:

1. $|B(x, y) \leq C\|x\| \, \|y\|$
2. $|B(x, x)| \geq c\|x\|^2 \qquad for\ all\ x \in X.$

Then for each $x \in X$ there exists a unique element x^* in X^* such that

$$x^*(y) = B(y, x).$$

Proof. Let X^* be the dual of X and define the mapping

$$T : X \to X^*$$

in the following manner: for each $x \in X$ we define $x^* \in X^*$ by the relation

$$x^*(y) = B(y, x)$$

and clearly this is a linear operator satisfying the relation

$$|\langle Tx, x \rangle| \geq c\|x\|^2$$

and thus T is a homeomorphism. Then for each $x^* \in X^*$ there exists an x in X (unique) such that

$$Tx = x^*.$$

This gives further that

$$x^*(y) = \langle Tx, y \rangle = B(y, y)$$

and the theorem is proved.

Remark 8.3.10.   For a constructive proof of the Lax-Milgram Lemma we refer to Petryshyn's paper (1965); and for a discussion and a new proof see the interesting paper of Sauer (1966).

For other results, as well as for applications to the Dirichlet problem for certain classes of partial differential equations, we refer to Browder (1966), (1976), Kachurovski (1968) as well as to the references quoted there.

8.4. SOME SURJECTIVITY THEOREMS IN
COMPLEX BANACH SPACES

Let X be a complex Banach space and let X^* be its dual. The monotonicity is defined as follows:

DEFINITION 8.4.1.   The mapping $T : X \to X^*$ is said to be monotone if, for all $x, y \in X$,

$$|\langle Tx - Ty, x - y \rangle| \geq f_M(\|x - y\|)$$

where $\|x\| \leq M$, $\|y\| \leq M$ and f_M is a continuously increasing function on $\mathbf{R}^+$ with $f_M(0) = 0$ for each M.

The following theorem represents the analogue of Theorem 8.3.3. for the case of complex Banach spaces.

THEOREM 8.4.2.   *Let X be a complex Banach space and let $T : X \to X^*$ be a mapping satisfying the following properties:*

1.   *T is complex monotone,*

2.   *T is hemicontinuous,*

3.   *T is coercive, i.e.*

$$\lim |\langle Tx, x \rangle| / \|x\| = \infty.$$

Then T is a homeomorphism.

Proof. We remark that the above proof of Theorem 8.3.3 works also in this case.

We mention without proof the following result of Browder (1966):

THEOREM 8.4.3.   *Let X be a complex space and $T : X \to X$ be a mapping satisfying the following properties:*

1.   *T is complex monotone,*

2.   *T is continuous on any finite dimensional subspace and X has the weak topology*

3. $\lim\limits_{\|x\| \to \infty} |\langle Tx, x \rangle| / \|x\| = \infty.$

Then T is a homeomorphism.

8.5. SOME SURJECTIVITY THEOREMS IN LOCALLY CONVEX SPACES.

Suppose that X and Y are two locally convex (Hausdorff) spaces satisfying the following conditions:

1.   there exists a mapping on $X \times Y$ with real values such that it is linear in both variables and (x, y) is continuous in x for each fixed y and it is also continuous in y for each fixed x.

2.   for each x there exists a y such that $(x, y) \neq 0$.

Using this mapping we can define the monotonicity for mappings $T : X \to Y$ as follows:

DEFINITION 8.5.1.   The mapping $T : X \to Y$ is said to be monotone if for any x, x' in X,

$$(x - x', Tx - Tx') \geqq 0$$

and T is called finitely continuous if it is continuous on each finite dimensional subspace of X.

THEOREM 8.5.2.   *Let X, Y be two locally convex (Hausdorff) spaces satisfying the above conditions 1 and 2. Then if $T : X \to Y$ is finitely continuous and*

$$(x - x_1, Tx - y_1) \geqq 0$$

for all $x \in X$, then $Tx_1 = y_1$.
   Proof. Suppose that the assertion is not true and thus we find $x_0 \in X$ such that

$$(x_0, Tx_1 - y_1) \neq 0.$$

Let $x_t = x_1 + tx_0$ with $t > 0$ and from the finite continuity of T we get,

$$(x_0, Tx_t - y_1) \geqq 0$$

and for $t \geqq 0$,

$$0 = \lim (x_0, Tx_t - y_1) \geqq (x_0, y_1 - Tx_1)$$

which is a contradiction. This gives us that $Tx_1 = y_1$ and the theorem is proved.

THEOREM 8.5.3.   *Let X, Y be two locally convex (Hausdorff) spaces of finite dimension satisfying the above two conditions 1 and 2.*

Suppose that M is a bounded set in X having the following properties:

1. $0 \in \overset{\circ}{\overline{\operatorname{conv}}} M$,

2. $\partial \operatorname{\overset{\circ}{conv}} M \subset M$,

3. $(x, Tx) \geq 0$ *for all* $x \in M$ *and where* $T: X \to Y$ *is supposed to be continuous.*

Then there exists an x_0 *such that*

1. $x_0 \in \operatorname{conv} M$,

2. $Tx_0 = 0$.

Proof. Since the spaces are finite-dimensional, $M' = \overline{\operatorname{conv}} M$ is a compact set in X. Thus, if $p_{M'}$ denotes the associated Minkowski functional then we have

$$\overline{\operatorname{conv}} M = (x, p_{M'}(x) < 1)$$

$$\partial \operatorname{conv} M = (x, p_{M'}(x) = 1).$$

We now define the mapping

$$T_1 : M' \to M'$$

by the relation

$$T_1 x = - Tx/p_{M'}(x)$$

and it is easy to see that we can apply Brouwer's fixed point theorem to obtain the existence of $x_0 \in M'$ such that $T_1 x_0 = x_0$. From condition 3 we get that $Tx_0 = 0$ and the theorem is proved.

For the following theorem we need the notion given in

DEFINITION 8.5.4.   Let V be a topologically linear space and let S be a subset of V. We say that $x_0 \in \overline{\operatorname{conv}} S \setminus S$ has the property (H) with respect to S if

$$\overline{\operatorname{conv}} S \cap F \subset S \cap F$$

for each finite-dimensional subspace F of X.

THEOREM 8.5.5.   *Let* X, Y *be two locally convex (Hausdorff) spaces satisfying the condition 1 and 2 mentioned above. Suppose that*

$$T: X \to Y$$

is of the form

$$T = T_0 + T_1$$

where T_0 is finitely continuous and monotone, T_1 is continuous and

$$x \rightarrow (x, T_1 x)$$

is continuous on the compact subsets of X. If S is a compact set in X and x_0 has the property (H) then if $y \in Y$ satisfies the property

$$(x - x_0, Tx - y) \geq 0$$

for all $x \in X$ then necessarily $y = Ts$ for some $s \in \overline{\text{conv } S}$.

Proof. If we set

$$S_1 = (x - x_0, x \in S)$$

$$T'(x - x_0) = T(x - x_0) - y$$

then to prove the theorem it suffices to prove that there exists $x_0' \in S_1$ such that $T'x_0' = 0$. Of course we can suppose that $x_0 = 0$ and then for any $x \in X$ we set

$$M_x = \overline{\text{conv } S} \cap (x, (x - z, T_0 x + T_1 z) \geq 0)$$

Since the function

$$x \rightarrow \langle x - z, T_0 x + T_1 z \rangle$$

is continuous we obtain that M_x is a compact set. Let $z_1, \ldots, z_m$ be in $X (m < \infty)$ and F be the finite-dimensional subspace of X generated by these elements. Let (e_i) be a basis of F and (f_i) be a dual basis, i.e. $f_j \in F$ and

$$f_j(e_i) = \delta_{i,j} = \begin{cases} 1 & i = j \\ 0 & i \neq j \end{cases}$$

We denote again by f_j the extension of f_j to X and we consider the operator

$$Px = \sum_{1}^{m} f_j(x) e_j$$

which is obviously a projection. It is also clear that $PX = F$ and that the operator

$$P^* : Y \rightarrow Y$$

defined by

$$(z, Pz') = (Pz, z')$$

for all $x \in X$ satisfies once again the relation $P^{*2} = P^*$.

Now, since T is finitely continuous, the mapping

$$r = P^*T$$

is continuous and we let

$$S_0 = S \cap F.$$

We have that

$$0 \in \overline{\text{conv } S_0}$$

and for any z in S_0,

$$\langle z, r(z) \rangle = \langle z, PTz \rangle = \langle Pz, Tz \rangle = \langle z, Tz \rangle \geq 0.$$

Then there exists $z_1' \in \overline{\text{conv } S_0} \subset \overline{\text{conv } S}$ such that

$$r(z_1') = PTz_1' = 0$$

and thus

$$P^*T_0 z_1' = -P^*T_1 z_1'.$$

Further we get,

$$0 \leq \langle z_i - z_1', T_0 z_i - T_0 z_1' \rangle$$
$$= \langle z_i - z_1', T_0 z_i \rangle - \langle P(z_i - z_1'), T_0 z_1' \rangle$$
$$= \langle z_i - z_1', T_0 z_i \rangle - \langle (z_i - z_1'), PT_0 z_1' \rangle$$
$$= \langle (z_i - z_1'), T_0 z_i + T_1' z_1' \rangle$$

and from this we obtain that the family of sets (M_x) has the finite intersection property. From the compactness of S we get that

$$\bigcap M_x$$

is nonempty. We note that the intersection is taken over $x \in X$.

Let x' be in the intersection and thus

$$\langle x - x', T_0 x + T_1 x' \rangle \geq 0$$

for all $x \in X$ and according to Theorem 8.5.2.

$$T_0 x' = -T_0 x'$$

and thus

$$(T_0 + T_1)x' = 0$$

and the theorem is proved.

From this theorem we obtain some results which we give as corollaries are for complex Banach spaces.

COROLLARY 8.5.6.   *Let X be a complex Banach space and $T: X \to X^*$ be a continuous mapping (X with the strong topology, X^* with the weak topology) satisfying the following properties:*

1. $\mathrm{Re} \langle x - y, Tx - Ty \rangle \geqq 0,$

2. $\mathrm{Re} \langle x, Tx \rangle \geqq f(\|x\|) \|x\|$

where f is a real valued continuous increasing function on $\mathbf{R}^+$ with the property

$$\lim_{x \to \infty} f(x) = \infty$$

Then T is onto.
Proof. We define

$$(x, y)_1 = \mathrm{Re}(x, y)$$

and with respect to this mapping on $X \times X \to \mathbf{R}$ the conditions of Theorem 8.5.5 are satisfied and thus the surjectivity follows.

COROLLARY 8.5.7.   *Let X be a reflexive Banach space and let B be a complex-valued function on $X \times X$ with the following properties:*

1.   *B is linear in the first variable and antilinear in the second variable,*

2.   *B is continuous in each variable separately,*

3. $\mathrm{Re}(B(x, x - y) - B(y, x - y)) \geqq 0,$

4. $\mathrm{Re}\, B(x, x) \geqq f(\|x\|) \|x\|$

where f is as given in the above corollary. Then for each $x^* \in X^*$ there exists a unique x in X such that

$$x^*(y) = B(y, x)$$

for all $y \in X$.
Proof. We define

$$T : X \to X^*$$

using the relation

$$(x, Ty) = B(x, y)$$

which clearly defined T uniquely. It is easy to see that T satisfies the above corollary and this implies the assertion of the theorem.

8.6. DUALITY MAPPINGS AND MONOTONICITY FOR SET-VALUED MAPPINGS

In what follows we give some results concerning monotonicity for set-valued mappings.

Let X be a Banach space, X^* be its dual and 2^{X^*} be the family of all subsets of X^*.

DEFINITION 8.6.1. Let C be a closed and convex subset of X and $T : C \to 2^{X^*}$ be a mapping. T is said to be monotone if, for any x, y in C and $u \in Tx$, $v \in Ty$,

$$\langle x - y, u - v \rangle \geqq 0$$

and T is called maximal monotone if T is monotone and there exists no monotone mapping $T_1 : C_1 \to 2^{X^*}$ such that

1. $G(T) \subset G(T_1)$,

2. $T \neq T_1$.

Here $G(T)$ denotes the graph of T, i.e. the set

$$((x, u), x \in C, u \in X^*, u \in Tx).$$

Remark 8.6.2.   In the family of mappings defined on C with values in 2^{X^*} we define a partial order as follows:

$$S \leqq S_1 \quad \text{if} \quad G(S) \subset G(S_1)$$

and we say that S is the restriction of S_1 or that S_1 is an extension of S. In this case the maximality says nothing other than that a monotone mapping has no proper extension.

The boundedness of a set-valued mapping is defined as follows:

DEFINITION 8.6.3.   The mapping $T : C \to 2^{X^*}$ is said to be bounded if, for each $r > 0$, there exists a $k(r) > 0$ such that for any x, $\|x\| \leqq r$ there exists $z \in Tx$ such that $\|z\| \leqq k(r)$.

We define now the finite continuity for set-valued mappings.

DEFINITION 8.6.4.   The mapping $T: C \to 2^{X^*}$ is called finitely continuous if:

1. $D_T = (x, \, Tx \neq \emptyset, \, x \in C) = C,$

2.   for each $x \in C$, Tx is closed and convex,

3.   for each finite set $(x_1, \ldots, x_n)$ in C, if $\overline{\text{conv}(x_1, \ldots, x_n)}$ is the convex closure of the set $(x_1, \ldots, x_n)$, then T is upper-semicontinuous on $\overline{\text{conv}(x_1, \ldots, x_n)}$ when X^* has the weak*-topology.

The coerciveness is defined as follows:

DEFINITION 8.6.5.   The mapping $T: C \to 2^{X^*}$ is called coercive with respect to the point $z_0 \in X^*$ if for some $r > 0$ and all x in C, $\|x\| > r$

$$\langle Tx, x \rangle - \langle z_0, x \rangle \geqq 0.$$

We mention, without proof, the following important result concerning the monotone set-valued mappings

THEOREM 8.6.6.   *Let X be a reflexive Banach space, C be a closed convex subset of X and*

$$T, T_0 : C \to 2^{X^*}$$

be two monotone mappings such that the following properties are satisfied:

1. $0 \in C$, $0 \in T(0)$,

2. T_0 *is finitely continuous, bounded and coercive with respect to a given point of X, say z_0.*

Then there exists a function g and $[u_0, z_0'] \subset G(T_0)$, $\|u_0\| \leqq g(\|z_0\|)\|z_0'\| \leqq g(\|z_0\|))$ such that for all $(x, y) < G(T)$,

$$\langle x - u_0, \, y + z_0' - z_0 \rangle \geqq 0.$$

For a proof as well as for some applications of this theorem we refer to Browder (1968), p. 79.

8.7. SOME APPLICATIONS

As we have already mentioned, the theory of monotone mappings has many interesting and useful applications. In what follows we give only two applications. The first refers to the existence of fixed points for nonexpansive mappings and the second to the existence of solutions of nonlinear integral equations.

First we give a new proof of Theorem 6.4.5. For the reader's convenience we recall the theorem: *if C is a closed, bounded and convex set in a real Hilbert space and $f : C \to C$ is nonexpansive then f has a fixed point in C.*

For this we need the following result:

LEMMA 8.7.1.   *If C is a closed, bounded and convex set in a Hilbert space H, then there exists a continuous mapping*

$$r : H \to C$$

with the following properties:

1. $r(x) = x$ if $x \in C$,

2. $\|r(x) - r(y)\| \leq \|x - y\|$.

Proof. For any x in H we define

$$\bar{r}(x) = \inf_{y \in C} \|x - y\|$$
$$r(x) = \bar{x}, \qquad \|x - \bar{x}\| = \bar{r}(x)$$

and it is known that r is well-defined (see also Theorem 2.5.8) and it is obvious that r satisfies the properties 1 and 2. Now from Theorem 8.2.1 the mapping

$$1 - r_0 f_0 r = 1 - f'$$

is monotone. We show now that the conditions of Theorem 8.3.3 are satisfied. Indeed, since

$$\langle (1 - f')x, x \rangle = \|x^2\| - \langle f'x, x \rangle$$

it simply follows that $(1 - f')$ is coercive and, since the other properties are easy to see, the mapping $1 - f'$ is surjective. This implies that there exists a point x_0 such that

$$f'x_0 = x_0$$

and this is just a fixed point of f.

For the second application we consider a measurable set in $\mathbf{R}^n$, Ω, and the kernel $K(x, y)$ is such that the operator

$$Ah = \int_\Omega K(x, y)H(y)\,dy$$

is a bounded Hermitian operator in $L^2(\Omega)$. We suppose that A is positive. We consider now the following integral equation

$$f(x) = \int_{\Omega} K(x, y)g(x, f(y)) \, dy$$

where g is a function satisfying the following properties:

1. $g(y, z) \in \mathbf{R}$ for all $y \in \Omega$ and $z \in R$; it is a continuous function in z for almost all $y \in \Omega$ and measurable for all $z \in \mathbf{R}$,

2. $|g(y, z)| \leqq \alpha(y) + v|z|$ with $b > 0$ and $\alpha(\,,) \in L^2(\Omega)$,

3. $|\partial g/\partial z (y, z)| \leqq M$ *for almost all* $y \in \Omega$.

4. $|zg(y, z)| \leqq \alpha_1 z^2$ for almost all $y \in \Omega$ and $M\|A\| \leqq 1, \alpha_1\|A\| < 1$.

Then we have the following result proved by Kachurovski (1964) (see also Kachurovski (1968) and Weinberg (1956)):

THEOREM 8.7.2.   *If K and g satisfy the above conditions then the equation*

$$f(x) = \int_{\Omega} K(x, y)g(x, f(y)) \, dy$$

has a solution in L^2.

   Proof. We consider in the space L^2 the mapping

$$h(f)(y) = g(y, f(y))$$

and, as we have supposed the operator A is positive, the square root $A^{1/2}$ exists (and is unique). Then we consider the equation

$$F_1(f) = f - A^{1/2}h(A^{1/2}f) = 0$$

and we remark that all the solutions of this equation are solutions of our original equation,

$$f = Ah(f).$$

   We prove now that the equation

$$F_1(f) = 0$$

has a solution.

   We remark that from conditions 2 and 3 we conclude that the mapping $f \rightarrow h(f)$ is continuous in L^2 and we show that it is monotone.

Indeed, for any f, g in L^2 we have,

$$\langle F_1(f) - F_1(g), f - g \rangle$$
$$= \langle f - A^{1/2}(h(f)) - g + A^{1/2}(h(g)), f - g \rangle$$
$$= \| f - g \|^2 - \langle A^{1/2} f - A^{1/2} g, h(A^{1/2} f) - h(A^{1/2} g) \rangle$$
$$\geqq \| f - g \|^2 - M \| A \| \, \| f - g \|^2$$
$$= (1 - M \| A \|) \| f - g \|^2$$

and the monotonicity of F_1 is proved.

We show now that F_1 is coercive. Indeed, if f is in L^2 we have,

$$\langle F_1(f), f \rangle = \| f \|^2 - \langle A^{1/2} h(A^{1/2} f), f \rangle$$
$$= \| f \|^2 - \langle h(A^{1/2} f), A^{1/2} f \rangle$$
$$= \| f \|^2 - \int_{\Omega} (A^{1/2} f)(y) g(y, A^{1/2}(y)) \, dy$$
$$\geqq \| f \|^2 - \alpha_1 \langle Af, f \rangle$$
$$\geqq (1 - \alpha_1 \| A \|) \| f \|^2$$

if $\alpha_1 > 0$. In the case $\alpha_1 < 0$ we obtain easily that $\langle F_1(f), f \rangle \geqq k \| f \|^2$ for some $k > 0$. Then by Theorem 8.3.3. $F_1(f) = 0$ has a solution and the theorem is proved.

For applications to differential equations in Banach spaces we refer to V. Barbu (1974), R. Kachurovski (1966), J. Lions and W. Strauss (1965), W. Strauss (1967), for applications to the Navier–Stokes equations see M. Shinbrot (1964) and S. Kaniel (1965).

Chapter 9

Families of Mappings and Fixed Points

9.0 INTRODUCTION

Let S be an arbitrary set and let $\mathscr{F}$ be a family of mappings $f : S \to S$. A point $s \in S$ is called a fixed point for the family $\mathscr{F}$ if, for any $f \in \mathscr{F}$, $f(s) = s$. Brouwer's and Schauder's theorems assert that for some S and $\mathscr{F}$ there exist fixed points. An interesting problem in this respect was noted by Isbel (1957): if $\mathscr{F}$ is a family of continuous commuting mappings on $[0, 1]$ into $[0, 1]$ then do there exist fixed points for $\mathscr{F}$? Recently this problem was solved in the negative, independently, by Boyce and Huneke (1969). The result is as follows:

THEOREM 9.0.1. *Let $I = [0, 1]$. Then there exist two continuous functions*

$$f : I \to I, \qquad g : I \to I \qquad f(g(x)) = g(f(x)), \quad \forall x \in I$$

which have no common fixed points.

The first result about fixed points for families of mappings was proved by A. Markov in 1936. S. Kakutani in 1938 found a direct proof for Markov's result and also proved a fixed point theorem for groups of affine equicontinuous mappings. An important extension of the Markov and Kakutani results was given by C. Ryll-Nardzewski in 1967. Another important result which greatly influenced the development of fixed point theory for some classes of mappings was the result of de Marr in 1963 about families of nonexpansive mappings. In what follows we give the above mentioned results as well as others.

9.1. MARKOV'S AND KAKUTANI'S RESULTS

Let X be a locally convex Hausdorff space and let C be a convex set in X.

DEFINITION 9.1.1. The mapping $f : C \to C$ is called an affine map if for all x, y in C and $t \in (0, 1)$,

$$f(tx + (1 - t)y) = tf(x) + (1 - t)f(y).$$

DEFINITION 9.1.2.   Let $F = (f)$ be a family of mappings. Suppose that each f is linear. Then the family is said to be equicontinuous on a set $M \subset X$ iff for every neighbourhood V of the origin of X there is a neighbourhood U of the origin of X such that for any x_1, x_2 in M, $x_1 - x_2$ is in U then, for any f, $f(x_1) - f(x_2)$ is in V.

Remark 9.1.3.   The same definition applies to arbitrary families of continuous functions (maps).

THEOREM 9.1.4   (Markov, 1936). *Let C be a compact and convex set in a locally convex Hausdorff space X and let F be a family of commuting affine continuous mappings. Then F has a fixed point in C.*

Proof. Let f be an arbitrary element in F and take n to be an integer. We consider the map $f_n : C \to C$, defined as follows:

$$f_n(x) = 1/n(x + f(x) + \cdots + f^{n-1}(x)).$$

Now we consider the family of sets of the form

$$(f_n(x), x \in C)$$

and we denote this family by F'. Clearly these sets are convex and compact. Now since the elements in F are commuting we get

$$f_n \circ g_m = g_m \circ f_n$$

for all f, g in F, since

$$f_n \circ g_m(C) \subset f_n(C) \cap g_m(C)$$

for arbitrary f, g in F. This implies that any finite subfamily in F' has a nonempty intersection. Since C is supposed to be compact we get that the intersection of all sets in F' is nonempty. Now we show that any point in this intersection is a fixed.

Denote the intersection by J and let $p \in J$ be an arbitrary point.

Let f again be an arbitrary element of F and suppose that $p \neq f(p)$. In this case there exists a neighbourhood of the origin of X, say U, such that $f(p) - p$ is not in U. Now since p is in $f_n(C)$, an element of F', we find $q \in C$ such that

$$p = 1/n(q + f(q) + \cdots + f^{n-1}(q))$$

for all integers n. Now,

$$f(p) - p = 1/n(f^n(q) - q)$$

is not in U. But $f^n(q) \in C$ and thus for any integer n

$$1/n(C - C)$$

is not a subset of U (because of the existence of q). Since $C - C$ is the image of C through the map

$$r(x, y) = x - y$$

we get that $C - C$ is compact. This clearly contradicts the fact that $1/n(C - C)$ is not a subset of U for all n. Thus $f(p) = p$ for all f. Since p was arbitrary the theorem is proved.

Remark 9.1.5.   Markov's proof uses the Tichonov generalization of the Schauder theorem.

Suppose that C is a compact convex set in a locally convex Hausdorff space X and $f_i : C \to C$, $i = 1, 2, 3, \ldots, N$ is a finite family of continuous affine mappings. In this case for any N positive numbers $a_1, \ldots, a_N$ such that $\sum_{i=1}^{N} a_i = 1$ we can consider the mapping

$$h(x) = \sum_{i=1}^{N} a_i f_i(x)$$

which is affine and continuous. It appears to be of interest to relate the fixed points of f_i, $F(f_i)$ with the fixed point of h, $F(h)$. The following result was proved by Anzai and Ishikawa and gives an answer to this problem.

THEOREM 9.1.6.   (Anzai and Ishikawa, 1977). *Let X, C, f_i, a_i and h be given as above. In this case $\bigcap_{i=1}^{N} F(f_i)$ is nonempty and is equal to $F(h)$.*

The proof will be given as a consequence of some results stated as lemmas.

LEMMA 9.1.7.   *Let X be a locally convex Hausdorff space; let C be a convex and compact subset in X; and let $f : C \to C$ be a continuous affine mapping. In this case the following assertions hold:*

1.   *for any $\varepsilon > 0$ there exists an integer N such that*

   $$(C - C) = \ni x_i - f(x_i)$$

   for all $x_0 \in C$, $i \geq N$ and where

   $$x_i = (1 - s)x_{i-1} + sf(x_i), \qquad s \in (0, 1).$$

2.   *a point of accumulation of $(x_i)_0^\infty$ is a fixed point of f.*

Proof. (1). Let I denote the identity mapping on C and thus we have,

$$x_i - f(x_i) = ((1-s)I + sf)^i x_0 - f(((1-s)I + sf)^i x_0)$$

$$= \sum_{h=0}^{i+1} (_iC_h(1-s)^{i-h}s^h -$$

$$- _iC_{h-1}(1-s)^{i-h+1}s^{h-1})f^h(x_0)$$

where

$$_iC_{i-1} = 0 = {}_iC_{i+1}.$$

Put

$$L_h(i) = {}_iC_h(1-s)^{i-h}s^h - {}_iC_{h-1}(1-s)^{i-h+1}s^{h-1}$$

for $0 \leq h \leq i+1$.

Thus $L_h(i) \geq 0$ if $0 \leq h \leq h_0$ and $L_h(i) < 0$ if $h_0 \leq h \leq i+1$ and h_0 is an integer satisfying the following relation

$$h_0 \leq (i+1)s < h_o + 1.$$

We have also the relation

$$\sum_{1}^{h_0} L_h(i) = \sum_{h=h_0+1}^{i+1} L_h(i) = {}_iC_h(1-s)^{i-h_0}s^{h_0}.$$

Now if

$$S(i) = {}_iC_{h_0}(1-s)^{i-h_0}S^{h_0}$$

by Stirling's formula

$$\lim_{i \to \infty} S(i) = 0$$

and from the convexity of C we get,

$$x_i - f(x_i) = \sum_{h=1}^{i+1} L_h(i)f^h(x_0)$$

$$= S(i)\sum_{h=0}^{h_0} (L_h(i)/S(i))f^h x_0 - S(i)\sum_{h=h_0+1}^{i+1} (L_h(i)/S(i))f^h(x_0)$$

$$\in S(i)(C - C).$$

From the convergence to zero of the sequence $(S(i))$ we obtain Assertion 1.

To prove (2): let p be an arbitrary accumulation point of the sequence (x_i). The continuity of f implies that for any neighbourhood $V \ni 0$ there exists an integer N_V such that

$$p - x_{i(k)} \in \tfrac{1}{3}V,$$
$$f(x_{i(k)}) - f(p) \in \tfrac{1}{3}V$$

for all $k \geq N$ and where the subsequence $(x_{i(k)})$ converges to p.

Now the set C is compact and thus $C - C$ is a compact set and thus we find an integer N with the property that

$$S(i(k))(C - C) \subset \tfrac{1}{2}V$$

for all $k \geq N$. In this case if $k \geq \max(N, N_V)$ we get

$$p - f(p) = (p - x_{i(k)}) + (x_{i(k)} - f(x_{i(k)})) +$$
$$+ (f(x_{i(k)}) - f(p))$$
$$\in \tfrac{1}{3}V + \tfrac{1}{3}V + \tfrac{1}{3}V \subset V$$

and this obviously implies that p is a fixed point of f. Thus the assertion (2) is proved.

LEMMA 9.1.8. *Let X, C, f be given as in Lemma 9.1.7. Then for any convex neighbourhood $V \ni 0$ there exists an integer N_V such that for any $i \geq N_V$ we find a $z_i \in F(f)$ such that $x_i - z_i \in V$ for any $x_0 \in C$, where x_i is defined as in Lemma 9.1.7.*

Proof. Since f is supposed to be continuous, and C being compact, for any convex neighbourhood $V \ni 0$ we find a convex neighbourhood $V' \ni 0$ such that

$$(x + V) \cap F(f) \neq \varnothing$$

for any x in C such that $x - f(x) \in V'$. Let N be an integer such that

$$S(i)(C - C) \subset V'$$

for all $i \geq N$, and this is possible, since $\lim_{i \to \infty} S(i) = 0$ and $C - C$ is compact.

In this case

$$x_i - f(x_i) \in V'$$

for all $x_0 \in C$ and thus z_i can be chosen in $(x_i + V) \cap F(f)$ for all $x_0 \in C$. This proves the Lemma.

Using these Lemmas we prove now Theorem 9.1.6. First we remark that we can suppose without loss of generality that $N = 2$.

Now it is obvious that

$$F(f_1) \cap F(f_2) \subset F(h).$$

Let p be arbitrary in $F(h)$. Clearly

$$h = sf_1 + (1 - s)f_2$$

and we set

$$A = sf_1 + (1 - s)I$$
$$B = (1 - s)f_2 + sI.$$

It is clear that

$$p = (A + B)/2 = ((A + B)/2)^i p$$

for all i. According to Lemma 9.1.8 for any convex neighbourhood $V \ni 0$ we find an integer N such that

$$A^i B^i p - z_i \in \tfrac{1}{2}V, \ i \geq N$$

and $z_i \in F(f_1)$.

Now for i in $[1, N]$ we put

$$z_i = z_N.$$

Let us consider

$$w_n = \sum_{i=0}^{n} 2^{-n}{}_nC_i z_i$$

and since f_1 is affine, $w_n \in F(f_1)$. But f_1 and f_2 are commuting and thus,

$$((A + B)/2)^n p - w_n = \sum_{i=0}^{n} 2^{-n}{}_nC_i(A^i B^{n-i}p - z_i)$$

$$= \sum_{i=0}^{N-1} 2^{-n}{}_nC_i(A^i B^{n-i}p - z_i) +$$

$$+ \sum_{N-1}^{n} 2^{-n}{}_nC_i(A^i B^{n-i}p - z_i)$$

$$\in \left(\sum_{i=0}^{N-1} 2^{-n}{}_nC_i\right)(C - C) +$$

$$+ \left(\sum_{N-1}^{n} 2^{-n}{}_nC_i\right)\tfrac{1}{2}V$$

and if we take n such that

$$\left(\sum_{i=0}^{N-1} 2^{-n}{}_nC_i\right)(C-C) \subset \tfrac{1}{2}V$$

we obtain

$$p - w_n = ((A+B)/2)^n p - w_n \in V.$$

Now since $w_n \in F(f_1)$ we obtain that $p \in F(f_1)$. Similarly we prove that $p \in F(f_2)$. Thus the proof of the Theorem is complete.

Remark 9.1.9.   From the finite intersection property it is clear that from Theorem 9.1.6. Markov's theorem follows.

Now we give Kakutani's theorem.

THEOREM 9.1.10   (Kakutani, 1938). *Let C be a convex and compact subset of a locally convex Hausdorff space X and let G be a group of affine maps $f : C \to C$. Suppose that G is equicontinuous on C.*
   Then G has a fixed point in C.

Remark 9.1.11.   We can suppose that the elements of G are defined on a larger set but with the properties stated in the theorem, i.e. C is invariant for each $f \in G$ and G is equicontinuous on C.

Proof. Since C is a compact set we consider the family of all closed subsets of C which are invariant for all $f \in G$. Clearly this set satisfies Zorn's Lemma, with the order by inclusion, and thus we obtain a set, say C_1, which is closed and invariant for all $f \in G$ and minimal.
   Now if C_1 contains only one point, the assertion of the theorem is proved. We suppose that this is not the case and we get a contradiction.
   In this case it is clear that the set $C_1 - C_1$ contains other points than $x = 0$. Also, since $C_1 - C_1$ is a compact set, we find a neighbourhood V of the origin which does not contain $C_1 - C_1$. Since X is a locally convex space we find a convex neighbourhood of the origin, say V_1, such that $tV_1 \subseteq V$ for all t in $(z, |z| \leqq 1)$. Now the group G is supposed to be equicontinuous on C and thus it is equicontinuous on C_1, and thus we find a neighbourhood U_1 of the origin of X such that for all x, y in C_1 and $x - y \in V_1$ then for all $g \in G$, $g(x) - g(y) \in V_1$.
   Let us consider the set $U_2 = \text{conv}\,(g(x) : g \in G, x \in C_1)$. From the fact that G is a group we obtain that $GU_2 = U_2$ and this is valid also for the closure of U_2, i.e.

$$G\bar{U}_2 = \bar{U}_2.$$

We suppose for a moment that this is true and the proof continues as follows: we set $d = \inf(s, s > 0, sU_2 \supseteq C_1 - C_1)$ and, from the above property of $C_1, d > 0$. Set further $U = dU_2$ and thus for each $t \in (0, 1)$, $C_1 - C_1$ is not contained in $(1 - t)\bar{U}$ and we have that $C_1 - C_1$ is in $(1 + t)U$. Since U is a neighbourhood of the origin, the open sets $(\frac{1}{2}U + z)_{z \in C_1}$ is a covering of C_1 and from compactness we find a finite covering, say C_2, corresponding to the elements $z_1, \ldots, z_m$.

Define the point $z = 1/m(z_1 + \cdots + z_m)$ and if c_1 is an arbitrary point of C_1 then for some i in $[1, m], z_i - c_1 \in 2^{-1}U$. But $z_i - c_1$ is in $(1 + t)U$ for all i in $[1, m]$. We obtain further,

$$z \in 1/m(2^{-1}U + (m - 1)(1 + t)U) + c_1$$

and for

$$t = \tfrac{1}{4}(m - 1)$$

we obtain that

$$z \in (1 - \tfrac{1}{4}m)U + c_1.$$

We now define the set C_2 as follows:

$$C_2 = C_1 \cap \bigcap_{c_1} c_1 \in C_1((1 - \tfrac{1}{4}m)U + c_1)$$

which is clearly nonempty and convex. Since, for each $T \in G, T(aU) \subseteq aU$ we obtain that $T(a\bar{U} + c_1) \subseteq a\bar{U} + Tc_1$.

Now the set $C_1 \neq C_2$ since $(1 - \tfrac{1}{4}m)\bar{U}$ does not include the set $C_1 - C_1$. Since G is a group we obtain that C_1 is an invariant set for G. This clearly implies that C_2 is an invariant set for G. Since C_2 is a proper subset of C_1 we have a contradiction with the minimality of C_1. This contradiction proves the theorem.

We prove now the assertion about $G\bar{U}_2$.

The assertion is an immediate consequence of the following well-known characterization of continuity: if X_1 and X_2 are two topological spaces then the function $f : X_1 \to X_2$ is continuous if and only if, for all parts S of X_1, $f(\bar{S}) \subseteq \overline{f(S)}$.

9.2. THE RYLL-NARDZEWSKI FIXED POINT THEOREM

An interesting extension of Kakutani's fixed point theorem for families of mappings was obtained by C. Ryll-Nardzewski.

It is interesting to note that the proof given by Nardzewski was probabilistic in nature. Later he found a proof which did not use probabilistic ideas, and Asplund and Namioka (1967) have found a simplified proof. In what follows we follow the presentation of Asplund–Namioka's proof as expounded by Burckell (1970).

Let G be a set. We say that G is a group if there exists a mapping ' $+$ ' of $G \times G$ with values in G such that the following properties hold: x, y and z arbitrary in G.

1. $(x + y) + z = x + (y + z),$

2.   there exists an element 0 such that $x + 0 = 0 + x = x,$

3.   for each x there exists a negative element, denoted by $- x$, such that $x + (- x) = (- x) + x = 0.$

The group G is called commutative (or Abelian) if

$$x + y = y + x$$

holds for all x, y in G.

A semigroup P in G is any subset with the property that

$$P + P \subset P.$$

Example 9.2.1.   If $\mathbf{Z}$ is the group of the integers then $\mathbf{Z}^+$, the set of positive integers, is a semi-group.

Example 9.2.2.   If $\mathbf{R}$ is the group of all real numbers then the set of all positive real numbers is a semi-group.

Let X be a Banach space and let Q be a subset of X and S be a semi-group of mappings defined on Q with values in Q.

DEFINITION 9.2.3.   The semi-group S is called non-contracting if 0 does not belong to the norm closure of $(Tx - Ty : T \in S, x, y \in Q, x \neq y)$.

Remark 9.2.4.   Sometimes this property of a semi-group is referred to as being distal.

THEOREM 9.2.5   (Ryll-Nardzewski). *Let X be a Banach space and let Q be a convex non-empty weak compact set in X and S be a semigroup of a mapping, defined on Q with values in Q, which is non-contracting. Then there exists a* *point $q \in Q$ such that*

$$Tp = p$$

for all $T \in S$.

For the proof we need the following

LEMMA 9.2.6. *Let X be a Banach space, let K be a non-empty, norm, separable, weak-compact and convex subset of X. Then, for every $\varepsilon > 0$ there is a norm closed convex proper subset C of K such that* diam $(K \setminus C) \leqq \varepsilon$.

Proof. Let $S = (x, x \in X, \|x\| \leqq \varepsilon/4)$ which is a weakly closed neighbourhood of zero of X. Let E be the weak closure of the set of all extreme points of K and, since K is supposed to be separable, a countable number of translations of S cover K and thus E. Since E is weak compact it is of second category in itself, with respect to the relative weak topology. This implies the existence of an element $k \in K$ and a weakly open set $W \subset X$ such that $(S + k) \cap E \supset W \cap E \neq \emptyset$. We consider now K_1, the closed convex hull of $E \setminus W = \{x, x \in E, x \notin W\}$, i.e. the smallest closed convex set containing $E \setminus W$ and let K_2 be the closed convex hull of $E \setminus W$. Since K_i, $i = 1, 2$ are in K they are weakly compact and the set $K_1 \cup K_2$ contains all the extreme points of K. In this case, by the Krein–Milman theorem about extreme points of sets, K is the closed convex hull of $K_1 \cup K_2$. But the closed convex hull of $K_1 \cup K_2$ is exactly the convex hull of $K_1 \cup K_2$, because it is the map

$$(x, y, s) \rightarrow sx + (1 - s)y$$

of $K_1 \times K_2 \times [0, 1] \rightarrow X$, conv $(K_1 \cup K_2)$ is already compact.

Now we remark that $K_1 \neq K_2$. Indeed, in the contrary case, the set $E \setminus W$, which is weakly compact, contains all the extreme points of K – which contradicts the fact that $W \cap E \neq \emptyset$.

It is obvious that diam $(K_2) \leqq \varepsilon/2$.

Let $r \in (0, 1]$ and define the map f_r by

$$f_r(x_1, x_2, t) = tx_1 + (1 - t)x_2$$

on $K_1 \times K_2 \times [r, 1]$ with values in K.

Clearly f_r is continuous and the image C_r of f_r is a weak compact set. It is easy to see that C_r is a convex set. Now we show that $C_r \neq K$. Indeed, in the contrary case, each extreme point is of the form

$$z = sx_1 + (1 - s)x_2, \qquad x_i \in K_i$$

and by extremality we obtain that $z = x_1$ which implies that K is a subset of K_1 and thus $K_1 = K$ which is not the case. Thus $C_r \neq K$.

Let z be in $K \backslash C_r$ and thus it has the form

$$z = sx_1 + (1-s)x_2, \qquad x_i \in K_i$$

and thus

$$\|z - x_2\| = s \qquad \|x_1 - x_2\| \leq rd, \qquad d = \operatorname{diam} K.$$

Now $\operatorname{diam} K < \infty$ and since $\operatorname{diam}(K_2) < \varepsilon/2$ we obtain

$$\operatorname{diam}(K \backslash C_r) \leq \varepsilon/2 + 2rd.$$

Now we take $r = \varepsilon/(4d+1)$ and $C = C_r$ clearly satisfies the lemma. Using this lemma we can prove Ryll-Nardzewski's fixed point theorem.

Proof of Theorem 9.2.5. First we remark that it suffices to consider only the case of finitely-generated semi-groups. Indeed, if (S_a) is the family of all finitely generated subsemi-groups of S and $F(S_a)$ is the fixed point set of the semi-group S_a, then clearly $F(S_a)$ is weakly compact and if $S_{a_{m+1}}$ is the semi-group generated by $S_{a_1}, \ldots, S_{a_m}$, we obviously have

$$F(S_{a_{m+1}}) \subset \bigcap_{i=1}^{m} F(S_{a_i}).$$

This obviously implies that the family of sets $(F(S_a))$ has the finite intersection property and, Q being a compact set, the assertion of the theorem follows.

Thus we consider only the case when S is finitely generated. Let $x_0 \in Q$ be arbitrary and denote by X_1 the closed linear space generated by $(Tx_0, T \in S)$ and consider the set

$$Q' = \overline{\operatorname{conv}(Tx_0, T \in S)} \subset Q.$$

We remark that Q' is compact with respect to the relative weak topology of X_1 (which is the weak topology of X_1). We also remark that X_1 is separable since S is countable. Using Zorn's lemma, we find a weak compact convex set invariant for S which is minimal (with respect to the partial order defined by inclusion). Let K be this set. If K is a single point then the assertion of the theorem is proved. Suppose that this is not the case, then we find two distinct points in K say x, y. Let

$$d_1 = \inf(\|Tx - Ty\| : T \in S)$$

which is positive since S is supposed to be non-contracting.

According to Lemma 9.2.6. we find a compact convex set C in K, not equal to K, such that $\operatorname{diam}(K \backslash C) < d_1/2$.

Now for every $T \in S$ we have

$$T(\tfrac{1}{2}(x + y)) \in C$$

since in the contrary case, C being convex, one of Tx, Ty is in $K \setminus C$. Say $Ty \in K \setminus C$. Thus we get

$$\|Ty - Tx\| = 2\|\tfrac{1}{2}Ty - \tfrac{1}{2}Tx\|$$
$$= 2\|Ty - T(\tfrac{1}{2}(x + y))\|$$
$$\leq 2 \operatorname{diam}(K \backslash C) < d_1$$

which contradicts the definition of d_1.

But if $\tfrac{1}{2}T(x + y) \in C$ then

$$\overline{\operatorname{conv}}(T(\tfrac{1}{2}(x + y)): T \in S)$$

is a weakly compact convex set invariant for S and clearly is a subset of K and not equal to K. But this contradicts the minimality of K. Thus K is a single point and the theorem is proved.

9.3. FIXED POINTS FOR FAMILIES OF NONEXPANSIVE MAPPINGS

Since the class of affine maps is in some sense very restrictive and in many applications we encounter other classes of mappings the problem of the existence of common fixed points appear naturally. The first result in this direction was obtained by Ralph de Marr in 1963 and refers to the class of nonexpansive mappings. First we give de Marr's result, since its proof contains many of the arguments which remain the same for subsequent extensions of the theorem: besides, this result was at the origin of a long series of papers dealing with common fixed points for families of mappings.

THEOREM 9.3.1 (de Marr, 1963). *Let X be a Banach space and let F be a family of nonexpansive mappings defined on a compact convex set C having values in C. If F has only commuting members then there exists a fixed point for the family F.*

For the proof we need some results which have their own intrinsic interest.

LEMMA 9.3.2. *Let X be a Banach space and let K be a compact set in X. Let*

$d = \text{diam } K$ *and suppose* $d > 0$. *In this case there exists* $u \in \text{conv } K$ *such that*

$$\sup(\|x - u\|, x \in K) < d.$$

Proof. Since K is a compact set we find x_1 and x_2 in K such that $d = \|x_1 - x_2\|$.

We define now a set K_0 as follows:

$$K_0 = (x_1, x_2, x_3, \ldots)$$

where the x_i's satisfy the relations

$$\|x_i - x_j\| = d \quad \text{or} \quad 0$$

and, since K_0 is a subset of a compact set K, it is necessarily a finite set, say

$$K_0 = (x_1, x_2, \ldots, x_m).$$

We further consider the point u defined as follows:

$$u = 1/m(x_1 + \cdots + x_m).$$

This is obviously in conv K and we show now that this point satisfies the condition of the lemma. Indeed, since K is compact we find a point z such that

$$\|z - u\| = \sup(\|x - u\|, x \in K)$$

and suppose now that $\|z - u\| = d$. In this case we get

$$d = \|z - u\| \leqq 1/m(\sum_1^m \|z - x_i\|) \leqq d$$

and thus for each i

$$\|z - x_i\| = d$$

i.e. z is in K_0', which is a contradiction. This implies that

$$\|z - u\| < d$$

and the lemma is proved.

LEMMA 9.3.3.   *Let* X *be a Banach space and* C *be a convex set in* X *and* $T : X \to X$ *be a nonexpansive mapping with the following properties:*

1.   *C is an invariant set for* T,
2.   *there exists a compact set* K *in* C *such that* $K = (Tx, x \in K)$ *and is not reduced to a point.*

Then there exists a compact convex set C_1 such that

1. $C_1 \subset K$ *is invariant for* T,

2. $K \cap C_{C_1}$ *is nonempty.*

Proof. According to Lemma 9.3.2. we find a u in conv K such that sup $(\|x - u\|) < d$, $d = \text{diam } K$. Now for each x in K we define the set $m(x)$ as follows:

$$m(x) = (y, \|y - x\| \leq d_1 = \sup(\|x - u\|, x \in K)).$$

Since u is in each $m(x)$ these sets are nonempty and

$$C_1 = \bigcap m(x)$$

is nonempty. Since each $m(x)$ is closed and convex the set C_1 is again a closed and convex set. We show now that this set satisfies the conditions of the lemma. Let $z \in C_1$ and thus z is in each $m(x)$ and p in K. Then we have

$$\|z - p\| \leq d_1$$

and from the property of K there exists $q \in K$ such that $Tq = p$ and from the nonexpansivity of T we get

$$\|Tz - p\| = \|Tz - Tq\| \leq d_1$$

and thus $Tz \in m(x)$ for each x. Then $Tx \in C_1 \cap C$ and property 1 is proved.

For the second property we remark that since K is compact we have, for some x_1 and x_2 in K, $\|x_1 - x_2\| = \text{diam } K$ and thus x_1 is not in $m(x_2)$.

Now we are ready to prove de Marr's Theorem.

First we remark that using Zorn's lemma we find a minimal set C_0 in C (with respect to the order induced by the inclusion) which is compact, convex and nonempty (and of course invariant for F). If this set C_0 reduces to a point the assertion of the theorem is proved. We show now that the contrary case, i.e. C_0, contains at least two different points leads us to a contradiction. Indeed, using Zorn's lemma again, we find a compact nonempty subset of C_0, say K which is invariant for all members of F. We show that K has the property

$$K(Tx, x \in K)$$

for each T in F. Indeed, in the contrary case we find an element S in F such that $SK \neq K$. We remark that the continuity of the mappings in F give that SK is compact. Let x be in SK and thus $x = Ly$ for some $L \in F$ and $y \in K$. The commutativity of F implies that

$$RLy = LRy \in SK$$

and thus

$$R(SK) \subset SK \subset K$$

for all $R \in F$. This clearly contradicts the minimality of K and thus K has the stated property.

Now the proof of the theorem can continue as follows: supposing K has at least two points (since in the contrary case the theorem is proved): then by Lemma 9.3.3 there exists a subset C_1 with the following properties:

1. $C_1 \subset C_0$,

2. $C_0 \cap C_{C_1} \neq \emptyset$.

In this case $C_1 \cap C_0$ is nonempty and invariant for the family F which is a contradiction. Thus C_0 reduces to a point and the theorem is proved.

The following theorem was proved by Belluce and Kirk (1966) and we give it here without proof.

THEOREM 9.3.4.   *Let X be a Banach space and let C be a bounded, closed and convex set in X. If F is a family of commuting nonexpansive mappings and M is a compact subset of C with the property that for all $x \in C$, $M \cap (T^n x) \neq \emptyset$. then F has a fixed point in C.*

We give now another result of Belluce and Kirk (1966) in the case of strictly convex spaces.

THEOREM 9.3.5.   *Let X be a strictly convex space and let C be a weakly compact subset of X. Suppose that F is a family of commuting nonexpansive mappings defined on C with values in X, with the property that for each $T \in F$, the set of fixed points is nonempty. Then F has a fixed point in C.*

Before the proof we remark that any uniformly convex space is strictly convex and for each nonexpansive mapping the set of fixed points is nonempty.

The proof will be given using two results and these are given as lemmas.

LEMMA 9.3.6.   *Let C be a weakly compact subset in a Banach space X and let $f : X \to \mathbf{R}$ be a lower weakly semicontinuous function. Then the infimum of f is achieved in C.*

Proof. First we remark that f is bounded from below in K. Indeed, in the contrary case we find a sequence (x_n) in K such that

$$f(x_n) \leqq -n.$$

We can suppose without loss of generality that the sequence (x_n) is convergent to a point, say x^*. In this case, f being lower semicontinuous we obtain,

$$f(x^*) \leqq \liminf f(x_n)$$

and this contradicts the fact that $f(x_n) \leqq -n$. Since f is thus bounded below, we set

$$a = \inf_{x \in C} f(x)$$

and thus there exists a sequence (x_n) such that $\lim f(x_n) = a$.

The compactness of C implies that some subsequence of (x_n) converges to a point x'. Then clearly $a = f(x')$.

LEMMA 9.3.7.   *Let X be a strictly convex Banach space and let K be a weakly compact convex set in X. In this case, for each $y \notin K$, there exists a unique $y_0 \in K$ such that*

$$\|y - y_0\| = \inf_{x \in C} \|y - x\|.$$

Proof. The function

$$f(x) = \|x - y\|$$

is obviously lower semicontinuous and, K being weakly compact, we can apply Lemma 9.3.6. Thus there exists $y_0 \in K$ such that

$$\min \|y - x\| = \|y - y_0\|$$

Now we use the strict convexity of X to prove that this point is unique. Indeed, suppose that there exists another point y_0' with the same property,

$$\min \|y - x\| = \|y - y_0\| = \|y - y_0'\|$$

and from the convexity of K we get further,

$$\|y - ty_0 - (1 - t)y_0'\| = \min \|y - x\|$$
$$= t\|y - y_0\| + (1 - t)\|y - y_0'\|$$

and this contradicts the strict convexity of X unless we have $y_0 = y_0'$.
The Lemma is proved.

Now the proof of Theorem 9.3.5 is as follows: For each T in the family F, $F(T)$ is convex (because X is strictly convex) and the continuity of T implies clearly that $F(T)$ is also closed. Then it is weakly closed and thus weakly compact as a subset of a weakly compact set. We prove now that the family of all fixed point sets of elements in F has the finite intersection property and the weak compactness of C implies the existence of a fixed point.

Let $T_1, \ldots, T_n$ be a finite number of elements of F such that $\bigcap_1^n F(T_i)$ is nonempty. If T is another element of F then

$$T\left(\bigcap_1^n F(T_i)\right) \subseteq \bigcap_1^n F(T_i)$$

and if $z \in F(T)$ we consider z_0, the element which is closest to z. Since $Tz_0 \in F(T_i)$ and $\|Tz_0 - z\| \leq \|z_0 - z\|$, we obtain that $Tz_0 = z_0$. This completes the proof.

9.4. INVARIANT MEANS ON SEMIGROUPS AND FIXED POINTS FOR FAMILIES OF MAPPINGS

One of interesting applications of the fixed point theorem of Kakutani concerns the existence of the mean value for almost periodic functions. We give now some results about the applications of the fixed point theorems for families of mappings on the one hand and, on the other hand, the application of some results about a class of semigroups to fixed point theorems for families of mappings.

Let G be a locally compact Abelain group and $f : G \rightarrow \mathbf{R}$ be a continuous function.

DEFINITION 9.4.1.   The function f is called almost periodic if

$$(f_g : g \in G) \qquad f_g(s) = f(gs)$$

is a precompact set in $C(G)$, the Banach space of all continuous functions on G.

Let $A(G)$ be the set of all almost periodic functions on the group G. We consider the set

$$\overline{\mathrm{conv}\,(f_s, \, s \in G)}$$

which is clearly a compact convex set. For each $g \in G$ we define the mapping

$$T_g f = f_g$$

for each f in $A(G)$. Clearly (T_g) is a semigroup of mappings having conv$(f_s,\ s \in G)$ as an invariant set. By Kakutani's theorem there exists a fixed point for this family and thus, as is easy to see, this is the constant function. The value of the constant is clearly the number satisfying the properties of the mean.

Now we give some results about invariant means and fixed points.

Suppose that S is a set and denote by $m(S)$ the set of all real-valued functions on S with the norm

$$f \rightarrow \|f\| = \sup(|f(s)|,\ s \in S).$$

DEFINITION 9.4.2.   An invariant mean μ on $m(S)$ is an element of $(m(S))^*$ such that for each $x \in m(S)$,

$$\underset{s \in S}{\text{glb}}\ x(s) \leq \mu(x) \leq \underset{s \in S}{\text{lub}}\ x(s).$$

Each mean μ on $m(s)$ has the following properties:

1. μ is in the unit sphere in $m(S)^*$,

2.   if e is the function whose values are 1 at every point of S then $\mu(e) = 1$,

3.   if $x(s) \geqq 0$ for all $s \in S$, then $\mu(x) \geqq 0$,

4.   the set of means on $m(S)$ is nonempty, convex and w^*-compact.

Suppose now that S is a semigroup. Then for each $s \in S$ we consider the mappings in $L(m(S))$ defined as follows:

$$R_s(f)(t) = f(ts)$$
$$L_s(f)(t) = f(st).$$

It is clear that

$$R_{ss'} = R_{s'}R_s$$
$$L_{ss'} = L_{s'}L_s.$$

DEFINITION 9.4.3.   A element $\mu \in m(S)^*$ is called left(right-)-invariant if

$$\mu(L_s x) = \mu(x), \qquad (\mu(Rs(x)) = \mu(x))$$

for all $x \in m(S)$ and $s \in S$.

DEFINITION 9.4.4.   A semigroup S is called amenable if there exists a mean μ on $m(S)$ which is left- and right-amenable.

If S is a group then the notion of amenability was considered by John von Neumann (1929).

The connection between amenability of a group of mappings of semi-groups and fixed points is indicated in the following theorem of M. M. Day:

THEOREM 9.4.5. (M. M. Day, 1961). *Let X be a locally convex topological linear space and let K be a convex and compact subset of X.*

Suppose that S is a semigroup of mappings acting on K with values in K. If S is amenable then S has a fixed point in K.

For a proof see Day's paper (1961). It is proved there that for the existence of a fixed point of S it is sufficient for S to be left-amenable (i.e. there exists a mean which is left-invariant).

For a generalization as well as for other related results we refer to papers of R. E. Huff (1970), T. Mitchell (1970) and N. W. Rickert (1967), F. P. Greenleaf (1969).

Chapter 10

Fixed Points and Set-Valued Mappings

10.0. Introduction

Let X be a topological space and $P(X)$ the family of all subsets of X. In many problems the following types of function are encountered:

$$f : X \to P(X).$$

For example, if we consider a family of bounded linear operators $\{T_z\}_{z \in C}$, C is the set of complex numbers, then for each T_z we can consider the spectrum, $\sigma(T_z)$ which is a closed nonempty set in C and thus we have a function

$$z \to \sigma(T_z)$$

which is of the above type.

Similarly we can consider the numerical range of the operators and thus we have a function

$$z \to W(T_z)$$

which in the case of Hilbert space operators has values in the family of convex sets of C.

For some applications it is useful to consider on $P(X)$ some topologies or at least on parts of $P(X)$. In the case of $X = C$ and the family of closed sets a metric on this family was defined by D. Pompeiu (1905) and later extended to all complete metric spaces by F. Hausdorff in his book (1962). (See also the book of K. Kuratowski (1962)).

The metric defined by Pompeiu and Hausdorff on the family of all closed set of a complete metric space is nowadays called 'the Pompeiu–Hausdorff metric'. In Section 10.1. We give the definition of the Pompeiu–Hausdorff metric as well as some properties which are useful for the applications which follow.

In the rest of the chapter we present various fixed point theorems for set-valued mappings. Among them we mention the generalization of the Brouwer fixed point theorem obtained by S. Kakutani (1941) and some theorems on contractive set-valued mappings obtained by S. Nadler and R. Fraser (1969), as well as other related results.

10.1. THE POMPEIU – HAUSDORFF METRIC

Let X be a metric space with the metric d.

For any two nonempty elements M and N of $P(X)$ we define the number $d(M, N)$ as follows: let $x \in M$ and set

$$d(x, N) = \inf\{d(x, y): y \in N\}$$

and

$$d(M, N) = \sup\{d(x, N): x \in M\}.$$

The number $d(M, N)$ has the following properties:

LEMMA 10.1.1.   *For any M, N, Q bounded elements in $P(X)$ (and of course nonempty),*

1. $d(M, N) = 0$ *if and only if $M \subseteq \bar{N}$,*

2. $d(M, N) \leq d(M, Q) + d(Q, N).$

Proof. The first assertion follows obviously from the definition of the number $d(M, N)$ and from the properties of the closure of a bounded set in a metric space.

The second assertion is proved as follows:

From the triangle inequality we have,

$$d(m, n) \leq d(m, q) + d(q, n)$$

for any $m \in M$, $n \in N$ and $q \in Q$. This implies that

$$\inf_{n}\{d(m, n)\} \leq d(m, q) + \inf_{n}\{d(q, n)\}$$

holds for all $q \in Q$. This gives further that,

$$\inf_{n}\{d(m, n)\} = d(m, N) \leq d(m, q) + d(q, N)$$

and since this is valid for all $q \in Q$ we obtain that

$$d(m, N) \leq d(m, Q) + d(Q, N)$$

and taking sup we obtain

$$d(M, N) \leq d(M, Q) + d(Q, N).$$

DEFINITION 10.1.2.   The function $d(,)$ defined on the family of all nonempty bounded sets of a metric space X, by the formula

$$d(M, N) = \sup_{m \in M} \{\inf_{n \in N} d(m, n)\}\}$$

is called the Pompeiu–Hausdorff semidistance.

Remark 10.1.3.   Simple examples show that in general

$$d(M, N) \neq d(N, M).$$

To avoid the difficulty posed by the above remark we define, following Pompeiu and Hausdorff, the following function on the family of all bounded sets of a metric space:

DEFINITION 10.1.4.   The Pompeiu–Hausdorff semi-metric on the family of all bounded sets of a metric space is the function defined by

$$\delta(M, N) = \max \{d(M, N), d(N, M)\}.$$

From the above lemma we have the following

THEOREM 10.1.5.   *On the family of all bounded and closed sets of a metric space the Pompeiu–Hausdorff semi-metric is a metric.*

For any metric space X we denote by 2^X the family of all closed and bounded sets in X, and $K(X)$ is the family of all nonempty compact subsets of X.

Theorem 10.1.5 says in this terminology that 2^X is a metric space with the Pompeiu–Hausdorff metric.

The properties of the set $K(X)$ as a metric subspace are given in the following theorem:

THEOREM 10.1.6.   *For any metric space X the following assertions hold:*

1.   *if X is a complete metric space then $K(X)$ is a complete metric space,*

2.   *if X is a separable metric space then $K(X)$ is a separable metric space.*

Remark 10.1.7.   We can mention also the following interesting property of the space $K(X)$: if X is a compact space then $K(X)$ is also a compact space.

Proof of Theorem 10.1.6. (1) Let $\{M_n\}$ be a sequence of bounded and closed sets in the metric space X which is supposed to be complete. Suppose that $\{M_n\}$ is a Cauchy sequence with respect to the Pompeiu–Hausdorff metric. Let M be the set defined as

$$M = \limsup M_n$$

and we prove that

$$\lim \delta(M_n, M) = 0.$$

That M is closed and bounded follows from the definition of lim sup and the properties of M_n.

Since $\{M_n\}$ is a Cauchy sequence, for any given $\varepsilon > 0$ there exists an N_ε such that

$$\delta(M_m, M_n) \leqq \varepsilon$$

if $n, m \geqq N_\varepsilon$. We prove that

$$\delta(M, M_{N_\varepsilon}) \leqq 2\varepsilon$$

and this clearly implies that

$$\delta(M, M_{n_\varepsilon}) \leqq \delta(M, M_{N_\varepsilon}) + \delta(M_{N_\varepsilon}, M_n) \leqq 2\varepsilon + \varepsilon = 3\varepsilon$$

for all $n \geqq N_\varepsilon$; and this clearly is property 1.

To prove the inequality

$$\delta(M, M_{N_\varepsilon}) \leqq 2\varepsilon$$

it is sufficient to prove that:

1.   for any $x \in M$, $d(x, M_{N_\varepsilon}) \leqq 2\varepsilon$

and

2.   for any $y \in M_{N_\varepsilon}$, $d(y, M) \leqq 2\varepsilon$.

We consider the following set

$$A_i = \{x, d(x, M_i) \leqq \varepsilon\}$$

and since the sequence $\{M_n\}$ is a Cauchy sequence, we obtain that for $n \geqq N_\varepsilon$, $M_n \subset A_{N_\varepsilon}$. From the definition of the set M it follows that

$$M \subseteq \bigcup_{i \geqq N_\varepsilon} A_i$$

and thus

$$M \subseteq A_{N_\varepsilon}$$

and this is equivalent to

$$d(x, M_{N_\varepsilon}) \leqq 2\varepsilon$$

for all $x \in M$.

To prove the assertion in 2 we consider the sequence of integers,

$$n_i = N_{\varepsilon/2^i}$$

for $\varepsilon > 0$ give. In each M_{n_i} we choose a point m_{n_i} such that

1. $\qquad m_{n_0} = y,$

2. $\qquad d(m_{n_i}, m_{n_{i-1}}) \leqq \varepsilon/2^{i-1}$

and we remark that this is possible since $\{M_n\}$ is a Cauchy sequence.

We now remark that the sequence $\{m_{n_i}\}$ is a Cauchy sequence since

$$d(m_{n_p}, m_{n_q}) \leqq \varepsilon/2^{q-1}$$

if $p > q$.

Since X is supposed to be a complete metric space we can consider $m = \lim m_{n_i}$ and from the definition of the set M we have that $m \in M$. Since

$$d(m_{n_p}, m_{n_0}) = d(m_{n_p}, y) \leqq 2\varepsilon$$

taking the limit, we get

$$d(m, y) \leqq 2\varepsilon.$$

and the completeness of $K(X)$ is proved.

We prove now the property that M is a compact set. For this it is sufficient to prove that M is totally bounded.

Let $\varepsilon > 0$ and choose N_ε such that it has the value 0 for all $n \geqq N_\varepsilon$ we have

$$\delta(M, M_n) \leqq \varepsilon.$$

Since the sets M_n are compact, we find a finite number of spheres of radius $\varepsilon/2$ which cover M_{N_ε}. It is clear that the spheres with the same centers and radius ε cover M. This gives M is a compact set.

10.2. CONTINUITY FOR SET-VALUED MAPPINGS

One of the most natural ways to define the continuity for set-valued mappings is to extend one of the many characterizations of continuity for single-valued mappings. While in some cases some alternatives give equivalent definitions, in general this is not the case.

Let P be a property of sets. We say that a set-valued mapping f is point P if and only if $f(x)$, for each x, has that property.

The properties which we shall use are closed, convex and compact.

DEFINITION 10.2.1.   Let X and Y be two topological spaces and $f: X \to P(Y)$ be a set-valued mapping such that for all $x \in X, f(x) \neq \varnothing$ ($\varnothing$ is the empty set). F is said to be upper semi-continuous at $x_0 \in X$ if for arbitrary neghbourhood $V \ni f(x_0)$ there exists a neighbourhood $x_0 \in U$ such that $f(x) \subset U$, for all $x \in U$.

DEFINITION 10.2.2.   Let X and Y be two topological spaces and $f: X \to P(Y)$ be a set-valued mapping such that for all $x \in X, f(x) \neq \varnothing$.

Then f is said to be lower semi-continuous at $x_0 \in X$ if for any open set V $f(x_0) \neq \varnothing$ there exists a neighbourhood $V_{x_0} \ni x_0$ such that $f(x) \subset V_{x_0}$ for all $x \in V_{x_0}$.

DEFINITION 10.2.3.   Let X and Y be two topological spaces and let f be a set-valued mapping defined on X with values in $P(Y)$ and $f(x) \neq \varnothing$ for all $x \in X$. Then f is said to be continuous at $x \in X$ if it is upper and lower semi-continuous at x. The set-valued mapping is called continuous on X if it is continuous at each point $x \in X$.

The following result gives a characterization of upper semi-continuity for set-valued mappings for the case of special spaces in terms of graphs.

THEOREM 10.2.4.   *Let X and Y be two compact spaces and $f: X \to P(Y)$ be a point, closed, set-valued mapping. Then f is upper semi-continuous iff*

$$G(f) = \{(x, y), \ x \in X, \ y \in Y, \ y \in f(x)\}$$

is closed in Y.

Proof. Suppose that X and Y are as in the theorem and $f: X \to P(Y)$ is upper semi-continuous. Let $(u, w) \in X \times Y$ which is not in $G(f)$.

In this case w is not in the closed set $f(u)$ and since Y is a compact space, we find two neighbourhoods $V_1 \ni w$, $V_2 \supset f(u)$ such that $V_1 \cap V_2 = \varnothing$. Now since f is upper semi-continuous, we find a neighbourhood $U \ni u$ such that $f(x) \subset V_2$ for all $x \in U$. Thus the neighbourhood $U \times V_1$ of (u, w) does not intersect $G(f)$ and this implies that $G(f)$ is closed.

Now, suppose that X and Y are as above and that $f: X \to P(Y)$ is a set-valued mapping its $G(f)$ closed. Since the space $X \times Y$ with the product topology is a compact space we have that $G(f)$ is compact.

Let u be an arbitrary point of X and V an arbitrary neighbourhood of $f(u)$.

If f is not upper semi-continuous at the point u, then for each neighbourhood $U \ni u$ we have

$$G(f) \cap (\bar{U} \cap (Y - V)) \neq \varnothing.$$

Now we remark that $G(f)$, $\bar{U}$, $Y - V$ are compact sets and thus they have the finite intersection property. In this case we have,

$$\bigcap G(f) \cap (\bar{U} \cap (Y - V)) \neq \varnothing$$

and if (a, b) is a point of this intersection then $a = u$ and $b \in f(u) \cap (Y - V)$ and this contradicts the choice of V as a neighbourhood of $f(u)$. This contradiction shows that f is upper semi-continuous.

For other properties of set-valued functions we refer to the interesting paper of R. E. Smithson (1972).

10.3. FIXED POINT THEOREMS FOR SOME CLASSES OF SET-VALUED MAPPINGS

Let X be a set.

DEFINITION 10.3.1.   Let $f : X \to P(X)$ be a set-valued mapping. We say that $x \in X$ is a fixed point of f if $x \in f(x)$.

More generally we give the following

DEFINITION 10.3.2.   A set X is said to have the fixed point property (abbreviated f.p.p.) for a family of set-valued mappings if each member of the family has a fixed point.

The first extension of topological fixed point theory of continuous mappings to the case of set-valued mappings was made by John von Neumann in 1937 in connection with the proof of the fundamental theorem of game theory.

Brouwer's fixed point theorem was extended in 1941 – by S. Kakutani – to point compact convex set-valued mappings on a conpact convex set in the Euclidean space. The extension of Schauder's fixed point theorem was given independently by H. Bohnenblust and S. Karlin in 1950, and by I. Glicksberg in 1952. The extension of Tychonoff's theorem was given by Ky Fan in 1952. Further results in this direction were obtained by F. Brouwder and others.

In what follows we follow Browder's treatment since it appears to cover the classical results in a simple and efficient manner.

Much recent work has been directed towards extending the contraction
principle to set-valued mappings. S. Nadler defined set-valued contractions
and proved that such mappings have fixed points in the case of complete
metric spaces. These results were extended by S. Nadler and R. Fraser, by
H. Covitz and S. Nadler.

R. E. Smithson defined contractive set-valued mappings and proved
fixed point theorem for such mappings. Using the Liapunov function
George V. Sehgal and R. E. Smithson were able to extend many results
from the single-valued mappings to set-valued mappings.

In the last part of the chapter we give these results.

For the fixed point theorems we give later we need a few results about a
partition subordinated to a covering. First we recall that a covering of a set
F is a family of sets $\{M_i\}_{i \in I}$ where I is an index set, such that for each $x \in F$
there exists an index $i_x \in I$ with the property that $x \in M_{i_x}$.

THEOREM 10.3.3.   *Let X be a compact space and $\{M_i\}_1^N$ be a finite open
covering of X (i.e. M_i are open sets). Then there exists the continuous functions*

$$f_n : X \to \mathbf{R}^+, \quad n = 1, 2, 3, \ldots, N$$

such that

1. $$\sum_{n=1}^N f_n(x) = 1,$$

2. $$0 \leq f_n(x) \leq 1,$$

3. $$f_n(x) = 0 \quad \text{for all } x \in X - M_n.$$

Proof. In the proof for the case of arbitrary compact spaces uses the
Lemma of Urysohn. However in the case of compact metric spaces a very
simple proof of the Urysohn Lemma can be given and this is left to the
reader.

We give now the Lemma of Urysohn.

LEMMA 10.3.4   (Urysohn). *Let F_0 and F_1 be disjoint closed sets in a normal
space X (i.e. a topological space X which is Hausdorff and for every pair of
closed sets F_1 and F_2 there exist disjoint open sets U_1 and U_2 such that
$F_1 \subset U_1$ and $F_2 \subset U_2$), then there exists a continuous real valued function
$f : X \to [0, 1]$ such that*

$$f(x) = \begin{cases} 0 & x \in F_0 \\ 1 & x \in F_1 \end{cases}.$$

Proof. Let $V_1 = X - F_1$ which is an open set. Since X is a normal space, there exists an open set $V_{1/2}$ such that

$$F_0 \subset V_{1/2}, \qquad \bar{V}_{1/2} \subset V_1.$$

Using again the fact that X is a normal space, we find the open sets $V_{1/4}$ and $V_{3/4}$ such that,

$$F_0 \subset V_{1/4}, \qquad \bar{V}_{1/4} \subset V_{1/2}$$

and

$$V_{1/2} \subset V_{3/4}, \qquad \bar{V}_{3/4} \subset V_1.$$

We can continue this process for all numbers of the form $m/2^n$ with $0 \leqq m \leqq 2^n$ by defining the sets V_r such that

$$F_0 \subset V_r, \qquad \bar{V}_r \subset V_1$$

and

$$\bar{V}_r \subset V_s$$

for $r < s$.

The function f is defined as follows:

$$f(x) = \begin{cases} 1 & \text{if } x \text{ is none of the sets } V_r, \\ \mathrm{glb}\{r, x \in V_r\} & \text{in the contrary case.} \end{cases}$$

From the definition of f it is clear that the range is in $[0, 1]$ and that on F_0 it takes the value zero and on F_1 it is 1. It remains to prove the continuity. For this purpose we consider a number $b \in (0, 1]$ and we remark that $f(x) < b$ if and only if $p \in V_r$ for some $r < b$ and thus $\{x, f(x) < b\} = \bigcup_{r < b} V_r$ which is clearly an open set. We consider now a number $a \in [0, 1)$ and we remark that $f(x) > a$ if and only if $x \notin V_r$ for some $r > a$. This gives that $\{x, f(x) > a\} = \bigcup_{r > a} \bar{V}_r'$ where for any set M, M' denotes the complement of the set M. Since the intervals of the form $[0, b)$ and $(a, 1]$ as well as their intersections form a basis for the topology of $[0, 1]$ the continuity of f is proved.

Remark 10.3.5.   It is not difficult to see that any compact Hausdorff space is normal.

Now we are in a position to prove Theorem 10.3.4. For this purpose we consider the closed sets

$$F_k = M'_k, \qquad k = 0, 1, \ldots, N$$

$$F_{\tilde{k}} = \left\{ \bigcup_{i=k+1}^{N} M_i \right\},$$

where

$$F_0 = \emptyset, \qquad F_{\tilde{N}} = X.$$

Clearly $F_1 \cap F_{\tilde{1}} = \emptyset$ and $F_k \cap F_{\tilde{k}} = F_{\widetilde{k-1}}$. Then by Urysohn's Lemma there exists a continuous real-valued function with the range in $[0, 1]$ such that

1. $\qquad f_1(x) = 0 \qquad x \in F_1$,

2. $\qquad f_1(x) = 0 \qquad x \in F_{\tilde{1}}$.

Suppose now that the functions $f_1, f_2, \ldots, f_{k-1}$ are chosen such that

1. $\qquad f_i(x) = 0 \qquad x \in F_i, \qquad i = 1, 2, \ldots, k - 1$,

2. $\qquad f_1(x) + \cdots + f_{k-1}(x) = 1 \qquad x \in F_{\widetilde{k-1}}$,

3. $\qquad f_1(x) + \cdots + f_{k-1}(x)$ is a number in $[0, 1]$.

In this case we consider the function

$$g(x) = 1 - f_1(x) - \cdots - f_{k-1}(x)$$

which is real-valued and continuous with the range in $[0, 1]$. Also the fact that $g(x) = 0$ on $F_{\widetilde{k-1}}$ is obvious. Now for the sets F_k and $F_{\tilde{k}}$ we find a real-valued continuous function f_k with the range in $[0, 1]$ such that

1. $\qquad f_k(x) = 0 \quad$ if $x \in F_k$,

2. $\qquad f_k(x) = g(x) \quad$ if $x \in F_{\tilde{k}}$,

3. $\qquad f_k(x) \leqq g(x)$.

Indeed, by the Urysohn Lemma we find a continuous real-valued function h which is zero on F_k and 1 on $F_{\tilde{k}}$, and with the range in $[0, 1]$. Clearly the function gh satisfies the above conditions.

Now the proof can be continued as follows: we consider the function $f_1(x) + \cdots + f_k(x)$ which has the range in $[0, 1]$ and is 1 on $F_{\tilde{k}}$. This process can be continued until $k = N$ and we obtain the functions $\{f_i\}$ having the stated properties.

Now we are in a position, using the above Theorem 10.3.4, and Brouwer's fixed point theorem, to obtain fixed point theorems for a class of set-valued mappings. These results, given after the manner of Browder, generalize and extend the results of Kakutani, Bohnenblust, Karlin, Glicksberg and Ky Fan.

THEOREM 10.3.6.   *Let X be a topological vector space and C be a nonempty compact convex set and let*

$$f : C \to P(C)$$

be a convex point set-valued mapping such that for each $x \in C$, $f(x) \neq \varnothing$.
   If for each $y \in C$, $f^{-1}(y) = \{x, x \in C, f(x) \ni y\}$ is open then f has a fixed point in C.

   Proof. Let $y \in C$ be arbitrary and, since $f^{-1}(y)$ is an open set, then it is clear that $\{f^{-1}(y)\}_{y \in C}$ is an open covering of C. But C is a compact set and thus we find a finite subcovering, say $f^{-1}(y_1) \cdots f^{-1}(y_n)$ such that

$$\bigcup_{i=1}^{n} f^{-1}(y_i) = C.$$

According to Theorem 10.3.3 we find a partition subordinated to this covering, say $\{\alpha_1, \ldots, \alpha_n\}$ such that

1. $\qquad \displaystyle\sum_{i=1}^{n} \alpha_i(x) = 1, \qquad 0 \leq \alpha_i(x) \leq 1,$

2. $\qquad \alpha_i(x) = 0 \qquad \text{for } x \in C - f^{-1}(y_i).$

Now we define the mapping

$$p : C \to C$$

by the formula

$$p(x) = \sum_{i=1}^{n} \alpha_i(x) y_i$$

which is clearly a continuous function. Let us consider the n-dimensional space generated by $y_1, \ldots, y_n$ and, since the induced topology in the Euclidean space is the Euclidean topology, we can apply Brouwer's fixed point theorem and we find an $x_0 \in C$ such that $p(x_0) = x_0$.
   It is now clear that x_0 is a fixed point of f and the theorem is proved.

From this theorem we obtain the following useful result given as

THEOREM 10.3.7.   *Let X be a topological space and let C be a nonempty compact convex set of X. Let $f : C \to X^*$ (X^* is the dual of X, i.e the set of all linear and continuous real-valued functions on X) be a continuous mapping. Then there exists a $u_0 \in C$ such that*

$$f(u_0)(u_0 - u) = \langle f(u_0), u_0 - u \rangle \geq 0.$$

Proof. Suppose that the assertion of the theorem is false. Then for each $u_0 \in C$ there exists an element $u \in C$ such that

$$\langle f(u_0), u - u_0 \rangle > 0.$$

For each $u_0 \in C$ we define

$$C(u_0) = \{u, u \in C, \langle f(u_0), u_0 - u \rangle < 0\}$$

and it is clear that this is a nonempty set for each $u_0 \in C$. Obviously $C(u_0)$ is convex. Since the mapping f is continuous, the function

$$v \in C \to \langle f(u), v - u \rangle$$

is again continuous. This gives us that the function $f^{-1}(u)$ is an open set of C. Then by Theorem 10.3.4. there exists an element $v_0 \in C$ such that $v_0 \in C(v_0)$. But for this element we have

$$0 > \langle f(v_0), v_0 - v_0 \rangle = 0$$

and this is a contradiction and the theorem is proved.

From this theorem we obtain a result which contains as a special case the theorems extending Tychonoff's theorem for the case of set-valued mappings as well as the results of Kakutani, Bohnenblust and Karlin and Gicksberg.

For this we consider a set associated with a convex set C.

DEFINITION 10.3.8.   Let C be a convex set in a locally convex topological space X and $x \in C$. Then x is said to be in $\delta(C)$ if there exists a finite-dimensional vector space F such that x lies in the boundary of $C \cap F$.

Now the result is as follows.

THEOREM 10.3.9.   *Let X be a locally convex linear Hausdorff space and C be a nonempty compact convex set in X. Let $f : C \to P(X)$ be an upper semi-continuous, point closed, and convex set-valued mapping.*

Suppose that for each $x \in \delta(C)$ there exists an element $w \in f(u)$ such that

$$w - u = \lambda(u_1 - u)$$

for some $u_1 \in C$ and $\lambda > 0$ Then there exists a fixed point of f.

Proof. Suppose that the assertion of the theorem is false. In this case, for each $u \in C$, 0 is not an element of $u - f(u)$ which is clearly a closed and convex set since $f(u)$ is convex and closed.

Then the theorem on separation of convex sets in locally convex spaces implies that there exist $\delta > 0$ and $w \in X^*$ such that

$$w(v) \geq \delta$$

for all $v \in u - f(u)$. Now for each $w \in X^*$, we set

$$N_w = \{u, u \in C, w(v) > 0, v \text{ arbitrary in } u - f(u)\}.$$

We consider now the interior of the sets N_w, $\{\mathring{N}_w\}$ and we remark that according to upper semi-continuity of f, these form a covering of C. But C is a compact subset in X and thus we find a finite subcovering, say that one given by the elements $w_1, \ldots, w_n$. Clearly

$$C = \bigcup_{i=1}^{n} \mathring{N}_{w_i}$$

and we consider now a partition subordinated to this covering, say $\alpha_1, \ldots, \alpha_n$, and we define the mapping

$$p(x) = \sum_{i=1}^{n} \alpha_i(x) w_i$$

which is a mapping on C with values in X^*. But for each $v \in u - f(u)$ we have

$$\langle p(u), v \rangle = \sum \alpha_i(u) \langle w_i, v \rangle > 0$$

since

$$\alpha_i(u) \neq 0 \rightarrow \langle w_i, v \rangle > 0.$$

In this case, Theorem 10.3.6 assures the existence of a point $u_0 \in C$ with the property

$$\langle p(u_0), u_0 - u \rangle \geq 0$$

for all $u \in C$.

We have now to discuss two cases:

Case 1. $u_0 \notin \delta(C)$. In this case for each $v \in X$ we find $\varepsilon > 0$ such that $u = u_0 - \varepsilon v \in C$ and thus

$$0 \leq \langle p(u_0), \varepsilon v \rangle = \varepsilon \langle p(u_0), v \rangle$$

which implies that

$$\langle p(u_0), v \rangle \geq 0$$

for all $v \in X$. Replacing v by $-v$ we have that $\langle p(u_0), v \rangle = 0$ for all $x \in X$.

But $\langle p(u_0), v \rangle > 0$ for each v in the nonempty set $u_0 - f(u_0)$. This contradiction proves the theorem in this case.

Case 2. $u_0 \in \delta(C)$. In this case by the definition of the set $\delta(C)$ there exists $w_0 \in f(u_0)$ and $u_1 \in C$, $\lambda > 0$ such that

$$w_0 - u_0 = \lambda(u_1 - u_0)$$

and

$$\langle p(u_0), u_1 - u_0 \rangle > 0.$$

This implies that

$$0 < \lambda^{-1} \langle p(u_0), w_0 - u_0 \rangle = -\lambda^{-1} \langle p(u_0), u_0 - w_0 \rangle$$

and since

$$\langle p(u_0), u_0 - w_0 \rangle > 0$$

we obtain that

$$\lambda^{-1} \langle p(u_0), u_0 - w_0 \rangle > 0$$

which is a contradiction and the theorem is completely proved.

As a corollary of this theorem we obtain, as we have mentioned above, the results of Kakutani, Bohnenblust and Karlin and Glicksberg and Ky Fan.

COROLLARY 10.3.10 *Let X be a locally convex topological space and C be a nonempty compact convex set in X. Let $f : C \to P(C)$ be an upper semicontinuous point closed and point convex set-valued mapping. Then f has a fixed point.*

Proof. We take in the above theorem w in $f(x) \subseteq C$ and for each $x \in \delta(C)$, $u_1 = w$ and $\lambda = 1$.

Also it is clear that Corollary 10.3.10 contains the case of finite-dimensional spaces as a special case.

As we have mentioned, John von Neumann has proved the first result about fixed points for the case of set-valued mappings and has used this result to prove some deep results in the theory of games. In what follows we show that Theorem 10.3.6 implies the result of J. von Neumann as well as the results of Nash and Ky Fan.

The following theorem was proved by Ky Fan [1966]

THEOREM 10.3.11. *Let* X *be a real topological space and let* $K_1, \ldots, K_n (n \geq 2)$ *be nonempty compact convex sets in* X *and let* $K = \bigcup K_n$.
Suppose that $S_1, \ldots, S_n$ *are subsets of* K *which have the following properties*:

1.   *if* $K_{\tilde{j}} = \bigcup_{k \neq j} K_k$ *and the generic point of* $K_0^{\tilde{}}$ *is* $x_{\tilde{j}}$ *then for each* j *and for each point* x_j *in* K_j *the set*

$$S_j(x_j) = \{x_{\tilde{j}}, x_{\tilde{j}} \in K_{\tilde{j}}, [x_j, x_{\tilde{j}}] \in S_j\}$$

is an open subset of $K_{\tilde{j}}$,

2.   *for each* j *and for each* $x_{\tilde{j}}$ *in* $K_{\tilde{j}}$ *the set*

$$S_j(x_{\tilde{j}}) = \{x_j, x_j \in K_j, [x_j, x_{\tilde{j}}] \in S_j\}$$

is a nonempty convex subset of K_j.

Then

$$\bigcap_{j=1}^{n} S_j$$

is nonempty.

Proof. We define a mapping

$$T : K \rightarrow P(K)$$

as follows: let $x \in K$ and $y \in K$. We say that $y \in T(x)$ if and only if for each $j \in [1, n]$,

$$[y_j, x_{\tilde{j}}] \in S_j$$

where $x_{\tilde{j}}$ represents the natural projection of x on $K_{\tilde{j}}$. Property 2 in the theorem asserts that $T(x)$ is a nonempty convex subset of K.

Let $y \in K$ and consider the set

$$T^{-1}(y) = \{x, x \in K, y \in T(x)\}$$

and we see that $x \in T^{-1}(y)$ if and only if $x_{\bar{j}} \in S_j(y_j)$ and this is exactly the set

$$T^{-1}(y) = \bigcap_{j=1}^{n} \{S_j(y_j) \times K_j\}$$

and these are open sets. Thus we can apply Theorem 10.3.6 and we obtain that T has a fixed point in K. Let u be that fixed point. From the definition of the mapping T we see that

$$u = [u_j, u_{\bar{j}}] \in S_j, \qquad 1 \leq j \leq n$$

which is equivalent to

$$u \in \bigcap_{j=1}^{n} S_j.$$

In his paper [1966] Ky Fan gave an analytic formulation of Theorem 10.3.11 using the so-called quasi-concave functions.

DEFINITION 10.3.12.   A real-valued function defined on a convex subset of a topological linear space X is said to be quasi-concave if for every real number r, $\{x, f(x) < r\}$ is a convex set.

The function f defined on a convex subset of X is said to be quasi-convex if the function $-f$ is quasi-concave.

Now the analytical form of Theorem 10.3.11 is as follows:

THEOREM 10.3.13.   Let $K_1, \ldots, K_n$ be $n(n \geq 2)$ compact convex sets in a real topological space X and let $K = \bigcup K_n$. Suppose that $f_1, \ldots, f_n$ are n real-valued functions on K with the following properties:

1.   for each j and for each point $x_j \in K_j$, the function $f_j(x_j, x_{\bar{j}})$ as a function of $x_{\bar{j}}$ is lower semi-continuous on $K_{\bar{j}}$,

2.   for each j and for each point $x_{\bar{j}} \in K_{\bar{j}}$, the function $f_j(x_j, x_{\bar{j}})$ as a function of x_j is quasi-convex on K_j.

Let $t_1, \ldots, t_n$ be n real numbers and suppose that for each j and for each point $x_{\bar{j}} \in K_{\bar{j}}$ there exists a point $y_j \in K_j$ such that $f_j(y_j, x_{\bar{j}})$ is greater than t_j then there exists a point $u \in K$ such that

$$f_j(u) > t_j, \qquad j = 1, 2, 3, \ldots, n.$$

Proof.   For each $j \in [1, n]$ we define the subsets S_j of K as follows:

$$S_j = \{u, u \in K, f_j(u) > t_j\}$$

and property 1 in the theorem asserts that for fixed $x_j \in K_j$ the set

$$S_j(x_j) = \{x_j^-, x_j^- \in K_j, [x_j, x_j^-] \in S_j\}$$

is an open subset of K_j. Property 2 asserts that for each fixed $x_j^- \in K_j^-$,

$$S_j(x_j^-) = \{x_j, x_j \in K_j, [x_j, x_j^-] \in S_j\}$$

is a convex subset of K_j and the condition about the numbers $t_1, t_2, \ldots, t_n$ asserts that the convex sets are nonempty. Thus we can apply Theorem 10.3.11. and we obtain that there exists a common fixed point of the sets S_j and this obviously is the conclusion of the theorem.

John von Neumann (1937) has proved the following

THEOREM 10.3.14.   *Let X and Y be two nonempty compact convex sets each in a finite-dimensional Euclidean space. Let E, F be two subsets of $X \times Y$. Suppose that for each $y \in Y$, $E(y) = \{x, x \in X, (x, y) \in E\}$ is nonempty and convex and also for each $x \in X$, $F(x) = \{y, y \in Y, (x, y) \in F\}$ is nonempty and convex. In this case $E \cap F \neq \varnothing$.*

This important result, which has applications in game theory was extended by Ky Fan (1966) to the following theorem:

THEOREM 10.3.15.   *Let $\{X_\tau\}_{\tau \in I}$ be a family (finite or infinite) of nonempty compact convex sets each in a real Hausdorff locally convex linear space. Let $X = \Pi_\tau X_\tau$ and $Y_\mu = \Pi_{\mu \neq \tau} X_\tau$ for each $\mu \in I$.*

Let $\{E_\tau\}_{\tau \in I}$ be a family (indexed with the same set I) of closed subsets of X. Suppose that for each point $x \in X$ and for any $\mu \in I$, the set

$$\mathrm{pr}_{X_\mu}(E_\mu(X_\mu \mathrm{pr}_{Y_\mu} x)) = R_\mu(x)$$

in X_μ is nonempty and convex. Then $\bigcap_\tau E_\tau \neq \varnothing$.

(Here pr_{X_τ} and pr_{Y_τ} denote the projections on X_τ and Y_τ respectively).

It is clear that Theorem 10.3.15 reduces to Theorem 10.3.14 when the set I has only two elements.

Proof. We consider the set X which, by the well known theorem of Tychonoff, is a compact set. We define now a mapping on X with values in $P(X)$ as follows: for $x \in X$, we define the set $T(x)$ as the set of all elements $y = (y_\tau)_{\tau \in I}$ for which $y_\tau \in R_\tau$ and we remark that it is nonempty and convex. Since X_τ are compact and convex it follows that $T(x)$ is also compact and convex. We show now that it is upper semi-continuous. Indeed, let $G(T)$ be the graph of T and we show that it is a closed subset in $X \times X$.

Let (x, y) be an element not in $G(T)$. Then there exists an index $\tau \in I$ such that $y_\tau \in R_\tau(x)$. Since X_τ is compact there exists a neighbourhood N_1 of y_τ in X_τ and a neighbourhood N_2 of x_τ^- where $x_\tau^- = \mathrm{pr}_{Y_\tau} x$, such that $N_1 \times N_2$ does not intersect E_τ.

We construct the neighbourhoods N_1' and N_2' such that

$$N_1' = N_1 \times Y_\tau, \qquad N_2' = X_\tau \times N_2$$

and we remark that for any $x \in N_2'$ and any $y \in N_1'$, we have $y_\tau \notin R_\tau(x)$ and thus $N_2' \times N_1'$ does not intersect $G(T)$ and thus $G(T)$ is closed in $X \times X$. Thus by Theorem 10.2.4 T is upper semi-continuous and we can apply Corollary 10.3.10 to obtain a fixed point of T. This implies the assertion of the theorem.

From this theorem we obtain the following famous result of Nash about the existence of equilibrium points in many-person games.

This theorem is as follows.

THEOREM 10.3.16. *Let $K_1, \ldots, K_n$ be ($n \geq 2$) nonempty compact convex sets in a real topological linear space. Further let $f_1, \ldots, f_n$ be n real-valued continuous functions defined on $K = \bigcap K_j$. Suppose that for each $j \in [1, n]$ and for each $x_j \in K_j^-$, $f_j(x_j, x_j^-)$ is quasi-concave as a function of x_j on K_j. Then there exists a point $u \in K$ such that for $j \in [1, n]$,*

$$f_j(u) = \max_{y_j \in K_j} f_j(y_j, u_j^-)$$

and where $u_j^- = \mathrm{pr}_{K_j^-} u$.

Proof. For each $x_j^- \in K_j$ we define

$$q_j(x_j^-) = \max_{y_j \in K_j} f_j(y_j, x_j^-)$$

and we remark that since f_j is continuous on a compact set, it is in fact uniformly continuous, and thus q_j is a real-valued continuous function on K_j^-. Let $\varepsilon > 0$ and consider the set

$$H_\varepsilon = \{u, u \in K, f_j(u) \geq q_j(u_j) - \varepsilon\}, \qquad u_j = \mathrm{pr}_{K_j} u.$$

It is clear that H_ε is a compact subset of K and decreases as ε decreases to zero. Let $H_0 = \bigcap_\varepsilon H_\varepsilon$ and we remark that H_0 is exactly the set of all points $u \in K$ for which the assertion of the theorem holds. Thus, in order to prove the theorem it will suffice to prove that for each ε, H_ε is a nonempty set, which of course implies that H_0 is a nonempty set.

Let $\varepsilon > 0$ and consider the set

$$S_j = \{u, u \in K, f_j(u_j, u_{\bar{j}}) > q_j(u_j) - \varepsilon\}$$

and since f_j is a quasi-concave function in u_j it follows that for each $u_{\bar{j}} \in K_j$. $S_j(u_{\bar{j}}) = \{y_j, y_j \in K_j, f_j(y_j, u_{\bar{j}}) > q_j(u_{\bar{j}}) - \varepsilon\}$ is convex and from the definition of q_j is nonempty. Since for each $y_j \in K_j$,

$$S_j^{-1}(y_j) = \{u_{\bar{j}}, u_{\bar{j}} \in K_{\bar{j}}, f_j(y_j, u_{\bar{j}}) > q_j(u_{\bar{j}}) - \varepsilon\}$$

is open by the continuity of the functions involved, then all the conditions to apply Theorem 10.3.11 obtain. Then S_j is nonempty and any point which is in this intersection is necessarily in H_ε.

The theorem of Nash is thus proved.

In his important paper Ky Fan [1966] gives other applications of the above theorems. We mention some of them; the interested reader may consult Ky Fan's paper.

Among the applications we mention here the Hardy–Littlewood theorem about doubly stochastic matrices, Sion's minimax theorem, Iohvidov's fixed point theorem; also a theorem about the extension of monotone sets (i.e. sets M in a product of two linear spaces E and F such that

$$\langle x_1 - x_2, y_1 - y_2 \rangle \geqq 0$$

for any $(x_i, y_i) \in M, i = 1, 2$.) is obtained, as well as an analogue of a theorem of Helly about the intersection of convex sets.

10.4. SET-VALUED CONTRACTION MAPPINGS

Let (X, d) be a complete metric space and δ be the Pompeiu–Hausdorff metric on the family of all bounded sets in X. We denote by $CB(X)$ the family of all nonempty closed and bounded sets in X. As we know this family forms a metric space with the Pompeiu–Hausdorff metric. Sometimes, we denote the Pompeiu–Hausdorff metric, associated with the space X, by δ_X. Now, following S. Nadler, we define the notion of set-valued Lipschitz mapping as follows.

DEFINITION 10.4.1. Let (X, d_X) and (Y, d_Y) be two complete metric spaces and $f : X \to CB(Y)$ be a set-valued mapping. The set-valued mapping f is said to be a Lipschitz mapping if for all $x, y \in X$ there exists a constant k such that

$$\delta_Y(f(x), f(y)) \leqq k d_X(x, y).$$

We say that f is a set-valued contraction if $k < 1$.

The following result easily follows from the definition of Lipschitz mappings and the compactness.

LEMMA 10.4.2.   *Let $f : X \to K(Y)$ be a Lipschitz mapping. Then for any compact subset of X, say K,*

$$\bigcup_{x \in K} f(x)$$

is a compact subset of Y, i.e. it is in $K(Y)$.

Also, from the definition of Pompeiu–Hausdorff metric we have the following

LEMMA 10.4.3.   *Let F_1, F_2 lie in $K(X)$ and let $f : X \to K(Y)$ be a set-valued Lipschitz mapping with the Lipschitz constant k. In this case*

$$\delta\left(\bigcup_{x \in F_1} f(x), \bigcup_{y \in F_2} f(y)\right) \leq k\delta_X(F_1, F_2).$$

The following example shows that the intersection of two set-valued contraction mappings is not necessarily a continuous function (and where the intersection is defined as the pointwise intersection of the values, supposing, of course, that at each point, the intersection is nonempty).

Example 10.4.4.   Let $I^2 = \{(x, y), 0 \leq x, y \leq 1\}$ and consider

$$F : I^2 \to CB(I^2)$$

defined as follows: $F(x, y)$ is the line segment in I^2 from the point $(x/2, 0)$ to the point $(x/2, 1)$ and

$$G : I^2 \to CB(I^2)$$

is defined as follows: $G(x, y)$ is the line segment in I^2 from the point $(x/2, 0)$ to $(x/3, 1)$.

In this case F and G are set-valued contraction mappings and for each (x, y) in I^2 the set-valued mapping $F \cap G$ is defined as follows:

$$F \cap G(x, y) = \begin{cases} (x/2, 0) & x \neq 0 \\ (x, y) & x = 0. \end{cases}$$

Then $F \cap G$ is not continuous.

The following lemma gives a method to construct Lipschitz mappings.

LEMMA 10.4.5. *Let X be a Banach space and let*

$$f : \mathrm{CB}(X) \to \mathrm{CB}(X)$$

be defined as follows : for each $F \in \mathrm{CB}(X)$, $f(F)$ is the convex hull of F. In this case,

$$\delta_X(F_1, F_2) \geqq \delta_X(f(F_1), f(F_2)).$$

Proof. Let $\varepsilon > 0$ and $p \in \mathrm{conv}\, F_1$. Then there exist $f_1, \ldots, f_n$ and $t_1, \ldots, t_n$ in $[0, 1]$ such that

1. $\qquad \sum t_i = 1,$

2. $\qquad \|p - \sum f_i t_i\| \leqq \varepsilon.$

From the definition of Pompeiu–Hausdorff metric it follows that for each $i \in [1, n]$ there exists a $g_i \in F_2$ such that

$$\|f_i - g_i\| \leqq \delta_X(F_1, F_2) + \varepsilon/2.$$

The point

$$q = \sum t_i g_i$$

is obviously in $\mathrm{conv}\, F_2$ and thus

$$\|p - q\| \leqq \|p - \sum f_i t_i\| + \|\sum t_i(f_i - g_i)\| + \|q - \sum t_i g_i\|$$
$$\leqq \delta_X(F_1, F_2) + \varepsilon.$$

This gives that

$$d_X(\mathrm{conv}\, F_1, \mathrm{conv}\, F_2) \leqq \delta_X(F_1, F_2) + \varepsilon$$

and similarly we can prove that

$$d_X(\mathrm{conv}\, F_2, \mathrm{conv}\, F_1) \leqq \delta_X(F_1, F_2) + \varepsilon$$

which gives that, since ε is arbitrary,

$$\delta_X(\mathrm{conv}\, F_1, \mathrm{conv}\, F_2) \leqq \delta_X(F_1, F_2).$$

As a consequence of this Lemma we have the following

THEOREM 10.4.6. *Let X be a Banach space and let C be a closed convex subset of X. Further let $f : C \to \mathrm{CB}(C)$ be a set-valued Lipschitz mapping with the Lipschitz constant k. We define a new set-valued mapping $f^{\sim}$ by the relation*

$$f^{\sim}(x) = \mathrm{conv}\, f(x).$$

In this case $f^{\sim}$ is a set-valued Lipschitz mapping with the Lipschitz constant k.

The following theorem of Nadler (1969) represent the extension of the contraction principle to the case of set-valued contraction mappings.

THEOREM 10.4.7. Let (X, d) be a complete metric space and let $f : X \rightarrow CB(X)$ be a set-valued contraction mapping. Then f has a fixed point.
Proof. Suppose that the Lipschitz constant of f is denoted by k.
Let p_0 be an arbitrary point in X and choose $p_1 \in f(p_0)$. Now, since $f(p_0)$ and $f(p_1)$ are in $CB(X)$ and $p_1 \in f(p_0)$ we find a point $p_2 \in f(p_1)$ such that

$$d(p_1, p_2) \leq k + \delta_X(f(p_0), f(p_1)).$$

Similarly, since $f(p_1), f(p_2)$ are in $CB(X)$ we find a point $p_3 \in f(p_2)$ such that

$$d(p_2, p_3) \leq k^2 + \delta_X(f(p_1), f(p_2)).$$

We can continue this process to obtain a sequence $\{p_i\}$ of points in X such that

$$d(p_i, p_{i+1}) \leq k^i + \delta_X(f(p_{i-1}), f(p_i))$$

for all $i \geq 1$.
We show now that the sequence $\{p_i\}$ is a Cauchy sequence. Indeed, for any i we have,

$$\begin{aligned}
d(p_i, p_{i+1}) &\leq \delta_X(f(p_{i-1}), f(p_i)) + k^i \\
&\leq k d(p_{i-1}, p_i)) + k^i \\
&\leq k(\delta_X(f(p_{i-2}), f(p_{i-1})) + k^{i-1}) + k^i \\
&\leq k^2 d(p_{i-2}, p_{i-1}) + 2k^i \leq \cdots \\
&\leq k^i d(p_0, p_1) + ik^i.
\end{aligned}$$

This implies that for any i and j,

$$\begin{aligned}
d(p_i, p_{i+j}) &\leq d(p_i, p_{i+1}) + \cdots + d(p_{i+j-1}, p_{i+j}) \\
&\leq k^i d(p_0, p_1) + ik^i + k^{i+1} d(p_0, p_1) + \\
&\quad + (i+1)k^{i+1} + \cdots + k^{i+j-1} d(p_0, p_1) + \\
&\quad + (i+j-1)k^{i+j-1} \\
&= \left(\sum_{n=i}^{i+j-1} (k^n d(p_0, p_1) + nk^n) \right)
\end{aligned}$$

and this gives obviously that $\{p_i\}$ is a Cauchy sequence. Since X is complete, let $p = \lim p_i$. In this case, from the continuity of set-valued contraction mappings, it follows that $\{f(p_i)\}$ converges to $f(p)$. Since $p_{i+1} \in f(p_i)$, we have that $p \in f(p)$ and the existence of fixed points for set-contraction mappings is proved.

In the case of metric spaces which are ε-chainable we can define the (ε, λ)-uniformly local contractive set-valued mappings as follows.

DEFINITION 10.4.8. Let (X, d_X) be a complete metric space. A set-valued mapping $f : X \to CB(X)$ is called an (ε, λ)-uniformly local contraction if, whenever $x, y \in X$ such that $d(x, y) < \varepsilon$, then $\delta_X(f(x), f(y)) \leq \lambda d(x, y)$.

THEOREM 10.4.9. Let (X, d_X) be a complete ε-chainable space and $f : X \to CB(X)$ be a set-valued mapping which is an (ε, λ)-uniformly local contraction. Then f has a fixed point.

Proof. We define now a new metric on X by the following formula: for any $x, y \in X$, we set

$$d\tilde{\,}(x, y) = \inf \left\{ \sum_{i=1}^{n} d(x_{i-1}, x_i) \right\}$$

where $x_0 = x$ and $x_n = y$, $(x_0, \ldots, x_n)$ is an ε-chain.

The fact that this is a metric on X is easy to see. Also we remark that

1. $d(x, y) \leq d\tilde{\,}(x, y)$,

2. $d(x, y) = d\tilde{\,}(x, y)$ if $d(x, y) < \varepsilon$.

In this case it follows that $(X, d\tilde{\,})$ is a complete metric space.
We denote by $\delta_{\varepsilon, X}$ the Pompeiu–Hausdorff metric induced by $d\tilde{\,}$.
We show now that f is a set-valued contraction mapping with respect to the metrices $d\tilde{\,}$ and $\delta_{\varepsilon, X}$.
Let $x, y \in X$ and consider an ε-chain,

$$x_0 = x, x_1, \ldots, x_{n-1}, x_n = y.$$

Since this is an ε-chain we have $d(x_i, x_{i+1}) < \varepsilon$ for $i = 0, 1, \ldots, n-1$, and since f is an (ε, λ)-uniformly locally contraction we have $\delta_X(f(x_i), f(x_{i+1})) \leq \lambda d(x_i, x_{i+1})$. In this case we obtain,

$$\delta_{\varepsilon,x}(f(x), f(y)) \leqq \sum_{i=0}^{n} \delta_{\varepsilon,x}(f(x_i), f(x_{i+1}))$$

$$= \sum_{i=0}^{n-1} \delta_x(f(x_i), f(x_{i+1}))$$

$$\leqq \lambda \sum_{i=0}^{n-1} d(x_i, x_{i+1})$$

and since this is an arbitrary ε-chain we have

$$\delta_{\varepsilon,x}(f(x), f(y)) \leqq \lambda d^{\sim}(x, y).$$

Now we can apply Theorem 10.4.7 and the existence of fixed points is proved.

As we know there exists a great number of generalizations of the notion of contraction mapping. It is natural to ask about the classes of set-valued mappings containing as particular cases the set-valued contraction mappings. In what follows we propose some classes.

DEFINITION 10.4.10. Let (X, d) be a complete metric space and $f : X \to CB(X)$. We say that f is said to be K-contractive if there exist $\alpha, \beta \in [0, 1]$, $\alpha + \beta \leqq 1$, such that,

$$\delta_x(f(x), f(y)) \leqq \alpha \delta_x(\{x\}, f(x)) + \beta \delta_x(\{y\}, f(y))$$

for all $x, y \in X$.

DEFINITION 10.4.11. Let (X, d) be a complete metric space and $f : X \to CB(X)$ be a continuous set-valued mapping. We say that f is densifying if for any bounded noncompact set $A \subset X$,

$$\alpha_{\delta_x}\left(\bigcup_{x \in A} f(x)\right) < \alpha_d(A)$$

and if

$$\alpha_{\delta_x}\left(\bigcup_{x \in A} f(x)\right) \leqq k\alpha_d(A)$$

f is said to be α-set Pompeiu–Hausdorff contraction.

Another class of set-valued mappings can be considered using the notion of orbit.

DEFINITION 10.4.12. Let X be a metric space and $f : X \to P(X)$ be a set-valued mapping. Let $x \in X$ and an orbit of x is any sequence $\{x_n\}$ such that $x_n \in f(x_{n-1})$, $n \geq 1$, $x_0 = x$.

It is clear from the definition that a point $x \in X$ may have many orbits.

DEFINITION 10.4.13. The set-valued mapping f is said to have diminishing orbital diameters if, whenever each orbit of x has a positive diameter, then

$$\lim_{n \to \infty} \operatorname{diam}(x_k, k \geq n) < \inf \{\operatorname{diam}(O(x)), \text{ for each } O(x)\}$$

where $\operatorname{diam}(A)$ means the diameter of the set A and $O(x)$ is an orbit of x.

Another class of set-valued mappings may be defined as follows.

DEFINITION 10.4.14. Let X be a complete metric space and f be a set-valued mapping $f : X \to P(X)$. Suppose that a real lower semicontinuous function F is defined on $\operatorname{CB}(X)^2$. The mapping f is said to be Pompeiu–Hausdorff F-contractive if the condition $F(f(x), f(y)) < F(\{x\}, \{y\})$ holds for all $x, y \in X$. If the function F is equal to Pompeiu–Hausdorff metric then we say that f is a weak-Pompeiu–Hausdorff contraction.

We give now some results about the fixed points for set-valued mappings using the so-called Liapunov functions

First we give the following

DEFINITION 10.4.15. A set-valued mapping f is called sequentially continuous on the orbit $O(x)$ if and only if there is a convergent sequence $x_{n_i} \to y_0$ of $O(x)$ such that for each k there is an element $y_k \in f(y_{k-1})$ such that $x_{n_i+k} \to y_k$.

The following lemmas follow from the definition:

LEMMA 10.4.16. *If f is a set-valued mapping which is point closed and upper semi-continuous and $x_n \to x_0$, $y_n \to y_0$ with $y_n \in f(x_n)$ then $y_0 \in f(x_0)$.*

LEMMA 10.4.17. *If f is a set-valued Lipschitz and point compact mapping then f is continuous.*

LEMMA 10.4.18. *Let f be a point compact set-valued mapping which is also upper semi-continuous and $x_n \to x_0$, $y_n \in f(x_n)$ for each n. Then there is a $y_0 \in f(x)$ and a subsequence $x_{n_i} \to y_0$.*

LEMMA 10.4.19. *Let f be a lower semi-continuous mapping and suppose either that $O(x)$ contains a convergent subsequence or that f is point closed, upper semi-continuous and X compact. Then f is sequentially continuous.*

We define now an important notion, which plays a fundamental role in the proofs of the fixed point theorems for set-valued mappings.

DEFINITION 10.4.20.   Let $x \in X$ and $O(x)$ be an orbit of the set-valued mapping f. Then a Liapunov function for f on $O(x)$ is a continuous single-valued function $V : X \to \mathbf{R}$ with the following properties:

1. $V(x_{n+1}) < V(x_n)$, $x_n \in O(x)$,

2. V is bounded below.

We give now an example of a Liapunov function.

Example 10.4.21.   Let $X = [0, 1]$ and define the set-valued mapping f as follows: $f(x) = [0, kx]$ where $k \in (0, 1)$. Let x_0 be an arbitrary point of $(0, 1)$ and $O(x) = \{x_n, x_n = k^n x_0\}$. As a Liapunov function we can take $V(x) = x$ and it is easy to see that it satisfies all the above properties.

The following lemma gives an important and useful property of Liapunov functions:

LEMMA 10.4.22.   *Let V be a Liapunov function on the orbit $O(x)$ associated with a set-valued mapping f. If there is a convergent subsequence $x_{n_i} \to y_0$ of $O(x)$ such that $x_{n_i+1} \to y_1$ then $V(y_0) = V(y_1)$.*
    Also, if $x_{n_i+k} \to y_k$ then $V(y_0) = V(y_k)$.
    Proof. Since $x_{n_i} \to y_0$, and since V is continuous, we have $V(x_{n_i}) \to V(y_0)$ and, since $x_{n_i+1} \to y_1$, $V(x_{n_i+1}) \to V(y_1)$. But $V(x_{n_i+1}) \leq V(x_{n_i})$ and thus $V(y_1) \leq V(y_0)$. On the other hand, $V(x_{n_j}) \leq V(x_{n_i+1})$ for $j \geq i$ and this implies that $V(y_0) \leq V(y_1)$. Thus $V(y_0) = V(y_1)$. An induction argument proves the last assertion.

The next theorem is an easy consequence of this lemma and has important corollaries.

THEOREM 10.4.23.   *Let $f : X \to P(X)$ and $O(x)$ be an orbit of f at x with convergent subsequences $x_{n_i} \to y_0$, $x_{n_i+k} \to y_k$ for $k = 1, 2, \ldots, m$. If there is a Liapunov function V on $O(x)$ such that $y_0 \neq y_k$ implies $V(y_0) \neq V(y_k)$ for some $k \in [1, m]$ then $y_0 = y_k$. Further, if $k = 1$ and $y_1 \in f(x_0)$ then y_0 is a fixed point of f.*

    Proof. From the lemma we have $V(y_0) = V(y_k)$ and the hypothesis of the theorem implies that $y_0 = y_1$.

In the following theorem we give a method to construct Liapunov functions.

THEOREM 10.4.24. *Let $f : X \to P(X)$ be a point compact set-valued mapping which is also continuous. Then the function*

$$V(x) = d(x, f(x))$$

is a continuous function.

Proof. Since $f(x)$ is a compact set we find $y \in X$ such that $d(x, f(x)) = d(x, y)$. Suppose that $x_n \to x_0$ and let $y_n \in f(x_n)$ such that $d(x_n, f(x_n)) = d(x_n, y_n)$. Since f is upper semi-continuous and point compact, some subsequence $y_{n_i} \to y_0 \in f(x_0)$. In this case, since $d(x_0, f(x_0)) = d(x_0, y_0)$ the continuity of the metric d implies that $V(x_n) \to V(x_0)$. Suppose now that there is a $y' \in f(y_0)$ such that $d(x_0, y') < d(x_0, y_0)$ and let $2\varepsilon = d(x_0, y_0) - d(x_0, y')$. Since f is lower semi-continuous, we find $z_n \in f(x_n)$ such that $d(y', z_n) < \varepsilon$ and this clearly implies that

$$d(x_n, f(x_n)) \leqq d(x_n, z_n) < d(x_n, y_n) = d(x_n, f(x_n))$$

which is a contradiction and the theorem is proved.

From this theorem we obtain the following result, given as

COROLLARY 10.4.25. *Let $O(x)$ be an orbit of a set-valued point compact and continuous function. Suppose that $O(x)$ contains a convergent subsequence. If the function $V(x) = d(x, f(x))$ is decreasing on $O(x)$ and $V(x_n) \to 0$ then f has a fixed point.*

Proof. By the above result V is a Liapunov function on $O(x)$ and if $x_{n_i} \to y_0$ we may assume that $x_{n_i+1} \to y_1 \in f(x_0)$. Then the theorem gives the assertion.

The following fixed point theorem refers to the case when the orbit has a special feature.

DEFINITION 10.4.26. An orbit $O(x)$ of a set-valued mapping is called regular if and only if

$$d(x_n, x_{n+1}) = d(x_n, f(x_n))$$

for all n.

THEOREM 10.4.27. *Let $f : X \to P(X)$ be a point compact contractive set-valued mapping and let there exist an orbit $O(x)$ such that the mapping contains a convergent subsequence. Then f has a fixed point.*

Proof. Let $O(x)$ be the orbit and suppose that $x_{n_i} \to y_0$ and $x_{n_i+1} \to y_1 \in f(x_0)$. In this case the function $V(x) = d(x, f(x))$ is continuous and, moreover, it is a Liapunov function. Now if $y_0 \neq y_1$ then, since f is contractive, $V(y_0) < V(y_1)$ and thus we have $y_0 = y_1$ and f has a fixed point by the above result.

10.5. SEQUENCES OF SET-VALUED MAPPINGS AND FIXED POINTS

It is a natural problem to ask about the convergence of sequences of set-valued mappings and the fixed points of the limit set-valued mapping. In what follows we give some results concerning this problem. For the case when different metrics are considered we refer to the paper of Fraser and Nadler (1969). Also we note that the result which we give for sequences of set-valued mappings was obtained by S. Nadler.

First we give an example to show that, without some assumptions about the ranges of the set-valued mappings, we have no convergence for any subsequence.

Indeed, let $X = \mathbf{R}$ and for each $x \in X$,

$$f_n(x) = \mathbf{R}$$

for $n = 0, 1, 2, 3, \ldots$, It is clear that $x_n = n$ is a fixed point of f_n for each $n = 0, 1, 2, \ldots$. Now it is obvious that no subsequence of the sequence $\{x_n\}_1^\infty$ has a subsequence converging to $x_0 = 0$.

For this reason we suppose that the values of the set-valued mapping are compact sets, i.e., we consider only point compact set-valued mappings.

THEOREM 10.5.1.   *Let (X, d) be a complete metric space and let δ_X be the Pompeiu–Hausdorff metric induced by d. Suppose that $\{f_n\}_1^\infty$ is a sequence of set-valued mappings defined on X, and that $f_n(x)$ form a compact subset of X and are contraction mappings.*

Let f_0 be another point compact set-valued contraction. Suppose that one of the following conditions are satisfied:

1.   *all the mappings f_n, $n \geq 1$ have the same Lipschitz constant $k > 1$ and the sequence $\{f_n\}_1^\infty$ converges pointwise to f_0,*

2.   *the sequence $\{f_n\}_1^\infty$ converges uniformly to f_0,*

3.   *the space (X, d) is locally compact and the sequence $\{f_n\}_1^\infty$ converges pointwise to f_0.*

In this case there is a subsequence $\{x_{n_i}\}$ of $\{x_n\}$ such that $x_{n_i} \to x_0$, x_i is a fixed point for f_i, $i = 0, 1, 2, \ldots$.

For the proof we need the following

LEMMA 10.5.2.   *Let (X, d) be a complete metric space and let $f_n : X \to P(X)$ be a sequence of contraction mappings with the fixed point x_n, $n = 1, 2, 3, \ldots$ and let $f_0 : X \to P(X)$ be another contraction mapping. Suppose that $\{f_n\}_1^\infty$ converges pointwise to f_0 and that $\{x_{n_i}\}$ is a convergent subsequence $\{x_n\}$. In this case $\{x_{n_i}\}$ converges to a fixed point of f_0.*

Proof. Let $x_0 = \lim x_{n_i}$ and choose an arbitrary $\varepsilon > 0$. We then choose an integer N_ε such that for all $n \geq N_\varepsilon$,

$$\delta_X(f_n(x_0), f_0(x_0)) \leq \varepsilon/2$$

which is possible since the sequence $\{f_n\}$ converges pointwise to f_0. Also we can suppose that $d(x_0, x_{n_i}) \leq \varepsilon/2$ for all $i \geq N_\varepsilon$.

Now if $i \geq N_\varepsilon$ we have

$$\delta_X(f_{n_i}(x_{n_i}), f_0(x_0)) \leq \delta_X(f_{n_i}(x_{n_i}), f_{n_i}(x_0)) +$$
$$+ \delta_X(f_{n_i}(x_0), f_0(x_0))$$
$$\leq \varepsilon/2 + \varepsilon/2 = \varepsilon$$

and this gives us that

$$\lim f_{n_i}(x_{n_i}) = f_0(x_0).$$

Since $x_{n_i} \in f_{n_i}(x_{n_i})$, for all $i = 1, 2, 3, \ldots$, it follows that x_0 is in $f_0(x_0)$ and the lemma is proved.

Using this lemma and other results proved for the case of single-valued contractions we can prove Theorem 10.5.1. as follows: For each n we define the mapping $\tilde{f_n}$ as follows:

$$\tilde{f_n}(A) = \bigcup \{f_n(x) : x \in A\}$$

and thus $\tilde{f_n}$ is defined on $K(X)$ with values in $K(X)$ and it is a contraction mapping. Then it has a fixed point, say A_n, which is in $K(X)$ (the family of all nonempty compact subsets of X). Suppose now that conditions 1 or 3 are satisfied. In this case the sequence $\{f_n\}$ converges uniformly on the compact subsets of X and thus $\{\tilde{f_n}\}$ converges pointwise to $\tilde{f_0}$. Now if we have uniform convergence then we have uniform convergence for the sequence $\{\tilde{f_n}\}$, i.e. if condition 2 is satisfied then we have again uniform convergence for the mappings $\tilde{f}$.

Then the sequence of compact sets $\{A_n\}_1^\infty$ converges to the compact set

A_0. In this case the set $S = \bigcup_{n=0}^{\infty} A_n$ is a compact subset of X and

$$\{f_n^{\tilde{}\ m}(x_n)\}$$

converges to A_n for $m \to \infty$. Since $x_n \in f_n^{\tilde{}\ m}(x_n)$ it follows that x_n is in A_n and thus $\{x_n\}$ is a compact set. Then there exists a convergent subsequence converging to a fixed point of f_0 and the theorem is proved.

Remark 10.5.3. The example given for the case of single-valued mappings, in the case of local compactness, shows that the local compactness in condition 2 is necessary.

Chapter 11

Fixed Point Theorems for Mappings on PM-Spaces

11.0. INTRODUCTION

The notion of metric space was introduced by M. Fréchet in 1906 and is a natural setting for many problems.

An essential feature is the fact that, for any two points in the space, there is defined a positive number called the distance between the points. However, in practice we find very often that this association of a single number for each pair of points is, strictly speaking, an over-idealization.

The idea thus appears that, instead of a single positive number, we should associate a distribution function with the point pairs. Thus, for any p, q, elements in the space, we have a distribution function $F_{p,q}(x)$ and we interpret $F_{p,q}(x)$ as the probability that the distance between p and q is less than x.

This concept was first introduced by K. Menger in 1942 under the name 'statistical metric spaces'. However, we use the name 'probabilistic metric spaces' or 'PM-spaces' for short.

The important paper of B. Schweizer and A. Sklar (1960) has given a new impulse to the theory of PM-spaces. For a detailed discussion of PM-spaces we refer to V. Istrățescu (1974).

For the reader's convenience, we give here some fundamental notions from the theory of PM-spaces, since these are used in the fixed point theorems, which we shall also give.

11.1. PM-SPACES

First we recall that a real-valued function defined on the set of real numbers is a distribution function if it is nondecreasing, left continuous and $\inf f(x) = 0$, $\sup f(x) = 1$. In what follows, $H(x)$ denotes the distribution function defined as follows:

$$H(x) = \begin{cases} 0 & x \le 0 \\ 1 & x > 0 \end{cases}.$$

DEFINITION 11.1.1.   A probabilistic metric space or PM-space, is a pair (S, F) where S is a set and F is a function defined on $S \times S$ into the set of distribution functions, such that if p, q and r are points of S, then

1. $F_{pq}(0) = 0$,   (F_{pq} denotes $F(p, q)$);

2. $F_{pq}(x) = H(x)$ if and only if $p = q$;

3. $F_{pq} = F_{qp}$;

4.   if $F_{pq}(x) = 1$ and $F_{qr}(y) = 1$,   then $F_{pr}(x + y) = 1$.

For each p and q in S and each real number x, $F_{pq}(x)$ is to be thought of as the probability that the distance from p to q is less than x.

We remark that any metric space may be regarded as a PM-space. Indeed, if (X, d) is a metric space, then we define, for each p and q in X, the distribution function F_{pq} as follows:

$$F_{pq}(x) = H(x - d(p, q))$$

It is not difficult to prove that we have in fact a PM-space.

Axiom 4 is the so called 'triangle inequality' for the PM-spaces. An important class of PM-spaces is the class which was originally considered by K. Menger.

DEFINITION 11.1.2.   A Menger PM-space is a triple (S, F, T), where S is a set, F is a function defined on the set $S \times S$ with values in the set of distribution functions such that:

1. $F_{pq}(0) = 0$ (F_{pq} denotes $F(p, q)$);

2. $F_{pq}(x) = H(x)$ if and only if $p = q$;

3. $F_{pq} = F_{qp}$;

4_m. $F_{pr}(x + y) \geqq T(F_{pq}(x), F_{qr}(y))$

and where T is a 2-place function on the unit square satisfying:

1. $T(0, 0) = 0$, $T(a, 1) = a$,

2. $T(a, b) = T(b, a)$,

3.   if $a \leqq c, b \leqq d$ then $T(a, b) \leqq T(c, d)$,

4. $T(T(a, b), c) = T(a, T(b, c))$.

Let us consider Δ^+ to be the set of all distribution functions and we consider now a class of functions defined on $\Delta^+ \times \Delta^+$ with values in Δ^+ which gives us the possibility of obtaining a generalization of the notion of PM-spaces. This method is due to Šerstnev.

DEFINITION 11.1.3.   A two place function $\tau : \Delta^+ \times \Delta^+ \to \Delta^+$ is called a triangle function if, for all F, G, K in Δ^+,

1. $\tau(F, H) = F$,

2. $\tau(F, G) \geq \tau(K, G)$ if $F \geq K$,

3. $\tau(F, G) = \tau(G, F)$,

4. $\tau(F, \tau(G, K)) = \tau(\tau(F, G), K)$.

DEFINITION 11.1.4.   A triangle function τ is continuous if it is a continuous function from $\Delta^+ \times \Delta^+ \to \Delta^+$ where Δ^+ is endowed with the topology induced by the modified Levy metric and if we take the product topology on Δ^+.

The modified Levy metric, introduced by D. Sibley, is defined as follows: for any F and G in Δ^+ and $h > 0$, we set

$$A(F, G, h) \leftrightarrow F(x - h) - h \leq G(x) \quad \text{for} \quad x \in [0, 1/h + h)$$

$$B(F, G, h) \leftrightarrow F(x + h) + h \geq G(x) \quad \text{for} \quad x \in [0, 1/h)$$

and the modified Levy metric, $L(F, G)$ is defined as

$$L(F, G) = \inf\{h, A(F, G, h) \text{ and } B(F, G, h) \text{ hold}\}.$$

Using the triangle function τ we can introduce the notion of PM-space under τ as follows:

DEFINITION 11.1.5.   A probabilistic metric space under τ is a triple (S, F, τ), where S is a set, F is a function defined on $S \times S$ with values in Δ^+, satisfying Axioms 1–3 of Definition 11.1.2 and the axiom

4$_s$. $F_{pr} \geq \tau(F_{pq}, F_{qr})$.

We mention that if T is as in Definition 11.1.2, then

$$\tau_T(F, G)(x) = \sup_{u + v = x} \{T(F(u), G(v))\}$$

and

$$\pi_T(F, G)(x) = T(F(x), G(x))$$

are examples of triangle functions. As the functions T we note the following examples:

1. $T(a, b) = \min(a, b)$.

2. $T(a, b) = ab$,

3. $T_m(a, b) = \max\{a + b - 1, 0\}$.

We use the notion of convergence in PM-spaces as introduced by Schweizer and Sklar (1960).

DEFINITION 11.1.6.   A sequence of points $\{p_n\}$ in a PM-space, converges to p if for every $\varepsilon > 0$ and $\lambda > 0$ there exists an integer $N_{\varepsilon, \lambda}$ such that

$$F_{p p_n}(\varepsilon) > 1 - \lambda$$

for all $n \geq N_{\varepsilon, \lambda}$.
  It is not difficult to see that this is equivalent to the following assertion:

THEOREM 11.1.7.   *A sequence $\{p_n\}$ converges to p if and only if for every $x \in \mathbf{R}$*

$$\lim F_{p p_n}(x) = H(x).$$

The notion of a Cauchy sequence is introduced as follows:

DEFINITION 11.1.8.   A sequence of points in a PM-space is said to be a Cauchy sequence if for any $\varepsilon > 0$ and $\lambda > 0$ there exists an integer $N_{\varepsilon, \lambda}$ such that, for all $n, m \geq N_{\varepsilon, \lambda}$,

$$F_{p_n p_m}(\varepsilon) \geq 1 - \lambda.$$

The notion of complete PM-spaces is defined as follows:

DEFINITION 11.1.9.   A PM-space is called complete if every Cauchy sequence is convergent to some point of the space.

11.2. CONTRACTION MAPPINGS IN PM-SPACES

In what follows we suppose that all the PM-spaces are complete.
  The notion of a contraction mapping defined on a PM-space was first defined by V. M. Sehgal who has also proved that every contraction mapping in a complete Menger space (i.e. a space in the sense of Definition 11.1.2) has a unique fixed point.

DEFINITION 11.2.1.   Let (S, F) be a PM-space and $f : S \to S$ be an arbitrary mapping defined on S. Then f is called a contraction if there exists $k \in (0, 1)$ such that for all p and q in S the following relation holds:

$$F_{f(p)f(q)}(x) \geq F_{pq}(x/k).$$

We can prove the following

THEOREM 11.2.2. *Every contraction mapping has at most one fixed point.*

Proof. Suppose that f is a contraction mapping and has two fixed points, p and q, $p \neq q$.

In this case we get

$$F_{f(p)f(q)}(x) \geq F_{pq}(x/k) = F_{f(p)f(q)}(x/k)$$
$$\geq F_{pq}(x/k^2) \cdots$$

and thus, for each $x \in \mathbf{R}$,

$$F_{pq}(x) \geq \lim F_{pq}(x/k^n) = H(x)$$

and this contradiction proves the assertion.

Let f be a contraction mapping defined on a PM-space and let p_0 be an arbitrary but fixed point in S. Following Sherwood, we define a function which plays a fundamental role in the following fixed point theorems.

Given p_0 in S we consider the sequence of iterations $\{p_m\}$, $p_{m+1} = f(p_m)$. $m = 0, 1, 2, \ldots$, and we set

$$G_{p_0}(x) = \inf \{F_{p_0 p_m}(x), m = 0, 1, 2, \ldots \}.$$

THEOREM 11.2.3. *Let (S, F, τ) be a complete PM-space under τ, where τ satisfies the condition*

$$\sup \{\tau(F, F), F < H\} = H$$

and let $f : S \to S$ be a contraction mapping. In this case, either

1. \ *f has a unique fixed point,*

or

2. *for every $p_0 \in S$, $\sup \{G_{p_0}(x)\} < 1$.*

Proof. Suppose that we find $p_0 \in S$ such that 2 does not hold.

In this case, the sequence of iterations, $\{p_m\}$ of p_0, has the property that

$$F_{p_n p_{n+m}}(x) \geq F_{p_0 p_m}(x/k^n) \geq G_{p_0}(x/k^n)$$

and since G_{p_0} is nondecreasing, we get

$$\lim F_{p_n p_{n+m}}(x) = 1$$

for all $x \in \mathbf{R}$. This gives us that $\{p_m\}$ is a Cauchy sequence. Since (S, F, τ) is complete, let $p = \lim p_m$.

We show that $f(p) = p$. Indeed, for any integer m, we have

$$F_{f(p)\,p}(x) \geq \tau(F_{f(p)\,p_m}, F_{p_m\,p})(x)$$
$$\geq \tau(F_{p\,p_{m-1}}(j/k), F_{p_m\,p})(x)$$

where $j(x) = x$, for all $x \in \mathbf{R}$. Thus we get,

$$F_{f(p)\,p}(x) \geq \lim \tau(F_{p\,p_{m-1}}(j/k), F_{p_m\,p})(x)$$

and using the continuity of τ, we obtain,

$$F_{f(p)\,p} = H$$

and this gives us that p is a fixed point of f.

From this we obtain the following result for the case of Menger spaces:

THEOREM 11.2.4. *Let (S, F, T) be a complete Menger space and let f be a contraction mapping $f : S \to S$. Then f has a unique fixed point when $T(a, b) = \min(a, b)$.*

Proof. Since f is a contraction mapping there exists a $k \in (0, 1)$ such that

$$F_{f(p)\,f(q)}(x) \geq F_{pq}(x/k)$$

for all p and q in S. Thus, if p_0 is given, then for the sequence of iterations $\{p_m\}$ we get,

$$F_{p_p\,p_m}(x) \geq F_{p_0\,p_m}((1-k)(1+k+\cdots+k^{m-1})x)$$
$$\geq \min \{F_{p_0\,p_1}((1-k)x), \ldots, F_{p_{m-1}\,p_m}(k^{m-1}(1-k)x))$$
$$= F_{p_0\,p_1}((1-k)x)$$

and thus

$$G_{p_0}(x) \geq F_{p_0\,p_1}((1-k)x)$$

and the assertion follows now from Theorem 11.2.3.

We consider now a class of mappings which, for a particular value of the parameter, reduce to the contraction mappings.

DEFINITION 11.2.5. Let (S, F) be a PM-space and $f : S \to S$ be a mapping.

The mapping f is called a power contraction mapping if there exists an integer n_0 such that the iteration f^{n_0} is a contraction mapping.

We show now that Theorem 11.2.3. holds also for this class.

THEOREM 11.2.6.   *Let (S, F, τ) be a complete PM-space under τ, where τ is as in Theorem 11.2.3, and let $f : S \to S$ be a power contraction mapping.*
In this case either:

1.   *f has a unique fixed point;*

or

2.   *for every p_0, $\sup \{G_{p_0}(x)\} < 1$.*

Proof. Let n_0 be the integer for which f^{n_0} is a contraction mapping. In this case according to Theorem 11.2.3, we have either conclusions 1 or 2 of that theorem. In the case when conclusion 1 holds, let p be the fixed point of f^{n_0}. Since

$$f(f^{n_0}(p)) = f^{n_0}(f(p)) = f(p)$$

and the uniqueness gives us that $f(p) = p$.

Suppose now that conclusion 2 of the Theorem 11.2.3 holds for f^{n_0}. In this case, we have,

$$G_{p_0}(x) \leqq G_{p_0}^{\sim}(x)$$

where $G_{p_0}^{\sim}$ denotes the function G_{p_0} which is defined with the mapping f^{n_0}. This clearly implies our assertion and thus the theorem is proved.

Remark 11.2.7.   Theorem 11.2.6 is true for pairs of commuting mappings, say f and g, such that either f or g is a contraction mapping.

We consider now new classes of mappings defined on PM-spaces.
For this we recall that if p is a point in a PM-space, then an (ε, λ)-neighbourhood of p is the set

$$U_p(\varepsilon, \lambda) = \{q, F_{pq}(\varepsilon) > 1 - \lambda\}.$$

As is well known an important class of metric spaces, introduced also by M. Fréchet, is the class of ε-chainable spaces.

DEFINITION 11.2.8.   A metric space (X, d) is ε-chainable if for any p and q in X there exists a finite sequence of points, $p_1, \ldots, p_n$, such that

$$d(p, p_1) \leqq \varepsilon, \quad d(p_1, p_2) \leqq \varepsilon, \ldots, d(p_{n-1}, p_n) \leqq \varepsilon,$$
$$d(p_n, q) \leqq \varepsilon.$$

The notion of ε-chainability for PM-spaces is defined similarly and this was first considered by Sehgal and Bharucha-Reid.

DEFINITION 11.2.9.   Let (S, F) be a PM-space. Then we say that this PM-space is (ε, λ)-chainable if for any p and q in S there exists the finite sequence $p_1 \ldots, p_n$ such that

$$p_1 \in U_{p_0}(\varepsilon, \lambda), \ldots, p_i \in U_{p_{i-1}}(\varepsilon, \lambda), \ldots, q \in U_{p_n}(\varepsilon, \lambda).$$

The following class of mappings represents a localization of the notion of contraction mapping.

DEFINITION 11.2.10.   Let (S, F) be a PM-space and let $f : S \rightarrow S$ be an arbitrary mapping. The mapping f is said to be an (ε, λ)-contraction if there exists $k \in (0, 1)$ such that for any $p \in S$,

$$F_{f(p)\, f(q)}(x) \geq F_{p\, q}(x/k)$$

for all q in $U_p(\varepsilon, \lambda)$.

THEOREM 11.2.11.   Let (S, F, T) be a complete Menger space with T satisfying the property that $T(x, x) \geq x$; and let $f : S \rightarrow S$ be an (ε, λ)-contraction. Then if (S, F, T) is (ε, λ)-chainable then f has a unique fixed point.

   Proof.   First we prove that for each $p \in S$ and for $x \in \mathbf{R}^+$ there exists an integer $N(p, x)$ such that

$$F_{f^m(p)\, f^{m+1}(p)}(x) > 1 - \lambda$$

for all $m \geq N(p, x)$.

   Since (S, F, T) is (ε, λ)-chainable, for $p = p_0$ and $q = f(p)$ we find the points

$$p_1, \ldots, p_n$$

such that

$$F_{p_{i+1} p_i}(\varepsilon) > 1 - \lambda, \qquad p_{n+1} = q.$$

Now since f is an (ε, λ)-contraction

$$F_{f(p_{i+1})\, f(p_i)}(\varepsilon) \geq F_{p_{i+1} p_i}(\varepsilon/k) > 1 - \lambda$$

and an induction argument shows that for any integer s, for the points $f^s(p_0)$ and $f^{s+1}(p)$ the sequence $\{f^s(p_1), \ldots, f^s(p_n)\}$ satisfies the definition of (ε, λ)-chainability.

   In this case for any $x > 0$ we get,

$$F_{f^s(p_{i+1})\, f^s(p_i)}(x) \geq F_{f^{s-1}(p_{i+1})\, f^{s-1}(p_i)}(x/k)$$
$$\geq \cdots \geq F_{p_{i+1} p_i}(x/k^s)$$

and now using Axiom 4_m we get,

$$F_{f^s(p_0) f^s(p_n)}(x) \geq T(F_{f^s(p_0)f^s(p_1)}(x/2),\, F_{f^s(p_1)f^s(p_n)}(x/2))$$

$$\geq T(F_{p_0 p_1}(x/2\, k^s),\, F_{f^s(p_1)\, f^s(p_n)}(x/2))$$

and from the property of T,

$$F_{f^s(p_0)\, f^s(p_n)}(x) \geq T(F_{p_0\, p_1}(x/2\, k^s),\, T(F_{p_1 p_2}(x/2^2\, k^s),$$

$$F_{f^s p_2\, f^s p_n}(x/2^2))).$$

If we set $d = x/2^n\, k^s$, we obtain,

$$F_{f^s(p_0)\, f^s(p_n)}(x) \geq T(F_{p_0 p_1}(d),\, T(F_{p_1 p_2}(d),$$

$$F_{f^s(p_2)f^s(p_n)}(x/2^2))).$$

We can repeat the above argument and we get,

$$F_{f^s(p_0)f^s(p_n)}(x) \geq T(F_{p_0 p_1}(d),\, T(F_{p_1 p_2}(d), \ldots$$

$$\ldots,\, T(F_{p_{n-2}p_{n-1}}(d),\, F_{p_{n-1}p_n}(d))))).$$

Now since n is a fixed integer, there exists $m_i > 0$ such that $F_{p_i p_{i+1}}(x/2^n k^s) > 1 - \lambda$ for each $s \geq m_i$, $i = 0, 1, \ldots, n-1$.

We define $N(p, x)$ as the number

$$\max\{m_0, \ldots, m_{n-1}\}$$

and it is clear that this number satisfies our assertion.

Using this fact the proof of the theorem is as follows: we consider an arbitrary point p in S and we find an integer $N(p, \varepsilon)$ such that

$$F_{f^n(p)\, f^{n+1}(p)}(\varepsilon) > 1 - \lambda$$

for all $n \geq N(p, \varepsilon)$. Since f is an (ε, λ)-contraction mapping, we get

$$F_{f^n(q)\, f^{n+1}(q)}(x) \geq F_{q\, f(q)}(x/k^n), \qquad q = f^{N(p,\varepsilon)}(p).$$

for $n = 0, 1, 2, \ldots$. Now, exactly as in the case of contraction mappings, we prove that the sequence $\{f^n(q)\}$ is a Cauchy sequence.

Since the space is complete, we can find a point such that $p^\sim$ such that $p^\sim = \lim f^n(p)$. We consider now the sequence of iterations of the point p, $\{f^n(p)\}$ and we remark that this is also a Cauchy sequence and we show now that $\lim f^n(p) = f(p^\sim)$. Indeed, if $U_{f(p^\sim)}(a, b)$ is a neighbourhood of $f(p^\sim)$, we obtain that there exists an integer N such that

$$f^n(p) \in U_p(a, b) \qquad U_p(\varepsilon, \lambda)$$

for all $n > N$, since $\{f^n(p)\}$ converges to $p^\sim$. Now using the fact that f is an (ε, λ)-contraction mapping we get

$$F_{f^{n+1}(p) f(p^\sim)}(a) \geq F_{f^n(p) p^\sim}(a/k) > 1 - b$$

and thus

$$\lim f^n(p) = f(p^\sim),$$

and thus $f(p^\sim) = p^\sim$.

The uniqueness follows easily from the fact that the space is (ε, λ)-chainable and we omit this part of the proof.

We consider now another generalization of the notion of contraction mapping.

DEFINITION 11.2.12.   Let (S, F) be a PM-space and consider $f_i : S \to S$, $i = 1, 2$. The pair $\{f_1, f_2\}$ is called a contractive pair of mappings if there exists $k \in (0, 1)$ such that for any p, q in S,

$$F_{f_1(p) f_2(q)}(x) \geq F_{pq}(x/k).$$

Clearly this reduces to contraction mappings when $f_1 = f_2$. The problem connected with the pairs of contractive mappings is about the existence of common fixed points, i.e. points $p \in S$ such that

$$f_1(p) = f_2(p) = p.$$

Let p_0 be an arbitrary point in S and let f_1 and f_2 be a pair of contraction mappings. We define now a sequence of points in S as follows:

$$p_{2m} \quad = f_2(p_{2m-1}), \qquad m = 1, 2, \ldots$$
$$p_{2m+1} = f_1(p_{2m}), \qquad m = 0, 1, 2, \ldots.$$

We also consider the function

$$G_{p_0}(x) = \inf \{F_{p_0 p_m}(x), m = 0, 1, 2, \ldots \}.$$

We have the following result about the fixed points for pairs of contraction mappings:

THEOREM 11.2.13.   *Let (S, F, T) be a complete Menger space, or let (S, F, τ) be a complete PM-space under τ, where τ satisfies the condition*

$$\sup \{\tau(F, F), F < H\} = H$$

and let (f_1, f_2) be a pair of contraction mappings. Then, either

1.　　　*the pair (f_1, f_2) has a unique fixed point*

or

2.　　　*for every $p_0 \in S$, $\sup\limits_{x} \{G_{p_0}(x)\} < 1$.*

Proof. Suppose that there exists $p_0 \in S$ such that

$$\sup_{x} \{G_{p_0}(x)\} = 1.$$

In this case we get,

$$\begin{aligned}
F_{p_{2m+1} p_{2n}}(x) &= F_{f_1(p_{2m}) f_2(p_{2n-1})}(x) \\
&\geqq F_{p_{2m} p_{2n-1}}(x/k) \geqq \cdots \\
&\geqq F_{p_0 p_{2n-2m-1}}(x/k^{2n-2m-1}) \\
&\geqq G_{p_0}(s/k^{2n-2m-1})
\end{aligned}$$

and similarly,

$$F_{p_{2n} p_{2n+2m}}(x) \geqq T(F_{p_{2n} p_{2n+1}}(x - \varepsilon), F_{p_{2n+1} p_{2n+2m}}(\varepsilon))$$

$$F_{p_{2n+1} p_{2n+2m+1}}(x) \geqq T(F_{p_{2n+1} p_{2n+2m}}(x - \varepsilon), F_{p_{2n+2m} p_{2n+2m+1}}(\varepsilon))$$

Using the relation obtained above and the property of G_{p_0} we get that $\{p_m\}$ is a Cauchy sequence. If $p^{\sim} = \lim p_m$ then, as in the case of contraction mappings, we show that this is a common fixed point of our pair.

From this result we obtain the following theorem in the case of Menger spaces.

THEOREM 11.2.14. *Let (S, F, T) be a complete Menger space and let (f_1, f_2) be a pair of contraction mappings. The pair has a common fixed point when $T(a, b) = \min(a, b)$.*

Another generalization of the notion of contraction mappings can be obtained using the notion of the 'probabilistic diameter' of a nonvoid set in a PM-space.

We note that the notion of probabilistic diameter of a set in a PM-space was defined by Egbert, as well as the notion of distance between sets in a PM-space.

DEFINITION 11.2.15. Let A be a nonvoid set in a PM-space (S, F). The 'probabilistic diameter' of the set A is the function in Δ^+ defined as

$$D_A(x) = \sup_{t < x} \{\inf F_{pq}(t): p, q \in A\}.$$

For properties of this function connected with the sets in a PM-space we refer to V. Istrăţescu (1974).

Suppose now that (S, F) is a PM-space and that $f : S \to S$ is a mapping. For each $p \in S$ we consider the sets

$$O_p(f)^i = \bigcup_{n=i}^{\infty} f^n(p)$$

and clearly we have the inclusions

$$O_p(f)^0 \supseteq O_p(f)^1 \supseteq O_p(f)^2 \supseteq \cdots$$

and thus, for the probabilistic diameters, we have the inequalities,

$$D_{O_p(f)^0}(x) \leq D_{O_p(f)^1}(x) \leq \cdots$$

and we set

$$\delta_p(x) = \lim \{D_{O_p(f)^i}(x)\}$$

and we call this function 'the probabilistic orbital diameter of the function f at the point p', or briefly 'the probabilistic orbital diameter at p'. The name is justified because the function $\delta_p(x)$ is in Δ^+.

We consider now a class of mappings suggested by the class of mappings with diminishing diameter in metric spaces.

DEFINITION 11.2.16.   Let (S, F) be a PM-space and let $f : S \to S$ be an arbitrary mapping. We say that f has diminishing probabilistic diameter at the point p if

$$\delta_p(x) > \delta_{f(p)}(x).$$

for all $x \in \mathbf{R}$.

In this case we have

THEOREM 11.2.17.   Let (S, F, T) be a Menger space and let $f : S \to S$ be a contraction mapping. Then f has diminishing probabilistic orbital diameter at each point.

Remark 11.2.18.   The assertion is true also for local (ε, λ)-contractions.

Another class of mappings generalizing the class of contraction mappings are defined as follows:

DEFINITION 11.2.19.   Let (S, F) be a PM-space. A mapping $f : S \to S$ is called locally power contracting if there exists a $k \in (0, 1)$ such that for each

$p \in S$ there exists an integer $n = n(x)$ such that for all $q \in S$,

$$F_{f^n(p)\,f^n(q)}(x) \geq F_{pq}(x/k).$$

An interesting problem is to give results about the existence of fixed points for functions in the above classes.

11.3. PROBABILISTIC MEASURES OF NONCOMPACTNESS

As we have seen, the measures of noncompactness play a fundamental role in the problem of the extension of various fixed point theorems. A natural problem appears in the case of PM-spaces, namely, to define some functions which are, in some sense, the analogue of the classical measures of noncompactness of Kuratowski or Hausdorff as well as for other measures of noncompactness.

The notion of Kuratowski probabilistic measure of noncompactness was introduced by Bocşan and Constantin (1973), under the name of Kuratowski functions.

In what follows we give the definition and properties for the Kuratowski probabilistic measure of noncompactness as well as for the Hausdorff probabilistic measure of noncompactness.

Let (S, F) be a PM-space and A be a nonvoid set in S. We say that the set A is bounded if

$$\sup_x \{D_A(x), x \in \mathbf{R}\} = 1$$

The Kuratowski probabilistic measure of noncompactness is defined as follows:

DEFINITION 11.3.1.   Let (S, F) be a PM-space and let A be a bounded set in S. The Kuratowski probabilistic measure of noncompactness of the set A is

$$\alpha_A(x) = \sup \{\varepsilon > 0, \text{ there exists a finite cover}$$

$$\text{of } A, \{A_i\}_1^n, \text{ such that } D_{A_i}(x) \geq \varepsilon,$$

$$i = 1, 2, 3, \ldots, n\}.$$

The basic properties of the Kuratowski measure of noncompactness are given in

THEOREM 11.3.2.   *The function*

$$A \to \alpha_A(x)$$

has the following properties:

1. $\alpha_A \in \Delta^+$,

2. $\alpha_A \geq D_A$,

3.   if A is nonvoid and $A \subseteq B$, then $\alpha_A \geq \alpha_B$,

4. $\alpha_{A \cup B}(x) = \min\{\alpha_A(x), \alpha_B(x)\}$,

5. $\alpha_A = \alpha_{\bar{A}}$ (where $\bar{A}$ denotes the closure of A).

Proof. 1. It is clear that $\alpha_A(0) = 0$ since $D_A(0) = 0$. Suppose now that $x_1 \leq x_2$. Since for each i, $D_{A_i}(x_1) \leq D_{A_i}(x_2)$ we get that α_A is nondecreasing.

Since A is bounded (as a set in a PM-space) there exists a cover of A such that

$$\sup\{D_{A_i}(x)\} = 1$$

and thus for $a \in (0, 1)$ we find x_i such that $D_{A_i}(x_i) > 1 - a$. In this case for $x^\sim = \max\{x_i\}$ we get $\alpha_A(x^\sim) > 1 - a$. This clearly implies that

$$\lim \alpha_A(x) = 1$$

We show now that α_A is left continuous. For this we consider

$$K_{A,x} = \{\varepsilon \geq 0, \text{ there exists a finite cover of } A, \{A_i\}_i^n,$$
$$\text{such that } D_{A_i}(x) \geq \varepsilon\}.$$

Let $x_0 \in (0, \infty)$ and ε in K_{A, x_0}. If $a > 0$ since D_{A_i} are left continuous functions, we find $\delta_i > 0$ such that $D_{A_i}(x_0) - D_{A_i}(x) < a$ for $0 < x_0 - x < \delta$, Let $\delta = \min_i \{\delta_i\}$ and $a < \varepsilon$. In this case, if $x - x_0 < \delta$ then $D_{A_i}(x) > \varepsilon - a$, and since $\varepsilon - a$ is in $K_{A,x}$, we get $\varepsilon - a \leq \alpha_A(x)$ which implies that

$$\alpha_A(x) - \alpha_A(x_0) > -a$$

which proves the left continuity of α_A.

2. This property is clear from the definition of the function α_A, as is property 3.

4. From the definition of the function $\alpha_A(,)$ it follows that for any $\delta > 0$ there exists $\varepsilon_\delta \in K_{Ax}$ and a finite covering of A, say $\{A_i\}_1^n$, such that

$$\alpha_A(x) - \delta < \varepsilon_\delta < D_{A_i}(\delta)$$

and similarly there exists $\eta_\delta \in K_{Bx}$ and a finite covering of B, say $\{B_j\}_1^m$, such that

$$\alpha_B(x) - \delta < \eta_\delta < D_{B_j}(\delta).$$

Now it is clear that $\{A_i \cup B_j\}$ is a finite cover of $A \cup B$ and for this we have

$$D_{A_i \cup B_j}(x) > \alpha_A(x) - \delta$$

i.e.

$$\alpha_A(x) - \delta < \varepsilon K_{A \cup B}(x)$$

and thus

$$\alpha_{A \cup B}(x) \geqq \alpha_A(x).$$

Similarly we have

$$\alpha_{A \cup B}(x) \geqq \alpha_B(x)$$

and the assertion in 4 is proved.

5. This property follows from the fact that for any set A,

$$D_{\bar{A}}(x) = D_A(x).$$

As we know using the Kuratowski measure of noncompactness we can characterize the sets whose closure is compact (i.e. the precompact sets).

This can be done also in the case of PM-spaces. For this we need the notion of precompact set in a PM-space.

DEFINITION 11.3.3.   Let (S, F) be a PM-space and A be a bounded set in S. We say that A is precompact if for any $\varepsilon > 0$, $\lambda > 0$ there is a finite cover of A, say $\{A_i\}_1^n$ such that

$$D_{A_i}(\varepsilon) > 1 - \lambda.$$

Then we can prove,

THEOREM 11.3.4.   Let (S, F) be a PM-space which is precompact. In this case for any $\varepsilon > 0$ and $\lambda > 0$ there exists a finite set $A_{\varepsilon, \lambda}$ in S such that for any $p \in S$ there exists $q \in A_{\varepsilon, \lambda}$ such that

$$F_{pq}(\varepsilon) > 1 - \lambda.$$

Proof. For any set A in s we define

$$D'_A(x) = \inf\{F_{pq}(x); \ p, q \in A\}$$

and we remark that

$$D_A(x_0) = \lim_{\substack{x < x_0 \\ x \to x_0}} D'_A(x) \leqq D'_A(x_0) \leqq F_{pq}(x_0).$$

We take $\varepsilon = 1/n$ and $\lambda = 1/m$ and we obtain a finite cover of S, say $\{A_i^{n,m}\}$ such that

$$D_{nm}(1/n) > (m - 1)/m$$

which implies that

$$F_{pq}(1/n) > 1 - 1/m$$

if $p, q \in A_m^{n,m}$. Let $p_i^{n,m} \in A_i^{n,m}$ and the set $\{p_i^{n,m}\}$ satisfies the condition of the theorem. If $\varepsilon > 1/n$ and $\lambda > 1/m$ the same argument shows that we find the finite cover of S, and the existence of the finite set is proved as above.

The definition of precompact sets in a PM-space is as follows:

DEFINITION 11.3.5. A bounded set in a PM-space is said to be precompact if for any $\varepsilon > 0$ and $\lambda > 0$ there exists a finite cover of A with sets $A_i, i = 1, \ldots, n$ such that

$$D_{A_i}(\varepsilon) > 1 - \lambda$$

for all $i = 1, \ldots, n$.

THEOREM 11.3.6.   *A bounded set in a PM-space (S, F) is precompact if and only if*

$$\alpha_A = H.$$

Proof. If A is a bounded set which is precompact then from the definition, for (ε, λ), ε near to 1, and λ near to zero, we get

$$\alpha_A(x) = 1.$$

The converse assertion follows from the definition of the function α_A.

If A and B are two subsets of a PM-space then, following Egbert, we define the probabilistic distance between A and B as follows:

$$F_{AB}(x) = \sup_{t \leqq x} T\left(\inf_{p \in A} \left[\sup_{q \in B} F_{pq}(t) \right], \inf_{q \in B} \left[\sup_{p \in A} F_{pq}(t) \right] \right)$$

and where T is a triangular t-norm as in Definition 11.1.2.

Using this notion we can define the probabilistic Hausdorff measure of noncompactness as follows:

DEFINITION 11.3.7.   Let A be a bounded set in a Menger space (S, F, T).

Then the probabilistic Hausdorff measure of noncompactness of A is defined as the function

$$\chi_A(x) = \sup \{\varepsilon > 0, \text{ there exists a finite set } F_\varepsilon \text{ in } S$$

$$\text{such that } F_{AF_\varepsilon}(x) > \varepsilon\}.$$

The properties proved for the Kuratowski probabilistic measure of noncompactness in Theorem 11.3.2 can also be proved for this function.

Using these measures of noncompactness we can introduce some classes of mappings which contain the classes discussed above.

DEFINITION 11.3.8. Let (S, F, T) be a complete Menger space and $f : S \to S$ be a mapping. The mapping f is called an α-set probabilistic contraction mapping if there exists $k \in (0, 1)$ such that for any bounded set A in S,

$$\alpha_{f(A)}(x) \geqq \alpha_A(x/k)$$

and the mapping f is called probabilistic densifying mapping if for any bounded set A in S,

$$\alpha_{f(A)}(x) > \alpha_A(x).$$

Remark 11.3.9. Using the Hausdorff probabilistic measure of noncompactness we can introduce similar classes of mappings.

It is not difficult to see that every contraction mapping is an α-set probabilistic contraction mapping.

11.4. SEQUENCES OF MAPPINGS AND FIXED POINTS

In what follows we consider sequences of mappings which have fixed points and which converge (in some sense) to a mapping.

The problem is to relate the fixed point of the limit mapping with those of the members of the sequence.

First we introduce a notion of convergence for mappings on a PM-space.

DEFINITION 11.4.1. Let (S, F) be a PM-space and $\{f_n\}$ be a sequence of mappings. We say that the sequence converges probabilistically uniformly to the mapping $f : S \to S$ if for any $\varepsilon > 0$ and $\lambda > 0$ there exists an integer $N_{\varepsilon, \lambda}$ such that for all $n > N_{\varepsilon, \lambda}$,

$$F_{f_n(p) f(p)}(\varepsilon) > 1 - \lambda$$

holds for all $p \in S$.

We can prove the following

THEOREM 11.4.2.   *Let (S, F, T) be a Menger space and let $\{f_n\}$ be a sequence of mappings converging uniformly to a contraction mapping f. Suppose that f has a unique fixed point and that each f_n has a fixed point p_n, then the sequence $\{p_n\}$ converges to the fixed point of f, say p_0.*

Proof. Let $x > 0$ and we show that

$$F_{p_n p_0}(x) \to H(x).$$

Indeed, we have,

$$F_{p_n p_0}(x) \geq T(F_{f^s_{p_n}(p_n) f^s_{p_n}}(x), F_{f^s_{np_n} f(p_0)}(x))$$
$$\geq T(1 - \lambda, F_{p_n p_0}(x/k^s))$$

for any integer s. Since $k \in (0, 1)$ this clearly implies that

$$F_{p_n p_0}(x) > 1 - \lambda$$

if $n \geq N$. This clearly gives our theorem.

The following theorem gives a result connected with pointwise convergence.

DEFINITION 11.4.3. Let (S, F, T) be a Menger space and let $\{f_n\}$ be a sequence of mappings, $f_n : S \to S$. The sequence converges pointwise to $f : S \to S$ if for each $p \in S$, $\{f_n(p)\}$ converges to $f(p)$.

We can prove the following

THEOREM 11.4.4.   *Let (S, F, T) be a Menger space and let $\{f_n\}$ be a sequence converging pointwise to f and for each n there exists p_n such that $f_n(p_n) = p_n$. Suppose further that f is a contraction mapping and that there exists a subsequence of $\{p_n\}$ which converges to a point p. In this case p is a fixed point for f.*

Proof. Let $\{n_k\}$ be the sequence of integers such that

$$\lim p_{n_k} = p.$$

We consider $F_{f(p) p}$ and we show that this is exactly H.
Indeed, we have further,

$$F_{p_{n_k} f(p)}(x) = F_{f_{n_k}(p_{n_k}) f(p)}(x)$$
$$\geq T(F_{f_{n_k}(p_{n_k}) f_{n_k}(p)}, F_{f_{n_k}(p) f(p)})(x)$$
$$\geq T(F_{p_{n_k} f_{n_k} f_{n_k}(p)}, F_{f_{n_k}(p) p})(x)$$

and using the fact that the subsequence $\{p_{n_k}\}$ is convergent to p, we get, applying the pointwise convergence, that

$$F_{f(p)\,p} = H$$

and the theorem is proved.

We close this chapter with a problem which appears of interest in connection with the results discussed above.

Let (S, F, T) be a Menger space then, using the convergence of points in a PM-space, we introduce the notion of continuous mapping on a PM-space in a natural way.

DEFINITION 11.4.5.   Let (S, F, T) be a complete Menger space. We say that this space has the fixed point property if for each continuous mapping $f : S \to S$ there exists a point $p, f(p) = p$.

Our problem is as follows:

Problem 11.4.6.   Characterize those Menger spaces or other classes of PM-spaces with the fixed point property.

Chapter 12

The Topological Degree

12.0. INTRODUCTION.

One way to think of the degree of a mapping $f : M \to N$ is as a number which measures the number of solutions of the equation $f(x) = y$. Of course we want this number to depend continuously on f as well as on y: i.e., for small perturbations of f this number remains unchanged in a small neighbourhood V of y. Of course, one way to do this is to consider the number of f-preimages of a point in the target space N and to try and define an 'appropriate counting' of preimages.

There exist several approaches to the definition of the topological degree, and in what follows we shall use the approach given in Milnor's book (1965) to define this notion in the case of finite-dimensional spaces. An interesting approach to the definition of this notion for the case of smooth maps was expounded by Eisenbud and Levine (1977).

This approach is presented in Eisenbud (1978).

Suppose now that we have a mapping

$$f : S^1 \to \mathbf{C}$$

i.e. a continuous function on the unit circle S^1 with values in the set of all nonzero complex numbers $\mathbf{C}$. Then we can define a number, $\deg f$, satisfying the following properties:

1. the number $\deg f$ is invariant under continuous deformations,

2. $\deg f$ is the only such invariant, this means that f can be deformed in g iff $\deg f = \deg g$,

3. there exists a map with a given degree, $\deg(,)$.

There exist several ways to define this number $\deg f$; for example, we approximate f by a finite Fourier series, $\sum_{-m}^{m} a_n z^n = P(z)$ and then we set

$$\deg f = N(f) - P(f)$$

where $N(,)$ represents the number of zeros and $P(,)$ the number of poles.

Of course we may ask about the extension to higher dimensions.

In this chapter we shall not present the deep and beautiful generalization to higher dimension of the above deg(,) obtained by Raoul Bott (1954). For the reader's convenience we give here the result of Bott:

THEOREM 12.0.1. *Let*

$$f : S^{n-1} \to GL(N, C) \qquad 2N \geqq 2$$

be a continuous mapping from the unit sphere S^{n-1} with values in the linear group $GL(N, C)$. Then if n is odd every map can be deformed to a constant map. In the case of n even, for any f we can define an integer, called $\deg f$, satisfying properties $1 - 3$ mentioned above.

For a proof we refer to Bott (1954), Atiyah-Bott (1964), Calderon (1966) and Bojarski (1963) and, for connections of Bott's deep results with elliptic operators, to the very interesting article of Atiyah (1967). For the readers who are interested in geometric topology an interesting exposition of the degree is given in the beautiful book of C. T. C. Wall (1972)

As is well-known the topological properties of the spheres in the case of infinite-dimensional spaces are very different from the properties of the spheres in the case of finite-dimensional spaces.

It was J. Leray and J. Schauder who remarked that we can extend the many properties known for finite-dimensional spaces to the case of infinite-dimensional spaces if we restrict our considerations to a suitable class of mappings; namely, maps which are perturbations of the identity by compact (continuous) maps.

Recently many results known for this class of mappings have further been extended to the following class of mappings: $I + f$, where f is an α-set contraction in the sense of Darbo.

12.1. THE TOPOLOGICAL DEGREE IN FINITE-DIMENSIONAL SPACES

The notion of the topological degree in the case of finite-dimensional spaces was first defined by L. E. J. Brouwer (1912).

In what follows we present an account of this theory and to this end we need some preliminary (well known notions) and for reader's convenience we recall them here.

Let X_n be a real finite-dimensional space of dimension n and let $((e_1, e_2, \ldots, e_n))$ be the family of all bases of X_n. First we define the notion of an orientation in X_n and this will be an equivalence class in the family of all bases of X_n.

DEFINITION 12.1.1. We say that two bases of X_n, $(e_1, \ldots, e_n)$ and $(e_1', \ldots, e_n')$ have the same orientation if

$$\det(a_{ij}) > 0$$

where

$$e_i' = \sum_{j=1}^{n} a_{ij} e_j.$$

Using the determinant properties it is easy to see that this defines an equivalence relation on the family of all bases of X_n.

We denote this equivalence by '$\sim$'.

DEFINITION 12.1.2. An equivalence class in the family of all bases of X_n is called an 'orientation'.

DEFINITION 12.1.3. If $X_n = \mathbf{R}^n$ then the orientation corresponding to the basis

$$e_1 = (1, 0, \ldots, 0)$$

$$\ldots\ldots\ldots\ldots$$

$$\ldots\ldots\ldots\ldots$$

$$\ldots\ldots\ldots\ldots$$

$$e_n = (0, \ldots, 0, 1)$$

is called the standard orientation.

For the definition of the topological degree we need some information about the set of critical values of a smooth map (i.e. if $f : M \to N$ is a smooth map, a point x is called a critical point of f if df_x is singular and then $f(x)$ is called a critical value

The following theorem of Sard (1942) gives the needed information.

THEOREM 12.1.4. *Let U be an open set in $\mathbf{R}^m$ and $f : U \to \mathbf{R}^n$ be a smooth map. Then the set*

$$M = (x, x \in U, \, df_x \text{ is singular})$$

has the Lebesgue measure zero (of course the n-dimensional Lebesgue measure).

Proof. We give the proof of this very important theorem by induction on

m. For an infinite dimensional version of this theorem see Smale (1965).

Of course we can suppose, without loss of generality, that $m \geq 0$ and $n \geq 1$. For each integer p we define the following set C_p:

$$C_p = (x, x \in U, \text{ the derivative of } f \text{ of order } \leq p \text{ vanishes at } x).$$

It is clear that we have the inclusions

$$C \supseteq C_1 \supseteq C_2 \supseteq C_3 \supseteq \cdots \supseteq C_p \supseteq C_{p+1} \supseteq \cdots .$$

We show now that the following assertions hold:

1. $\text{meas}(f(C - C_1) = 0$

2. $\text{meas}(f(C_p - C_{p+1})) = 0 \quad \text{for} \quad p \geq 1,$

3.   if k is sufficiently large

         $\text{meas}(f(C_k)) = 0$

(where meas means the Lebesgue measure).

For the assertions we remark that we can suppose $n \geq 2$ since if $n = 1$ we clearly have $C = C_1$. For the proof we need the following famous theorem of G. Fubini (for a proof of which we refer to the book by S. Sternberg on Differential Geometry (1964)): if S is a measurable set in $\mathbf{R}^n = \mathbf{R} \times \mathbf{R}^{n-1}$ then it has the Lebesgue measure zero if S intersects each hyperplane in a set of measure zero (here of course we refer to the $(n-1)$-dimensional Lebesgue measure).

To prove the first assertion we consider x', an arbitrary point which is not in C_1. Then, according to the definition of the set C_1 there exists a partial derivatives of f, say $\partial f_i/\partial x_j$ which is not zero at x'. We then define the following function

$$h(x) = (x, x, \ldots, x_{i-1}, f_i, x_{i+1}, \ldots, x_n)$$

and we remark that this function has a nonsingular derivative at x'. Then we get that the function h maps diffeomorphically a neighbourhood $V_{x'}$ of x' onto an open set W'. Then the mapping

$$g = f_0 h^{-1}$$

maps W into $\mathbf{R}^m$ and since the set of all critical points of g is exactly $h(V_{x'} \cap C)$, the set of its critical values is $f(V_{x'} \cap C)$.

If

$$(x_1, \ldots, x_{i-1}, t, x_{i+1}, \ldots, x_m) \cap V_{x'}$$

then $g(x_1, \ldots, x_{i-1}, t, x_{i+1}, \ldots, x_m)$ is in the hyperplane

$$\mathbf{R}^{i-1} \times (t) \times \mathbf{R}^{m-i}$$

and thus g preserves the hyperplanes.

Let g_t be the restriction of g to $\mathbf{R}^{i-1} \times (t) \times \mathbf{R}^{m-i} \cap V_{x'}$ and we remark that g and g_t have the same set of critical values.

Then according to the induction hypothesis, g_t has for its set of critical values the Lebesgue measure zero and this implies that the set of critical values of f intersects each hyperplane in a set of Lebesgue measure zero. Since it is easy to see that this is a measurable set, by Fubini's theorem it has the Lebesgue measure zero.

Thus the first assertion is proved.

Let us now prove the second assertion. Let us consider x' in $C_p - C_{p+1}$ and thus there exists a derivative, say,

$$\frac{\partial^{p+1} f_j}{\partial x_{i_1} \cdots \partial x_{i_{p+1}}}$$

which is not zero at x'. We consider then the following function

$$g_1(x) = \frac{\partial p_{f_j}}{\partial x_{i_2} \cdots \partial x_{i_{p+1}}}$$

and we remark that it vanishes at x' and $\partial g_1/\partial x_{i_1} \neq 0$. We now define another mapping, g_2, by the relation

$$g_2(x_1, \ldots, x_{i_1-1}, g_1(x), x_{i_1+1}, \ldots, x_m)$$

which maps diffeomorphically a neighbourhood V_x of x' into an open set W. It is easy to see that g_2 maps $C_p \cap V_x$ into a hyperplane in $\mathbf{R}^m$. We now define the mappings g_3 and g_4 by the relations:

$$g_3(x) = f_0 g_2^{-1}$$

$$g_4(x) = g_3/\text{hyperplane} \cap V_{x'}$$

and according to the induction hypothesis, the set of critical values of g_4 has a zero measure in $\mathbf{R}^m$. Since

$$g_{4,0} h(C_p \cap V_x) = f(C_p \cap V_x)$$

we obtain that $C_p \cap V_{x'}$ has the Lebesgue measure zero. But $C_p - C_{p+1}$ can be covered by a countable family of sets, as above, and we obtain that it has the Lebesgue measure zero.

Now for assertion 3 we consider the cube in $\mathbf{R}^m$, I_m, with edge l such that it is in U. Now, since the cube is a compact set, by the definition of C_p, we have, using Taylor's formula,

$$f(x + h) - f(x) = 0(x, h)$$

with

$$\|0(x, h)\| \leq c \|h\|^{p+1}$$

and x in $C_p \cap I_m$, $x + h \in I_m$. We decompose I_m into cubes with the edge l/s and we have s^m cubes. If $I_{m,1}$ is the cube which contains a point x in C_p then any point of $I_{m,1}$ is of the form

$$x + l'$$

with

$$\|l'\| \leq m^{1/2}(l/s).$$

Now from the difference $f(x + h) - f(x)$ we get that $f(I_{m,1})$ is in the cube with edge $2c(m^{1/2}l)^{p+1}/s^{p+1} = 2cm^{(p+1)/2} \cdot (l/s)^{p+1}$ and the centre of the cube is $f(x)$. Then we obtain further that

$$f(C_p \cap I_m) \subset \bigcap_1^{s^m} I_f^i$$

where I_f^i are cubes of edge $\leq 2cm^{(p+1)/2}(1/s)^{p+1}$ and thus the total volume is less than

$$s^m (a/s^{p+1})^m = a^m s^{m-(p+1)n}.$$

From this formula we obtain that if

$$1 + p > m/n$$

the volume tends to zero and this implies that $f(C_p \cap I_m)$ has the measure zero. Of course this implies that for the same p, $f(C_p)$ has the measure zero.

From assertions 1–3 it follows clearly that $f(C)$ has the Lebesgue measure zero and thus Sard's theorem is proved.

We also need the notion of homotopic mappings as well as of smooth homotopic mappings. In what follows we give these definitions and we prove an important property of smooth connected manifolds.

DEFINITION 12.1.5. Let X, Y be two topological spaces and consider the continuous mappings

$$f : X \to Y$$

$$g : X \to Y.$$

We say that f is homotopic to g if there exists a continuous mapping

$$F : [0, 1] \times X \to Y$$

such that

$$F(0, x) = f(x), \qquad F(1, x) = g(x).$$

Here, of course $[0, 1] \times X$ has the product topology and if f is homotopic with g we write this as $f \sim g$. The function F is called the homotopy between f and g.

Concerning the homotopy we can prove the following important property stated as

LEMMA 12.1.6. ' $\sim$ ' is an equivalence relation.

Proof. (1) For any $f : X \to Y$ clearly $f \sim f$. In fact we can take the homotopy F equal to f.

(2) If $f \sim g$ and F is the homotopy between f and g then we define a homotopy between g and f by

$$F'(t, x) = F(1 - t), x)$$

which satisfies the required property.

(3) Suppose now that $f \sim g, g \sim h$. We show now that we can define a homotopy between f and h as follows: if F is the homotopy between f and g, G is the homotopy between g and h then we set

$$H(t, x) = \begin{cases} F(2t, x) & 0 \leq t \leq \frac{1}{2} \\ G((2t - 1), x) & \frac{1}{2} \leq t \leq 1 \end{cases}$$

which is obviously a homotopy between f and h and the Lemma is proved.

Suppose now that X, Y are two smooth manifolds and f, g are smooth mappings. We define the smooth homotopy between f and g exactly as above but with the additional property that the homotopy F be smooth. Then we have the following

LEMMA 12.1.7. *The smooth homotopy is an equivalence relation.*

Proof. We consider the functions

$$\varphi(t) = \begin{cases} 0 & t \leq 0 \\ e^{-1/t} & t > 0 \end{cases}$$

$$\psi(t) = \varphi(t - \tfrac{1}{3})/(\varphi(t - \tfrac{1}{3}) + \varphi(\tfrac{2}{3} - t))$$

and we remark that $\varphi(t)$ has the property that

$$\psi(t) = 0$$

if $0 \leq t \leq \tfrac{1}{3}$ and

$$\psi(t) = 1$$

if $\tfrac{2}{3} \leq t \leq 1$. Suppose now that $f \sim g$ and if F is the corresponding homotopy we define

$$F_1(t, x) = F(\psi(1 - t), x)$$

which has the property that

$$F_1(0, x) = g(x)$$

$$F_1(1, x) = f(x)$$

and thus $g \sim f$. Clearly F_1 is smooth.

Suppose now that $f \sim g$, $g \sim h$ and F, G are the corresponding smooth homotopies. Then we define

$$H(t, x) = \begin{cases} F_1(2t, x) & 0 \leq t \leq \tfrac{1}{2} \\ G_1(2t - 1, x) & \tfrac{1}{2} \leq t \leq 1 \end{cases}$$

where F_1, G_1 are defined in an obvious way. It is easy to see that this is a smooth homotopy and thus $f \sim h$. Since, for any smooth f, $f \sim f$ the Lemma is proved.

THEOREM 12.1.8.   *Let V be a smooth connected manifold and let x, y be two arbitrary points in V. Then there exists the mapping*

$$f : V \to V$$

with the following properties:

1.   *f is a diffeomorphism,*

2.   *$f(x) = y$,*

3.   *there exists a smooth homotopy F between f and I(Ix = x) such
            that*

$$x \rightarrow F(t, x)$$

is a diffeomorphism of V onto V.

Proof. First we consider the following function on $\mathbf{R}^n$,

$$\varphi(x) = \begin{cases} \exp(1 - \|x\|^2) & \text{if } \|x\| < 1, \\ 0 & \text{if } \|x\| \geq 1. \end{cases}$$

We remark that this is a smooth function and we consider then the
function $t \rightarrow x(t)$ defined by the differential equations,

$$\mathrm{d}x_i/\mathrm{d}t = y_i^0 \varphi(x_1, x_2, \ldots, x_n)$$

$$x(0) = x_0.$$

From the theory of differential equations, it is known that this has a
unique solution which is also smooth. Now

$$(t, x_0) \rightarrow x(t) = F(t, x_0)$$

satisfies the relation

$$F(t + s, x_0) = F(t, F(s, x_0))$$

and this clearly implies that, for each t, $F(t,)$ is a diffeomorphism of $\mathbf{R}^n$ onto
$\mathbf{R}^n$. We remark that this diffeomorphism, for appropriate t and $y_0 = (y_1^0, \ldots, y_n^0)$, maps the origin into a given point of the unit sphere.

Suppose now that we have the manifold V with the stated properties and
that we introduce an equivalence relation on the points of V: two points p and
g of V are called isotopic if there exists a diffeomorphism $f : V \rightarrow V$ such that
$f(p) = q$ and f is smooth homotopic to the identity. It is obvious that this is
an equivalence relation and thus we obtain a partition of V into disjoint sets.
Now, each point in the interior of V has a neighbourhood diffeomorphic to
$\mathbf{R}^n$ and the argument given above shows that if r and s are sufficiently near
to each other then in fact they are isotopic. This implies that the interior of V
is in only one equivalence class and since the equivalence classes are open
sets, by the connectedness of V there exists only one equivalence class, and
this proves the assertion of Theorem 12.1.8.

The following assertion is perhaps well known, but we give it for
completeness.

THEOREM 12.1.9.   *Let $f : M \rightarrow N$ be a smooth map between smooth manifolds* *and where M is supposed to be compact. Then the function*

$$y \to N_y(f) = (\text{the number of solutions of the equation } f(x) = y)$$

is a locally constant function.

Proof. The fact that $y \to N_y(f)$ is locally constant means that for each $y \in N$ there exists a neighbourhood $V_y \ni y$ such that

$$N_z(f) = N_y(f)$$

for all $z \in V_y$. Now, as a consequence of the inverse theorem $N_y(f)$ is a finite number for each y. Let

$$\{x_1, x_2, \ldots, x_m\} = \{x, f(x) = y\}$$

and we choose the open neighbourhoods of these points, say $V_{x_1}, \ldots V_{x_m}$ such that

$$V_{x_i} \cap V_{x_j} = \varnothing \qquad i \neq j$$

and since these are mapped diffeomorphically onto the neighbourhoods $W_1, \ldots, W_m$ the required neighbourhood of y is

$$W_0 = W_1 \cap W_2 \cap \cdots \cap W_m - f\left(M - \bigcup_1^m V_{x_i}\right)$$

which satisfies the requirement of the theorem.

Let M be a smooth manifold and U be an open subset of M.

We define now the notion of diffeomorphism preserving orientation.

DEFINITION 12.1.10. Let M and U be as above suppose that $h : U \to \mathbf{R}^m$ (or H^m) a diffeomorphism. We say that h is orientation preserving if, for each $x \in U$, df_x maps the orientation of TM_x into the standard orientation of $\mathbf{R}^m$. An oriented smooth manifold is a smooth manifold with a given orientation in each tangent space such that for each point x in M there exists a neighbourhood V_x of x and a diffeomorphism which is orientation preserving.

For each x, df_x which is orientation preserving, we say that

$$\text{sgn} \, df_x = +1$$

and in the contrary case we say that

$$\text{sgn} \, df_x = -1.$$

We note that we have, from the above remarks, that $\text{sgn} \, df_x = \text{sgn} \det f_x$.

Let M, N be two oriented n-dimensional manifolds without boundaries, with M is supposed compact and connected.

The Brouwer degree of any smooth map

$$f : M \to N$$

is defined as follows:

DEFINITION 12.1.11.   Let M, N and f be as above. Then the Brouwer degree

$$\deg(f, y)$$

of f at y, y a regular value, is the number

$$\sum_{x \in f^{-1}(y)} \text{sgn } df_x.$$

First we remark that the definition has sense since the function

$$y \to N_y(f)$$

is locally constant and thus the sum in the above definition is finite. Another remark is that we have defined the degree on a dense subset of N.

To prove some properties of the function $\deg(f, y)$ we need some results given in the following lemmas.

LEMMA 12.1.12.   *Let M, N and f be as above and suppose that there exists a compact oriented manifold X such that M is oriented as the boundary of X. Then, if f extends to a smooth map $f_1 : X \to N$,*

$$\deg(f, y) = 0$$

for any regular value y.

Proof. We have to consider two cases:

Case 1. y is a regular value for both f and f_1. Then $f_1^{-1}(y)$ is a 1-dimensional compact manifold and thus is a finite union of arcs and circles (because every 1-dimensional compact and connected (smooth) manifold is diffeomorphic either to the circle S^1 (or to some interval of real numbers). The boundary points are on the arcs situated on M. Let γ be an arc in $f_1^{-1}(y)$ with the boundary $\partial\gamma = \{a\} \cup \{b\}$. The assertion of the lemma follows if we show that

$$\text{sgn } df_a + \text{sgn } df_b = 0.$$

We check now the orientation induced on our arc by the orientations in

M and X. Take x to be an arbitrary point in our arc and suppose that $(e_1, \ldots, e_{n+1})$ is a positively oriented basis for the tangent space TX_x with e_1 tangent to γ. In this case e_1 determines uniquely the orientation of TM_x iff df_{1x} maps the basis $(e_2, \ldots, e_{n+1})$ into a positively oriented basis of TN_y. Then we consider the function

$$x \rightarrow e_1(x)$$

which is a smooth function and clearly $e_1(x)$ points outward at one boundary point and inward to the other point. Then it follows that at one point of the boundary the sgn is $+1$ and at the other it is -1. This gives the assertion and thus the lemma is proved in this special case.

Case 2. y is a regular value for f but not for f_1.

We know that the function $\deg(f, y)$ is locally constant and thus we find a neighbourhood of y, say W such that

$$\deg(f, y) = \deg(f, z)$$

for all $z \in W$. Then, by Sard's theorem we find z_0 in W which is a regular value for both functions. Then by Case 1 the assertion of the lemma follows.

LEMMA 12.1.13.   *Let $f, g : M \rightarrow N$ and suppose that f, g are smooth homotopic. Then*

$$\deg(f, y) = \deg(g, y) \text{ for any common regular value } y.$$

Proof. We consider the manifold

$$[0, 1] \times M^n$$

which has a boundary consisting of $\{0\} \times M^n$ with the wrong orientation and $\{1\} \times M^n$ with the correct orientation. Then the $\deg(F, y)$ of $F/\partial([0, 1] \times M^n)$ is equal to

$$\deg(g, y) - \deg(f, y)$$

and since, by the Lemma 12.1.12, this is zero, the assertion is proved.

THEOREM 12.1.14.   *Let f, M and N be given as in the Definition 12.1.11. Then $\deg(f, y)$ is independent of the regular value y.*

Proof. Let y, z be two distinct regular values. Then we can choose a diffeomorphism

$$h : N \rightarrow N$$

such that $h(y) = z$ and h is (smooth) homotopic to the identity, by Theorem

12.1.8. Then $h_0 f$ is homotopic to f and thus

$$\deg(h_0 f, z) = \deg(f, z)$$

and since

$$(h_0 f)^{-1}(z) = f^{-1}(h^{-1}(z)) = f^{-1}(y)$$

we obtain that the $\deg(f, y)$ is independent of the regular value y.

THEOREM 12.1.15.   *Let f and g be smooth homotopic. Then*

$$\deg(f, y) = \deg(g, y) = \deg(f).$$

Proof. If f is given then by the above theorem $\deg(f, y)$ is independent of the regular value y and thus this justifies the name for the number $\deg(f, y)$, the degree of f, which is denoted by $\deg(f)$.

Suppose that f is smooth homotopic to g.

We know that

$$\deg(f, y) = \deg(g, y)$$

for any common regular value y. Now if y and z are two distinct regular values then using again Theorem 12.1.8 we can easily prove that

$$\deg(f, y) = \deg(g, z)$$

for any regular values.

Now we consider the case when M is a bounded open subset in a finite-dimensional space and we remark that all the above results apply. Thus for any smooth map $f : M \to X_n$ we can define the number

$$\deg(f, M, y)$$

called the degree of f at y satisfying the following properties:

1.   (The Additivity Property). If the solutions of

$$f(x) = y \qquad y \notin f(\partial M)$$

are contained in the open subsets $M_1, \ldots, M_m$ and $\cup_i M_i \subset M$ then

$$\deg(f, M, y) = \sum_{i=1}^{m} \deg(f/M_i, M_i, y), \qquad y \notin f(\partial M_i).$$

2.   (The Smooth Homotopy Property). If $f, g : M \to X_n$ are smooth

homotopic then

$$\deg(f, y) = \deg(g, y), \qquad F(t,)(\partial M) \not\ni y.$$

3.   (The Normalization Property).

$$\deg(I, M, y) = 1.$$

4.   (The Translation Invariance Property).
$$\deg(f, M, y) = \deg(f - y, M, 0).$$

5.   (The Excision Property). If $y \notin f(\partial M), k \subset \bar{M}, k$ closed and $y \notin f(k)$,

$$\deg(f, M, y) = d(f, M - k, y).$$

The last property 5 follows easily from the definition.

The topological character of the topological degree consists in the fact that it can be defined for the class of continuous functions on $\bar{M}$, M is a bounded open subset of a finite-dimensional subspace. The definition is as follows:

DEFINITION 12.1.16.   Let $f : \bar{M} \to X_n$ be any continuous function defined on the closure of M, with values in X_n. Let (f_n) be a sequence of smooth functions converging uniformly to f. Then

$$\deg(f, M, y) = \lim \deg(f_n, M, y)$$

where $y \notin f(\partial M)$.

The justification of the definitions is as follows: let

$$d = \inf\{d_n(y, z), z \in f(\partial M)\} \qquad (d_n \text{ is the metric on } X_n)$$

which is positive. Since the sequence (f_n) converges uniformly we find N such that for all $n \geq N$

$$|f_n - f| \leq \tfrac{1}{2}d.$$

and thus y is not a convex combination of the form

$$sf_n + (1 - s)f_m$$

for some $s \in [0, 1]$ and all $n, m \geq N$.

Then

$$\deg(f_n, M, y) = \deg(f_m, M, y)$$

and thus the above limit exists. The fact that it is independent of the sequence (f_n) which approximates f is easy to see and the justification is thus complete.

From the method of defining the degree for continuous mappings on $\bar{M}$ and the properties established for the case of smooth mappings we obtain that the deg (f, M, y) has the following properties:

1. (The Additivity Property). If the solutions of

    $$f(x) = y \qquad y \notin f(\partial M)$$

    are contained in the open subsets $\cup_{i=1}^{m} M_i$ (as above) then

    $$\deg(f, M, y) = \sum_{1}^{m} \deg(f_i, M_i, y), \qquad f_i = f/M_i.$$

2. (The Homotopy Property). If f, g are homotopic and

    $$y \notin F(t,)(\partial M)$$

    then

    $$\deg(f, M, y) = \deg(g, M, y).$$

3. (The Normalization Property).

    $$\deg(I, M, y) = 1.$$

4. (The Translation Invariance Property).

    $$\deg(f, M, y) = \deg(f - y, M, 0).$$

5. (The Excision Property). If M_1 is a closed subset in $\bar{M}$ and $y \notin f(M_1)$ then

    $$\deg(f, M, y) = \deg(f_1, M - M_1, y), \quad f_1 \text{ is the restriction of } f.$$

Proof. This follows from the fact that from smooth mappings this is immediate from the definition formula of $\deg(,)$.

6. (The Cartesian Product Property). Suppose that $M \subset X_n$ and $M' \subset X_m, f : M \to X_n, f' : M' \to X_m$ then

    $$\deg((f, f'), M \times M', (y, y')) = \deg(f, M, y) \cdot \deg(f', M', y')$$

    whenever each term makes sense.

Proof. This follows from the case of smooth mapping and for the case of smooth maps this follows from the determinant properties.

Now since the topological degree is locally constant it follows that it is constant on every component of $X_n - f(\partial M)$ and thus for each nonempty component C_M of $X_n - f(\partial M)$ we can define the number

$$\deg(f, C_M, y)$$

as the number

$$\deg(f, M, y), \qquad y \in C_M.$$

An important and powerful property of the topological degree was proved by Jean Leray (1950), presented in his famous paper at the Congress of Mathematicians in 1950 and known as the 'Multiplication Property' of the topological degree. The power of this theorem was illustrated by Leray by proving the Jordan separation theorem.

Suppose that M is given as above and thus, since $\bar{M}$ is a compact set, then it has at least one unbounded component. Since this component must contain the exterior of any ball it is unique and thus we can denote this by M_∞.

Now for any point y in $M_\infty - f(\bar{M})$ we have

$$\deg(f, M, y) = 0$$

and thus we can define the $\deg(f, M_\infty, y) = \deg(f, M) = 0$. Since the degree is constant for any component C_M then we define

$$\deg(f, C_M)$$

as this number.

The multiplication property of the topological degree is as follows:

THEOREM 12.1.17. *Let M be a bounded open set in X_n, a finite dimensional space, and let $f : M \to X_n$, $f' : X_n \to X_n$ be two continuous functions and C_{M_i} be the bounded components of $X_n - f(\partial M)$. If $y \notin f_0 f(\partial M)$ then we have*

$$\deg(f_0' f, M, y) = \sum_i \deg(f', C_{M_i}, y) \deg(f, C_{M_i}).$$

Proof. Of course we can suppose that the mappings are smooth and thus the proof of the assertion follows by inspection of the defining formula for the topological degree. Of course, we can suppose that y is not the image of critical points.

Then we have, by definition,

$$\deg(f'_0 f, M, y) = \sum_{x, (f'_0 f)(x) = y} \operatorname{sgn} d(f'_0 f)_x$$

$$= \sum_{x, (f'_0 f)(x) = y} \operatorname{sgn} d f'_{f(x)} \operatorname{sgn} d f_x$$

$$= \sum_x \deg(f, M, z) \operatorname{sgn} d f'_{f(x) = z}$$

$$z \in X_n - f(\partial M)$$

and this can be written, using the components of M, exactly as the formula in the theorem.

For the beautiful proof of Jordan's separation theorem using the multiplication property of the topological degree we refer to Leray's paper. Another interesting application is the proof of Brouwer's invariance of domains as well as a proof of Borsuk's theorem. For these we refer to Schwartz's book (1969). We give these on infinite-dimensional spaces. See Section 12.7.

12.2. THE LERAY–SCHAUDER TOPOLOGICAL DEGREE

In what follows we intend to extend the Brouwer degree to certain mappings defined on bounded, closed sets in Banach spaces. These mappings are perturbations of the identity by compact mappings. First we permit perturbations by finite-dimensional mappings (i.e. mappings whose range lie in a finite-dimensional space) and then, using an approximation argument, we extend to arbitrary compact mappings.

First we give the following result, which is an approximation theorem of Granas (1962); the second refers to the topological degree computed relative to subspaces and this will be used in conjunction with the approximation theorem of Granas to define the topological degree for arbitrary compact mappings.

LEMMA 12.2.1.   *Let S be a topological space and let X be a Banach space. Then the set of all continuous mappings defined on S with values in X such that $f(S)$ is relatively compact is contained in the closure of the following set: $(g : S \to X, g(S)$ is contained in a finite dimensional space) $= K_f(S, X)$.*

Proof.   Consider first the case when S is a subset in a Banach space X^-. Then there exists for each $\varepsilon > 0$ a mapping $p_\varepsilon : S \to X^-$ such that

$$\|p_\varepsilon(x) - x\| \leqq \varepsilon.$$

First we note that the existence of such a mapping is proved implicitly in the proof of the Schauder fixed point theorem.

Now the proof is as follows: since S is compact and $\varepsilon > 0$ is given, then there exist $x_1, \ldots, x_n$ in S such that S is covered by balls with the centres at x_i and radius at most ε. We define then for each i, $1 \leqq i \leqq n$ the function p_i by

$$p_i(x) = \max(0, \varepsilon - \|x - x_i\|)$$

and it is clear that for each x there exists at least one $i = i(x)$ such that $p_i(x) > 0$. We note also that from the definition the functions are continuous. We define now the functions

$$q_i(x) = p_i(x) \Big/ \left(\sum_1^n p_j(x) \right)$$

and we have that $\sum_1^n q_i(x) = 1$ for all x.

Now the construction of the function p is very simple. We set

$$p_\varepsilon(x) = \sum_1^n q_i(x) x_i$$

and it is easy to see that

$$\|p_\varepsilon(x) - x\| = \| \sum_1^n q_i(x)(x - x_i)\| \leqq \varepsilon$$

and in this case the lemma is proved.

The general case follows easy from this special case. Indeed, since f has the stated property, $\overline{f(S)}$ is a compact set in X and the mapping $(,)$ constructed with respect to this set has the property that the mapping $f_\varepsilon = p_\varepsilon \circ f$ satisfies the inequality

$$\|f_\varepsilon(x) - f(x)\| \leqq \varepsilon$$

for all $x \in S$ and the lemma is proved.

Remark 12.2.2.   We have from the construction of the mapping $p_\varepsilon(,)$ that the values are in $\overline{\operatorname{conv} f(S)}$.

Remark 12.2.3.   Since f_ε has values in a finite-dimensional space the fact proved in the above lemma implies that for f compact there exists a sequence (f_n) of finite dimensional mappings which converges to f and the values of f_n are in $\operatorname{conv} f(S)$ for each n. Indeed, we can take in the above lemma $\varepsilon = 1/n$ and the assertion follows.

Let $f : S \to S_1$ where S, S_1 are topological spaces. We recall that the function f is said to be proper if, whenever K_1 is a compact subset of $S_1, f^{-1}(K_1)$ is a compact set in S. The mapping f is called open (closed) if whenever F_1 is an open (closed) set in S_1, $f^{-1}(F_1)$ is an open (closed) set in S.

It is not difficult to see that any proper continuous mapping is closed when S and S_1 are metric spaces.

PROPOSITION 12.2.4.   *Let X be a Banach space and let $F : X \to X$ be a continuous k-set contraction where S is a closed subset of X. Then I-F is a proper mapping.*

Proof. Let K be any compact set of X and consider the set

$$M = (I - F)^{-1}(K)$$

and suppose the contrary, i.e. that this is not a compact set. In this case the Kuratowski measure of noncompactness of M is strictly positive and since for all $x \in M$ we have,

$$x = (x - F(x)) + F(x) \in K + F(M)$$

we obtain

$$0 < \alpha(M) \leq \alpha(K) + \alpha(f(M)) \leq k\alpha(M)$$

which is a contradiction. Thus M is compact and the proposition is proved.

COROLLARY 12.2.5.   *If F is a continuous k-set contraction then I-F is closed.*

PROPOSITION 12.2.6.   *Let E_n be a finite dimensional space and F_m be a subspace of E_n. Let $f : \bar{\Omega}_n \to E_n$ be continuous, $\Theta^n \notin f(\partial \Omega_n)$ and*

$$(I - f)(\bar{\Omega}_n) \subseteq F_m.$$

Then we have

$$d(f, \Omega_n, \theta^n) = d(f/\Omega_n \cap F_m, \Omega_n \cap F_m, \theta^m).$$

Proof. We can identify E^n with $\mathbf{R}^n$ and F_m with $\mathbf{R}^m \times \Theta^{n-m}$. We may assume that f is in C^∞ and that Θ^n is a regular value of f. Then we have, since $x = (y, \Theta^{n-m}) \in \mathbf{R}^m \times \Theta^{n-m} \subset \mathbf{R}^n$,

$$\operatorname{sgn} f'(x) = \operatorname{sgn} \det f'(x) = \det (f/\bar{\Omega}_n \cap \mathbf{R}^m)'(y)$$

and this is $a(f, \Omega_n)$ which is exactly $d(f, \Omega_n, \Theta^n)$. The assertion is proved.

Remark 12.2.7.   In order to justify the definition of the Leray–Schauder degree definition, the above proposition plays a fundamental role, since if we consider finite dimensional spaces, say E_n and E_m then we can consider these spaces, in a canonical way, as subspaces of $E_n \oplus E_m$, and this last space is again finite dimensional.

Now we show how we can define the degree for mappings of the form $I - f$, f is a continuous compact mapping.

THEOREM 12.2.8.   *Let X be a Banach space and Ω be a bounded open subset of X. If $f : \bar{\Omega} \to X$ is continuous and compact, $y \notin (I - f)(\partial\Omega)$ then there exists an integer-valued function denoted by $d(I - f, \Omega, y)$ satisfying the following properties*:

1.   Additivity Property. *If $(x, (I - f)(x) = y) \cap \Omega$ is contained in $\bigcup \Omega_i, i = 1, \ldots, m, \Omega_i$ are open and pairwise disjoint and $y \notin I - f)(\partial\Omega_i)$ then*

$$d(I - f, \Omega, y) = \sum_{i=1}^{m} d(I - f, \Omega_i, y).$$

2.   Homotopy Property. *If f, g are two compact continuous mappings, Ω, y as above and for each $t \in [0, 1]$ there exists f_t continuous and compact with $f_0 = f$, $f_1 = g$, $t \to f_t$ continuous, $(I - f_t)(\partial\Omega) \not\ni y$, then*

$$d(I - f, \Omega, y) = d(I - g, \Omega, y).$$

3.   Normalization Property. *If $y \in \Omega$ then*

$$d(I - f, \Omega, y) = 1.$$

4.   Translation Invariance Property. *For any $y \notin f(\partial\Omega)$,*

$$d(I - f, \Omega, y) = d(I - f - y, \Omega, \Theta).$$

Proof. First we remark that we can suppose without loss of generality that $y = \Theta$. Indeed, if we suppose that there exists a function having as values integers satisfying the properties 1, 2, 3 and 4 then we set

$$d(I - f, \Omega, y) = d(I - f - y, \Omega, \Theta)$$

by definition and this clearly satisfies all the required properties.

   Further we note that the mappings f with finite dimensional range satisfy the property that y is not in $(I - f)(\partial\Omega)$ is dense in the set of mappings as in the theorem. Thus if we have a mapping with the properties $1-5$, defined for mappings with finite dimensional range, then this extends uniquely to the class of mappings as in the theorem. We show now that we can define such an extension. As we noted we can suppose $y = \Theta$ and consider the family of all finite dimensional subspaces of X, say β_X. This family has a natural order relation, namely, the order defined by inclusion and it is easy to see that β_X with this order relation is a directed set. Let f and E be arbitrary in β_X and set

$$d(I - f, E) = \begin{cases} d_E((I - f)_{\bar{\Omega} \cap E}, \ \bar{\Omega} \cap E, \theta) & \text{if } (I - f)(\bar{\Omega}) \subset E, \\ 0 & \text{otherwise} \end{cases}$$

and where $d_E(. , . , .)$ means the degree computed with respect to the finite dimensional space E.

Now for f with finite dimensional range and $(I - f)/\partial\Omega \not\ni \Theta$ there exists an E such that

$$(x - f(x))_{x \in \Omega} \subset E$$

and for all $E_1 \supset E$ we have

$$(x - f(x))_{x \in \bar{\Omega} \cap E_1} \subset E_1$$

$$(+) \qquad 0 \notin (I - f)/\partial\Omega \cap E_1$$

and according to the Proposition 12.2.6,

$$d(I - f, E_1) = d(I - f, E).$$

In the case that we have a subspace E_2 for which a similar property holds as $(+)$ but it is not a subspace of E then we consider these as subspaces of $E \oplus E_2$ and then Proposition 12.2.6. justifies our definition.

Since β_X is a directed set we define

$$d(I - f, \Omega, \Theta) = \lim_{\beta_X} d(I - f, E)$$

and it is obvious that Properties 1 and 3 of the degree are satisfied. We prove now that the degree so defined has the property of being continuous with respect to the uniform topology, i.e., for f and y given, there is a neighbourhood of f such that for all g in that set and $(I - g)/\partial\Omega$ avoiding Θ, we have

$$d(I - f, \Omega, \Theta) = d(I - g, \Omega, \Theta).$$

Indeed, let U be a neighbourhood of f in the set of all mappings with finite dimensional range such that $I - h$ on $\partial\Omega$ avoids θ. There exists an $E \in \beta_X$

such that for all $E_1 \supseteq E$

1. $(I - f)(\bar{\Omega}) \subset E_1$

2. $(I - h)(\bar{\Omega}) \cap E_1 .$

Then the continuity of the Brouwer degree implies that

$$d(I - f, E_1) = d(I - h, E_1)$$

which implies further that

$$d(I - f, \Omega, \Theta) = d(I - h, \Omega, \Theta).$$

The homotopy property is a consequence of a rather more general property stated as

PROPOSITION 12.2.9.   *Let $I = [a, b]$ be a closed, bounded interval and*

$$f : \Omega \times [a, b] \to X$$

be a continuous compact map and suppose that $y : [a, b] \to X$ is continuous and has the property that

$$x - f(x, t) \neq y(t)$$

for all t in I and x in $\partial\Omega$. Then

$$d(I - f(\,.\,, t), \Omega, y(t)) = \text{const.}$$

Proof. Since $d(\,.\,,\,.\,,\,.\,)$ is translation invariant we may suppose without loss of generality that $y(t) = 0$ for all $t \in I$. Since $d(I - f(\,.\,, t), \,.\,, \Theta)$ is well defined for each t, we must show that it has the same value for different t. Now, the function f can be approximated uniformly by functions with finite dimensional range, as follows from Lemma 12.2.1. Thus, if t_0 and t_1 are two arbitrary points in I, the functions $f(\,.\,, t_0)$ and $f(\,.\,, t_1)$ are simultaneously approximated by finite-dimensional mappings and thus, by the definition of the degree, there exists $E \in \beta_X$ such that for all $E_1 \supset E$,

1. $f(x, t) \in E_1, \qquad x \in \bar{\Omega}$ and $t \in I$,

2. $d(I - f(\,.\,, t_0), \Omega, \Theta) = d(I - f(\,.\,, t_1), \Omega, \Theta)$
      $$= d_E((I - f(\,.\,, t_0))/(\bar{\Omega} \cap E_1), \Omega \cap E_1, \Theta)$$
      $$= d((I - f(\,.\,, t_1))/(\bar{\Omega} \cap E_1), \Omega E_1, \Theta).$$

If we consider $I - f$ restricted to $\bar{\Omega} \cap E_1 \times [t_0, t_1]$, we note that it is continuous and then, applying the continuity of the Brouwer topological degree, we get that the values of degree at t_0 and t_1 are the same. Since these were atbitrary in I Proposition 12.2.9 is proved.

In a similar way we prove that the degree as defined above satisfies the product formula as well as the excision property which is as follows:

PROPOSITION 12.2.10 (Excision Property). *Let f, Ω be as above and suppose that K is a closed subset of Ω and $y \notin f(\partial\Omega)$, $y \notin f(K)$ then*

$$d(I - f, \Omega, y) = d(I - f, \Omega - K, y).$$

We have now that the Brouwer degree is extended to certain classes of mappings on Banach spaces which have many properties similar to finite dimensional mappings, namely the perturbations by compact mappings of the identity. The problem of the further extension of the degree – possibly to all mappings (continuous) on a Banach space – then appears natural. However, in the following subsection we give Leray's example to show that this is, in fact, impossible. In subsection 12.4, we show how the topological degree can be defined for perturbations of the identity with α-set contractions.

We now present certain results in the case when the function f is supposed to have a derivative (Fréchet derivative). We prove further that in this case the derivative is compact if f is compact (See Theorem 12.5.24).

Suppose that we have on an open set of a Banach space X a continuous compact and (Fréchet) differentiable mapping with the Fréchet derivative df_x at the point x.

Suppose that x_0 is a fixed point of $f : \Omega \to X$ in Ω.

DEFINITION 12.2.11. The fixed point x_0 is said to be an isolated fixed point for f if there exists a neighbourhood U_{x_0} of x_0 such that f does not have other fixed points in this neighbourhood.

DEFINITION 12.2.12. The index of the isolated fixed point x_0 of f is by definition the following number:

$$i(f, x_0) = \lim_{\varepsilon \to 0} d(I - f, x_0 + B_\varepsilon, \Theta)$$

where

$$B_\varepsilon = (x, X \ni x, \|x\| < \varepsilon).$$

We remark that this number is well defined, which follows from the excision property of the topological degree. The following result gives an important and interesting conexion between the indexes associated with the isolated fixed point of f and the index associated with Θ for the (Fréchet) derivative at x_0 of f.

THEOREM 12.2.13. *Let Ω be an open subset of a Banach space X and $f : \Omega \to X$ be a continuous and compact mapping having a Fréchet derivative at x_0 which is a fixed point for f. If 1 is not an eigenvalue of $\mathrm{d}f_{x_0}$ then x_0 is an isolated fixed point of f and*

$$i(f, x_0) = i(\mathrm{d}f_{x_0}, \Theta).$$

Proof. We know that $\mathrm{d}f_{x_0}$ is compact and since 1 is not an eigenvalue of $\mathrm{d}f_{x_0}$, $(I - \mathrm{d}f_{x_0})$ is a bijection and this implies that there exists an $M > 0$ such that

$$\|x - \mathrm{d}f_{x_0}(x)\| \geqq 2M\|x\|$$

which implies that Θ is an isolated fixed point of $\mathrm{d}f_{x_0}$. Let $\varepsilon > 0$ such that for $x \in x_0 + B_\varepsilon$

$$\|f(x) - f(x_0) - \mathrm{d}f_{x_0}(x - x_0)\| \leqq M\|x - x_0\|$$

and thus

$$\begin{aligned}
\|(1 - t)(I - f)(x) &+ t\mathrm{d}f_{x_0}(x - x_0)\| \\
&\geqq \|\mathrm{d}(1 - f)_{x_0}(x - x_0)\| - (1 - t) \cdot \\
&\quad \cdot (1 - f)(x) - \mathrm{d}_{(I - f)x_0}(x - x_0)\| \\
&\geqq M\|x - x_0\|
\end{aligned}$$

which gives us that x_0 is an isolated fixed point of f.

We have further

$$d(I - f, x_0 + B_\varepsilon, \Theta) = d(\mathrm{d}_{(I - f)x_0} - \mathrm{d}_{(I - f)x_0}(x_0), x_0 + B_\varepsilon, \Theta)$$

and since

$$\mathrm{d}_{(I - f)x_0} - \mathrm{d}_{(I - f)x_0}(x_0) = I - \mathrm{d}f_{x_0} - \mathrm{d}f_{x_0}(x_0) - x_0$$

has its zeros only at x_0, applying the excision property of the degree, we have for $a > \|x_0\|$,

$$\begin{aligned}
d(\mathrm{d}_{(I - f)x_0} &- \mathrm{d}_{(I - f)x_0}(x_0), x_0 + B_\varepsilon, \Theta) \\
&= d(\mathrm{d}(I - f)_{x_0} - \mathrm{d}(I - f)_{x_0}(x_0), a/\varepsilon B_\varepsilon, \Theta).
\end{aligned}$$

Since for every $t \in [0, 1]$ and any x, $\|x\| = a'$,

$$\mathrm{a}' > \max (a, \|\mathrm{d}(I - f)_{x_0}^{-1}(\mathrm{d}(I - f)_{x_0}(x_0)\|)$$

we have

$$\Theta \neq d_{(I - f)_{x_0}}(x) - td_{(I - f)_{x_0}}(x_0)$$

and the homotopy invariance of the topological degree, implies that

$$d(d(I - f)_{x_0} - d(I - f)_{x_0}(x_0), a'/\varepsilon B_\varepsilon, \Theta)$$
$$= d(d(I - f)_{x_0}, a'/\varepsilon B_\varepsilon, \Theta)$$

and then we get the equality with $i(df_{x_0}, \Theta)$ and the theorem is proved.

The following famous result of Leray–Schauder gives a formula for the index in terms of multiplicity of some eigenvalues.

THEOREM 12.2.14   (Leray–Schauder, 1934). *Let* $X, \Omega, f, 1, x_0$ *be given as above. Then we have*

$$i(f, x_0) = (-1)^m$$

where m is the sum of multiplicities of eigenvalues in $(0, 1)$, *in the case of a real Banach space and* $i(f, x_0) = 1$ *if* X *is a complex Banach space.*

Proof. From the above arguments, if $\|x\|$ is sufficiently small,

$$f(x) = df_\Theta(x) + w(x)$$

(we suppose without loss of generality that $x_0 = \Theta$, since the general case reduces to this by a translation), and

$$\|w(x)\| \leq 2M/3 \|x\|$$

and where M satisfies the inequality $\|x - df(x)\| \geq M\|x\|$.

We consider now the homotopy

$$h(x, t) = x - tf(x) - (1 - t)df_\Theta(x)$$

and we get that

$$d(I - f, B_\varepsilon, \Theta) = d(I - df_\Theta, B_\varepsilon, \Theta).$$

If X_1 is the linear space generated by all eigenvectors of df_Θ corresponding to eigenvalues in $(0, 1)$ then this is finite dimensional and has a complement, say X_2. Clearly $m = \dim X_1$ and set further

$$B_{1,\varepsilon} = B_\varepsilon \cap X_1, \qquad B_{2,\varepsilon} = B_\varepsilon \cap X_2.$$

Using the product formula for the topological degree we get

$$d(I - df_\Theta, B_\varepsilon, \Theta) = d(I - df_\Theta, B_{1,\varepsilon}, \Theta) \cdot d(I - df_\Theta, B_{2,\varepsilon}, \Theta).$$

If
$$h_1(x, t) = (I - t\,\mathrm{d}f_\Theta)(x)$$

then for $x \in \partial B_{2,\varepsilon}$ and $t \in (0, 1)$ we have $h_1(x, t) = 0$ iff $\mathrm{d}f_\Theta$ has an eigenvalue in $(0, 1)$ and this is not possible in X_2, by the definition of X_1.
This implies that

$$\mathrm{d}(I - \mathrm{d}f_\Theta, B_{2,\varepsilon}, \Theta) = \mathrm{d}(I, B_{2,\varepsilon}, \Theta)$$

and considering

$$h^-(x, t) = ((2t - 1)I - t\,\mathrm{d}f_\Theta((x)$$

we have that on the boundary of $B_{2,\varepsilon}$ it avoids zero and thus

$$\mathrm{d}(I - \mathrm{d}f_\Theta, B_{2,\varepsilon}, \Theta) = \mathrm{d}(-I, B_{1,\varepsilon}, \Theta)$$

which is $(-1)^m$.

Now is X is a complex space then the real dimension of X_1 is even and this clearly implies the assertion.

We give now the notion of index associated with pairs (S, A) and $f : S \to A$ where S and A are topological spaces and $A \subset S$.

DEFINITION 12.2.15.   A triple (S, f, A) is called admissable if S is ANR (i.e. an absolute neighbourhood retract; which means that if we have any metric space M, a closed subset of M, say M_1 and a continuous function $f : M_1 \to S$, then there exists an open neighbourhood V of M_1 and a continuous mapping $F : V - X$ such that $F(m) = f(m)$ for all $m \in M_1$). (S is called an absolute retract if F can be defined on all of M), A is an open subset of S and $f : \bar{A} - S$ has the property that

$$F(f) \cap \partial A = \varnothing, \qquad F(f) = \{x, f(x) = x\}.$$

DEFINITION 12.2.16.   A fixed point index is a function defined on all admissable pairs and satisfies the following properties:

1. *Homotopy Property.* If $f_t : A \to X$ is a homotopy and (S, f_t, A) is admissable for all $t \in I = [0, 1]$ then

$$i(f_0, A) = i(f_1, A).$$

2. *The Additivity Property.* If (S, f, A) is an admissable triple and $A_1, \ldots, A_m$ are open sets in A such that $F(f) \cap (A - \cup_1^m A_i = \varnothing$ then

$$i(f, A) = \sum_{j=1}^{m} i(f, A_j).$$

3. *The Normalization Property.* If S is a compact ANR and $f : S \to S$ is a map then $i(f, X) = L(f)$, the Lefschetz number of f (See R. Brown (1971)).

4. *Commutativity.* If S and T are compact ANR's and A is an open subset of S and $f : A \to T$, $g : T \to S$ are maps such that (S, gf, A) is admissable then

$$i(gf, A) = i(fg, g^{-1}(A)).$$

If (S, f, A) is an admissable triple and x_0 is in A then we say that x_0 is an isolated fixed point of f if there is a neighbourhood V_{x_0} of x_0 with the property that $F(f/V_{x_0}) = (x_0)$.

Note 12.2.17.   Definition 12.2.11 is another formulation of this definition and conversely.

DEFINITION 12.2.18.   The index at x_0 is by definition $i(f, V_{x_0})$ for such a neighbourhood and is denoted, as in the case of Banach spaces, by $i(f, x_0)$.

Now using the properties of the topological degree an index function can be defined when the function is of the form $I - f$, f compact and continuous.

If G is a bounded open set in a Banach space X and $f : \bar{G} - X$ is compact then we set

$$i_x(f + y, G) = \mathrm{d}(I - f, G, y)$$

and this is an index function. Properties 1, 2 and 4 follow from the corresponding properties of the topological degree. For 4 we refer to R. Brown (1971), F. Browder (1960).

We note that the uniqueness of the fixed point index in the category of finite polyhedra satisfying a somewhat different list of axioms was first proved in 1953 by Barrett O'Neill. Further, an elementary and ingenious proof of the uniqueness of the index function was given by R. Brown in 1970; for the readers interested in the applications of algebraic topology to fixed point theory the book of Brown (1971) and his paper on Leray's Index Theory may be of interest as well as the original papers of Leray (1945), (1945), (1959) and Deleanu (1959), F. Browder (1960), and Thompson (1969).

12.3. LERAY'S EXAMPLE

The following problem appears naturally in connection with the topological degree: is it possible to define the local degree for arbitrary continuous mappings defined on a Banach space X with values in X?

J. Leray (1936) has given an interesting example which shows that such a definition is in fact not possible. In what follows we present Leray's example.

Let us consider the real Banach space $C_{[0,1]}$ of all real-valued functions (continuous) on $[0, 1]$ with the usual norm,

$$f \to \|f\| = \sup_{t \in [0, 1]} |f(t)|.$$

We consider the following set

$$D = (f, f \in C_{[0, 1]}, \|f - f_0\| < \tfrac{1}{2})$$

where

$$f_0(s) = \tfrac{1}{2}$$

for all $s \in [0, 1]$.

Let

$$h : [0, 1] \to [0, 1]$$

continuously and with the property that $h(0) = 0$, $h(1) = 1$.

We now define, using the function h, a mapping on $\bar{D}$ with values in $\bar{D}$ by

$$H(f)(s) = h(f(s)).$$

We remark that we can define a homotopy by the relation

$$H(x, t) = th(s) + (1 - t)x(s)$$

for any $x \in \bar{D}$.

If y is in $\bar{D} - D$ then for each t, $H(y, t)$ is in $\bar{D} - D$, as is easy to see. From the form of $H(x, t)$ we deduce that the mapping H is homotopic to I in $\bar{D} - (y_0)$ where $y_0 \in D$. In this case, if the local topological degree can be defined and if it has the standard properties, we get that the degree of H relative to $\bar{D}$ and y_0 is 1 and according to the properties of the degree, we obtain that the equation

$$H(x(s)) = y_0(s)$$

has a solution in D.

If we choose y_0 suitably, then the above equation, i.e.

$$H(x(s)) = y_0(s)$$

has no solutions.

An example of such a function is as follows:

$$y_0(s) = \tfrac{1}{4} + \tfrac{1}{2}s$$

and h is as follows:

$$h(s) = \begin{cases} s & 0 \leqq s \leqq \tfrac{1}{2} \\ 1 - s & \tfrac{1}{2} < s \leqq \tfrac{5}{8} \\ \tfrac{5}{8}(s - 1) + 1 & \tfrac{5}{8} < s \leqq 1. \end{cases}$$

12.4. THE TOPOLOGICAL DEGREE FOR k-SET CONTRACTIONS

In what follows we consider the problem of defining the topological degree for the case when we permit perturbations of the identity by k-set contractions.

Since many expositions and extensions of the degree to larger classes of mappings are based upon the index theory for compact metric absolute neighbourhood retracts we begin with a very short indication of this. For details and generalizations we refer to Browder's papers (1948), (1960).

First we recall that a topological space S is said to be an absolute neighbourhood retract if, given any metric space X and a closed subset of X, say A, a continuous map $f : A \rightarrow S$, then there exists an open neighbourhood U of A and a continuous map $F : U \rightarrow S$ such that $F(x) = f(x)$ for all $x \in A$.

If F can be defined on X then we say that S is an absolute retract. We use the notation ANR and AR to denote the fact that a space is an absolute neighbourhood retract, or an absolute retract, respectively.

We mention that from a result of Dugundji (1951) it follows that any convex set of a topological linear space is an AR.

In the abovementioned papers F. E. Browder has constructed an index function defined on the category of ANR which are compact metric spaces. Thus the objects of this category are these ANR and the morphisms are the continuous mappings. For details about category theory we refer the reader to any book on categories (since there are only few!). Let us denote this category of ANR-spaces which are compact metric spaces by Ω. If A is in Ω and G is an open subset of A and $f : \bar{G} \rightarrow A$, ($\bar{G}$ means the closure of G) has the property that has no fixed points in $\bar{G} \backslash G$ then the number $i_A(f, G)$ can be defined and satisfies the following properties:

(a) Suppose that $f : \bar{G} \rightarrow A$ has no fixed points in $\bar{G} \backslash G$ and there exist two disjoint open subsets of G, G_1 and G_2 such that the fixed points of f are in

$G_1 \cup G_2$ then

$$i_A(f, G) = i_A(f, G_1) + i_A(f, G_2).$$

(b) Let $I = [0, 1]$ and A, G be given as above. Suppose that there exists a function

$$F : \bar{G} \times I \to G$$

and for each $t \in I$,

$$F_t : \bar{G} \to A, \qquad F_t(x) = F(x, t)$$

has no fixed points in $\bar{G} \backslash G$. If F is continuous we have

$$i_A(F_0, G) = i_A(F_1, G).$$

(c) if $G = A$ then

$$i_A(f, G) = \Lambda(f), \qquad \text{where } \Lambda(f) \text{ is the Lefschetz number of } f.$$

(d) If A, B are in Ω and $f : A \to B$ is continuous and for V, an open subset of B, there exists $g : \bar{V} \to A$ continuous, and suppose that $f_0 g$ has no fixed points in $\bar{V} \backslash V$. Consider $U = f^{-1}(V)$. In this case the function $g_0 f$ has no fixed points in $\bar{U} \backslash U$ and

$$i_B(f_0 g, V) = i_A(g_0 f, U).$$

The property (a) is called the 'additivity' property, (b) is the 'homotopy' property, (c) is the so-called 'normalization' property and the last, (d), is called the 'commutativity' property.

It is worth noting the following particular case of the commutativity property: suppose that A, B are in Ω and $A \subseteq B$. Let $f : A \to B$ and let V be an open subset of B, g: let $V \to A$ be a continuous mapping with no fixed points in $\bar{V} \backslash V$. Then the following holds:

$$i_B(f_0 g, V) = i_B(g, V) = i_A(g_0 f, f^{-1}(V)) = i_A(g, V \cap A).$$

In his paper in 1971 R. Nussbaum has defined using this index function a generalized index as follows: Suppose that we have A in Ω and consider G to be an open subset of A. If $f : G \to A$ is a continuous mapping with the property that $(x, f(x) = x)$ is compact or empty, then we can choose an open neighbourhood of $F(f) = (x, f(x) = x)$ such that the closure will be in G. Let V be this neighbourhood. Then we define the index as follows:

DEFINITION 12.4.1 (R. Nussbaum, 1971). The generalized index $i_{A, gen}(f, G)$ is $i_A(f, V)$.

We note that $i_A(f, V)$ is well defined and the problem is to show that the number so defined is in fact independent of V satisfying the above requirements.

R. Nussbaum proves that this is indeed the case and that the generalized index so defined satisfies all the properties (a)–(d) listed above. We show now how we can use this index (generalized) to define a topological degree for k-set contractions.

Let us consider X to be a Banach space and G to be an arbitrary open subset of X. If $f : G \to X$ is continuous we consider the following sets:

$$G_1 = \overline{\text{conv}} \, f(G)$$

$$G_2 = \overline{\text{conv}} \, (f(G \cap G_1))$$

$$\cdots\cdots\cdots\cdots\cdots$$

$$G_n = \overline{\text{conv}} \, (f(G \cap G_{n-1}))$$

$$\cdots\cdots\cdots\cdots\cdots$$

and we note that, by definition, these are closed and convex. Define now the set

$$G_\infty = \bigcap_n G_n.$$

Now if f is any k-set contraction then we easily get that G_∞ has the Kuratowski measure of noncompactness equal to zero and thus it is a compact and convex set (the convexity property follows from the fact that G_n's of all the sets are convex). The index for k-set contractions is defined using the set G_∞ constructed from G and the given k-set contraction f.

DEFINITION 12.4.2. (R. Nussbaum, 1971). The index for $f, i_X(G, f) = i_{G\infty}(F, G \cap G_\infty)$ and the degree is defined by the formula

$$d(I - f, G, y) = i_X(f + y, G).$$

It has been proved in Nussbaum (1971) that the degree so defined coincides with the usual degree of Leray–Scauder for compact maps.

In his (1972) paper R. Nussbaum has extended the index (and thus the degree) to local condensing mappings. We conjecture that this method of defining the degree can be adapted to obtain an index (and thus a degree) for local power k-set contraction mappings as well as for certain classes of local power condensing mappings.

We note that the index (and thus the degree) is defined by restricting the map to a certain 'nice compact' set. This important remark permits the use of approximation theorems which will be used to prove the uniqueness.

12.5. THE UNIQUENESS PROBLEM FOR THE TOPOLOGICAL DEGREE

An important problem appears in connection with the extension of the topological degree theory: to what extent does the extended notion of 'degree' have properties in common with the classical theory of L. E. J. Brouwer? The first mention of this problem is in the famous paper of Nagumo (1951a). It is mentioned there that the uniqueness of the Brouwer degree can be proved by simplicial approximation. The case of the Leray–Schauder degree is treated in the interesting paper of O'Neill (1953) in which the uniqueness is obtained using the uniqueness of the fixed point index introduced via cohomology theory.

An elementary proof of the uniqueness of the fixed point index was obtained by Robert Brown (1970). For the reader's convenience we give, following Brown, the axiomatic definition of the index; the axioms we give are an easily modified list of the axioms proposed by Felix Browder (1960)

Let C be a collection of topological spaces and let C' denote the collection of all triples (X, f, U) where $X \in C$, $f: X \to X$ is a map and U is an open subset of X such that there are no fixed points of f on the boundary of U.

DEFINITION 12.5.1.   A fixed point index on C is a function

$$i: C' \to \mathbf{Z}$$

($\mathbf{Z}$ is the set of integers) such that the following properties are true:

1. (Localization). If $(X, f, U) \in C'$ and $g: X \to X$ is a map such that $f(x) = g(x)$ for all $x \in \mathrm{Cl}\, U$ then

$$i(X, f, U) = i(X, g, U).$$

2. (Homotopy property). If $f: X \to X$, $g: X \to X$ are two maps and for each $t \in [0, 1]$ there exists $f_t: X \to X$ and

$$f_0 = f, \qquad f_1 = g, \qquad (X, f_t, U) \in C'$$

then

$$i(X, f_0, U) = i(X, f_1, U).$$

3. (Additivity property). Let $U_1, \ldots, U_m$ be a finite collection of mutually disjoint open subsets of U and $(X, f, U) \in C'$. If $f(x) = x$ for all $x \in U - U_{i=1}^m U_i$, then

$$i(X, f, U) = \sum_{j=1}^m i(X, f, U_j).$$

4. (Commutativity). Suppose that X, X' are in C and $f : X \to X'$, $g : X' \to X$ are maps such that $(X, g_0 f, U) \in C'$. Then

$$i(X, g_0 f, U) = i(X, f_0 g, g^{-1}(U)).$$

5. (The Weak Normalization property). Suppose that $(X, f, U) \in C'$ and f is the constant map, i.e. $f(x) = x_0$ for some $x_0 \in U$. Then

$$i(X, f, U) = 1.$$

We now give a list of axioms for the topological degree in the case of Banach spaces.

Let C be the family of all Banach spaces and let $O(C)$ be the collection of all open subsets of members in C which are also bounded.

DEFINITION 12.5.2.   A degree on C is any function on the triples (X, f, U) where $X \in C$, $U \subset X$ and it is in $O(C)$, $I - f : \bar{G} \to X$ and f is an α-set contraction. The values of this function are integers and the properties of the function, which we denote by

$$\deg(I - f, G, y)$$

where $y \notin (I - f)(\partial G)$, are as follows:
1. (The Additivity Property). If

$$(x, (I - f)(x) = y) \cap G$$

contained in $\bigcup_1^m G_i \subset G$, with $(G_i)_1^m$ a finite collection of pairwise open disjoint sets and $y \notin (I - f)(\partial G_i)$ then

$$\deg(I - f, G, y) = \sum_{i=1}^m \deg(I - f, G_i, y).$$

2. (The Homotopy Property). Suppose that f, g are two α-set contractions, G is as above and for y,

$$\deg(I - f, G, y), \qquad \deg(I - g, G, y)$$

are defined. Suppose that for each $t \in [0, 1]$ there exists f_t an α-set contraction such that $f_0 = f, f_1 = g$ and $t \to f_t$ is continuous. If

$$y \notin (I - f_t)(\partial G)$$

then

$$\deg (I - f_t, G, y)$$

is constant for $t \in [0, 1]$.

3. (The Normalization Property):

$$\deg (I, G, y) = 1.$$

4. (The Product Property). If f_i, G_i, y_i are such that

$$\deg (I - f_i, G_i, y_i), \qquad i = 1, 2$$

are defined then

$$\deg ((I - f_1)(I - f_2), G_1 \times G_2, (y_1, y_2))$$

is also defined and is equal to

$$\deg (I - f_1, G_1, y_1) \cdot \deg (I - f_2, G_2, y_2).$$

In his interesting paper 'A Bifurcation Theorem for k-Set Contractions' in *Pacific J. Math* **44** (1973) J. W. Thomas has proved that if we consider only differentiable α-set contractions then the deg(,) satisfying the above four properties is unique. At the end of his paper, Thomas remarks that it is possible to remove the differentiability condition in the case of Hilbert spaces by using the Weierstrass approximation of P. M. Prenter (1970). The Weierstrass approximation theorem of Prenter refers to Hilbert spaces and to a special class of Banach spaces. The first general approximation theorem of this type was obtained by V. Istrățescu (1977) (1980) and in the lectures at the Centro Linceo di Scienze Matematiche ed loro Applicazioni it was used to prove the uniqueness of the topological degree for the case of Banach spaces. See also the author's book *Topics in Operator Theory*, Rome, 1978.

In what follows we first give these approximation theorems.

To this end we need some information about polynomial mappings, and for more details we refer to Hille and Phillips (1957).

Let X be a Banach space and for the integer k we consider the Banach space

$$X_k : \underbrace{X \times X \times X \times \ldots \times X}_{k-times}$$

and let $L : X_k \to X$ be a linear and bounded operator.

DEFINITION 12.5.3. A mapping $f : X_k \to X$ is called k-linear if it is linear in each of its arguments separately.

A 0-linear operator on X is any constant function on X; in what follows we shall identify the 0-linear maps with their range.

DEFINITION 12.5.4. The norm of a k-linear map f is the number

$$\|f\| = \sup \|f(x_1, \ldots, x_k)\| \qquad (\|x_1\| = 1, \ldots, \|x_k\| = 1).$$

We remark that if f is a k-linear map and g is an l-linear map then we can define in a natural way the maps $f_0 g = fg$ and gf which are clearly $(k + 1)$-linear maps.

DEFINITION 12.5.5. A k-linear map is said to be symmetric if

$$f(x_1, \ldots, x_k) = f(x_{p(1)}, \ldots, x_{p(k)})$$

for any permutation p of the integers $(1, 2, 3, \ldots, k)$.

DEFINITION 12.5.6. A polynomial on a Banach space X is any function P of the form

$$P(x) = a_0 + L_1(x) + L_2(x) + \cdots + L_n(x)$$

where $a_0 \in X$, L_i are symmetric i-linear mappings. We call n the degree of P.
Also we use the following standard notation:

$$L_i(x, x, \ldots, x) = L_i x^i = L_i(x).$$

Thus the polynomial P is a function of the form

$$P(x) = a_0 + L_1 x + L_2 x^2 + \cdots + L_n x^n.$$

Remark 12.5.7. We can consider the values of the functions in another Banach space, say Y, and in certain cases the above properties are preserved. This is the case when we consider $Y = \mathbf{R}$, the real numbers and $Y = \mathbf{C}$, the complex numbers.

Remark 12.5.8. In the cases considered in the above remark the set of polynomials forms an algebra with pointwise multiplication and this fact is of crucial importance for the approximation theorems which we prove.

In the case of complex Banach spaces we can consider the class of functions which represents, in some sense, the analogue of the polynomials in the conjugate variable. These are defined as follows:

DEFINITION 12.5.9. A k-antilinear map on a Banach space X is any map

$f : X_k \to X$ which is antilinear in each argument. The map is called symmeric if it is invariant for all permutations of the variables.

DEFINITION 12.5.10. A mapping f is called of (k, l)-type if it is a function on X_{k+l} which is linear in k-variables and anti-linear in l variables. If this is preserved for permutations we call such a map symmetric of (k, l)-type. If f is symmetric of (k, l)-type we write

$$f(x, x, \ldots, x) = f x^k x^{-l}$$

and the norm is defined as usual.

DEFINITION 12.5.11. Let X be a complex Banach space. A polynomial on X is any function of the form

$$P(x) = a_0 + L_1 x + \cdots + L_j x^k jx^{-l}j + \cdots + L_n x^k nx^{-l}n$$

where

$$l_j + k_j = j, \qquad j = 1, 2, 3, \ldots, n.$$

Remark 12.5.12. As in the cases considered above, the set of polynomials forms an algebra when the space of values is C and X.

In the expression of P given above we call n the degree of the polynomial P.

We give now some simple examples of polynomials.

Examples 12.5.13. Let X be a Hilbert space $L^2_{[0, 1]}$ and set

$$P(f_1, f_2, f_3) = \iiint K(t_1, t_2, t_3) \times$$
$$\times f_1(t_1) \bar{f}_2(t_2) \bar{f}_3(t_3), dt_1 \, dt_2 \, dt_3$$

which is clearly a polynomial of $(1, 2)$-type.

If X is the space of all operators on a Hilbert space H then

$$P(T_1, \ldots, T_n, S_1, \ldots, S_m) = T_1, \ldots, T_n \cdot S_1^*, \ldots, S_m^*$$

is clearly a polynomial of (n, m)-type: if the operators commute then this is obviously a symmetric polynomial of (n, m)-type.

For the proof of the approximation theorem we consider certain sets which plays a fundamental role. These are defined as follows:

Let X be a Banach space and K be a compact subset of X and we consider the following spaces of functions:

1. $C(K, X) = (f, f : K \to X, f$ continuous),

2. $P(K, X) = (p, p : K \rightarrow X$, p continuous and polynomial), and if
          $Y = \mathbf{R}$ or $\mathbf{C}$,

3. $C(K, Y) = (f, f : K \rightarrow Y$, f continuous),

4. $P(K, Y) = (p, p : K \rightarrow Y$, p continuous and polynomial).

It is clear that $C(K, Y)$ and $P(K, Y)$ are algebras under pointwise multiplication. Our first basic result relates these subspaces.

THEOREM 12.5.14.   *The algebra $P(K, \mathbf{R})$ is dense in $C(K, \mathbf{R})$.*

Here of course we consider the norm given by the usual sup-norm. Also we note that if $K = [0, 1]$ our theorem is exactly the Weierstrass theorem about the density of polynomials in the space of real-valued continuous functions of $[0, 1]$.

Proof of Theorem 12.5.14.   Our proof depends upon a basic idea used by de Branges (1959). This consists of remarking that the support of certain measures has a special property with respect to some subspaces and using this fact for commutative subalgebras of the algebras of continuous functions on a compact Hausdorff space. For this we need the following notion:

DEFINITION 12.5.15.   A level set for a family of functions $S \subset C(K, \mathbf{R})$ is a subset $K_0 \subset K$ such that S/K_0 contains only constant functions.

It is obvious that any point in K is a level set. Then using Zorn's Lemma we readily obtain that any point is contained in a maximal level set and clearly every maximal level set is closed. Of course this is valid for arbitrary level sets.

For the proof of the Theorem 12.5.14. we need the following

LEMMA 12.5.16.   *Let F be a continuous linear functional on $C(K, \mathbf{R})$ such that:*

1.   *F is an extreme point of the set of all linear continuous functionals Z on $C(K, \mathbf{R})$ such that $\|Z\| \leq 1$, Z annihilates $P(K, \mathbf{R})$,*

2.   *If μ is the measure determining F so that*

          $$F(f) = \int f(s) \, d\mu(s),$$

then the support of μ, supp μ, is a level set of $P(K, \mathbf{R})$.

Proof. Suppose that this is not so and thus we can find an element of

$P(K, \mathbf{R})$, say g_0, such that it is not constant on $\operatorname{supp}\mu$. We know that the $\operatorname{supp}\mu$ is equal to the $\operatorname{supp}|\mu|$. We can suppose without loss of generality that

$$0 \leqq g_0 \leqq 1$$

and then we consider the following measures

$$\mu_1 = g_0\mu, \qquad \mu_2 = (1 - g_0)\mu.$$

It is easy to see that

$$\|\mu_1\| + \|\mu_2\| = 1$$

and thus the measures

$$\mu_1^* = \mu_1/\|\mu_1\|, \qquad \mu_2^* = \mu_2/\|\mu_2\|$$

have the property that

$$\mu = \|\mu_1\|\mu_1 + \|\mu_2\|\mu_2$$

which contradicts the fact that F is an extreme point. Thus $\operatorname{supp}\mu$ is a level set for $P(K, \mathbf{R})$ and the lemma is proved.

Now we are ready to prove Theorem 12.5.14. Indeed, by the Hahn–Banach theorem, Krein–Milman theorem and Riesz–Kakutani theorem there exists a nonzero continuous linear functional F which satisfies the conditions of the lemma. The lemma gives us that the $\operatorname{supp}\mu$ of the measure determining F is a level set for $P(K, \mathbf{R})$ and thus it reduces to a point, say t_0. Then we have

$$F(f) = f(t_0)$$

which is clearly an impossibility. This completes the proof of the theorem.

The approximation theorem is as follows:

THEOREM 12.5.17. *The space $P(K, X)$ is dense in $C(K, X)$.*

Proof. First we consider the case of real Banach spaces. We need the notion of a normal function and some properties connected with normal functions.

DEFINITION 12.5.18. A function $f : K \to \mathbf{R}$ is called normal if it is continuous and the values are in $[0, 1]$.

The following result can be proved by induction: see also 10.3.3.

THEOREM 12.5.19. *For any open covering of K, $(v_1, \ldots, v_n)$ there exist normal functions $h_1, \ldots, h_n$ such that*:

1. $h_i = 0$ *in the exterior of v_i, $i = 1, 2, 3, \ldots, n$.*

2. $1 = \sum_1^n h_i(s),$ $s \in K.$

Using this and Theorem 12.5.17, the proof of the approximation theorem is as follows: for any $\varepsilon > 0$ we consider the sets

$$v_t = (s, \|f(t) - f(s)\| < \varepsilon)$$

for f fixed in $C(K, X)$. Then clearly this is an open set and $(v_t)_{t \in K}$ is an open covering of K. From the compactness we can find a finite sets $v_{t_1}, \ldots, v_{t_n}$ (we set $v_{t_i} = v_i$) and thus by Theorem 12.4.19 we find the normal functions (h_i) and this gives us the possibility of considering the function

$$f_1(t) = \sum_1^n f(t_i) h_i(t)$$

and we remark that

$$\|f(t) - f_1(t)\| < \varepsilon.$$

Since the functions h_i are approximatable by functions in $P(K, \mathbf{R})$ we obtain that f can be approximated by functions in $P(K, X)$ and the theorem is proved.

Now we consider the case of complex Banach spaces. To adapt the technique of the proof in the case of real Banach spaces to complex Banach spaces we need the following notion:

DEFINITION 12.5.20. *If S is a family of functions in $C(K, \mathbf{C})$ then K_0 is called an antisymmetry set for S if S/K_0 contains only constant real functions.*

Exactly as in the case of level sets, each point is contained in a maximal set of antisymmetry and each maximal set of antisymmetry is closed. The following analogue of Lemma 12.5.16 can be proved in a similar way and thus we omit the details.

LEMMA 12.5.21. *Let F be a continuous linear functional on $C(K, \mathbf{C})$ such that*

1.   *F is an extreme point of the set of all functionals Z, continuous on*
        $C(K, \mathbf{C})$ *such that* $\|Z\| \leq 1$, *Z annihilates* $P(K, \mathbf{C})$,

2.   *if* μ *is the measure determining F so that*

$$F(f) = \int f(s) \, d\mu(s)$$

then the support of μ, *supp* μ *is an antisymmetry set for* $P(K, \mathbf{C})$.

This Lemma permits us to obtain an analogue of Theorem 12.5.14 for the case of $P(K, \mathbf{C})$ and this is given as

THEOREM 12.5.22. $P(K, \mathbf{C})$ *is dense in* $C(K, \mathbf{C})$.

Since the proof using Lemma 12.5.21 is exactly as for that of Theorem 12.5.14., we omit the details.

The approximation theorem in the case of complex Banach spaces can be proved as for the real spaces and the result is as follows:

THEOREM 12.5.23. $P(K, X)$ *is dense in* $C(K, X)$.

For the uniqueness of the topological degree we need the following fact about α-set contractions proved by Daneš (1971) and Nussbaum (1971). This incidentally contains the fact about compact mappings used in Theorem 12.2.13.

THEOREM 12.5.24.   *Let G be an open subset of a Banach space X and* $f : G \to X$ *be an* α-*set contraction which is (Fréchet) differentiable at* $x_0 \in G$. *Then the Fréchet differential* df_{x_0} *of f at* x_0 *is an* α-*set contraction,* df_{x_0} *is a Friedholm operator of index zero and* $\sigma(df_{x_0}) \cap (z, |z| > k_{df_{x_0}})$ *is a finite set.*

Proof. Let $\varepsilon > 0$ and denote $df_{x_0} = T$. Consider S to be an arbitrary bounded set of diameter $d > 0$. It is clear that the assertion is proved if we show that

$$\alpha(A(S)) \leq kd + \varepsilon$$

where $\alpha(\,,)$ is the Kuratowski measure of noncompactness and $k = k_{df_{x_0}}$.
Since f is Fréchet differentiable at x_0 we have

$$f(x) = f(x_0) + T(x - x_0) + R(x - x_0)$$

with

$$R(z) = 0(\|z\|).$$

In this case we find $r > 0$ such that if $\|x - x_0\| < r$ then

$\|R(x - x_0)\| \leq (\varepsilon/2d)\|x - x_0\|$. Pick s in $(0, d)$ and set

$$sS = (sx, x \in S)$$

and we remark that we have

$$T(sS) = sT(S) \subset (f(x) - f(x_0) - R(x - x_0))$$

with

$$x \in (x_0 + sS).$$

Since

$$\|R(x - x_0)\| \leq (\varepsilon/2d)\|x - x_0\| \leq \varepsilon s/2$$

we get that the set is contained in the $\varepsilon s/2$ neighbourhood of the set $(f(x) - f(x_0): x \in sS)$ and since f is supposed to be an α-set contraction, the measure of noncompactness of this set is less than or equal to $\varepsilon s + ksd$. Since ε is arbitrary we get that

$$\alpha(sT(S)) \leq ksd.$$

This obviously implies the assertion about df_{x_0}.

The fact that the points in $\sigma(df_{x_0})$ satisfying the inequality in the theorem are contained in a finite set is valid for more general maps, namely for local power α-set contractions. For more details as well as a proof we refer to Constantin (1972) and the present author's book about operator theory.

Now the fact that df_{x_0} is a Fredholm operator of index zero follows from the fact that df_{x_0} is an α-set contraction.

Remark 12.5.25.   It is of interest to know if the above result is valid in the case of local power α-set contractions.

Suppose that f is an α-set contraction defined on G and we suppose that it is differentiable at $x_0 \in G$. We may assume without loss of generality that $0 \in G$ and that $f(0) = 0$. (Indeed, if $f(0) \neq 0$ then we make a translation which obviously not change the differentiability properties of f.)

By definition, the index of an isolated fixed point of f is the number $\mathrm{ind}(f, x_0) = \deg(I - f, S, x_0)$. In our case $x_0 = 0$ and S is open ball such that f has no other fixed points in S other than Θ. Then we have the following result, stated for the case of compact mapping by Krasnoselskii (1964):

PROPOSITION 12.5.26.   *If f is as above and z_0 is not a characteristic value for f' and where z_0 satisfies the inequality $|z_0| > k_f$. Then there exists a* *neighbourhood $V \ni \Theta$ such that no eigenvectors for eigenvalues close to z_0 are contained in V.*

Proof. Suppose on the contrary that such a neighbourhood does not exist and thus for each n, in the ball with radius $1/n$ there exist eigenvectors x_n and these are associated with the eigenvalues near z_0. Without loss of generality we suppose that these eigenvalues are (z_n) and that $z_n \to z_0$. Thus we have that

$$f(x_n) = z_n x_n$$

and since f is an α-set contraction we readily get that (x_n) has a convergent subsequence. Again we can suppose that the subsequence is just (x_n). This implies that z_0 is an eigenvalue and this contradiction proves the proposition.

PROPOSITION 12.5.27. *Let f be as above and suppose that 1 is not an eigenvalue for f'. Then Θ is an isolated solution of the equation*

$$x - f(x) = \Theta$$

and the index of Θ is equal to $(-1)^\beta$ and where β is the sum of multiplicities of the characteristic values of f' in $(0, 1)$.

Proof. According to the hypothesis, 1 is not an eigenvalue of f' and this implies that there exists a positive constant M such that

$$\|y - f'(y)\| \geq M \|y\|$$

for all y. But f is supposed to be Fréchet differentiable and thus there exists a function $w(x)$ such that in a neighbourhood of zero,

$$f(x) = f'(x) + w(x)$$

where w satisfies the property that

$$\lim_{\|x\| \to 0} w(x)/\|x\| = 0.$$

Let $d > 0$ such that for x in a ball with radius d, $\|w(x)\| \leq M\|x\|/3$ which is possible as follows from the property of w stated above. In this case we have

$$\|x - f(x)\| \geq \|x - f'(x)\| - \|w(x)\|$$
$$\geq M\|x\| - M\|x\|/3 = (2M)/3 \cdot \|x\|$$

and thus Θ is an isolated solution of the equation

$$x - f(x) = \Theta.$$

Let us consider the following homotopy

$$H(x, t) = x - tf(x) - (1 - t)f'(x)$$

which has the property that for any x, $\|x\| = d$ (d is as considered above) then

$$H(x, t) \neq \Theta$$

for all t, which can be easily verified.

In this case by the properties of the degree, listed above (Def. 12.5.2),

$$d(I - f, B_d, \Theta) = d(I - f', B_d, \Theta)$$

where B_d is the ball with the center in Θ and radius d. We note that $d(I - f', B_d, \Theta)$ exists since f' is an α-set contraction. Consider now X_1 the space generated by all the eigenvectors corresponding to all eigenvalues of f' situated in $(0, 1)$. Then according to Theorem 12.5.24, this space is finite-dimensional, and we let β be its dimension. Let us consider X_2 to be the complementary subspace of X_1 and further set $B_{i,d} = B_d \cap X_i$ and let P_1 be the projection of X onto X_1.

Since the degree satisfies the product formula, we get,

$$d(I - f', B_d, \Theta) = d(I - f, B_{1,d}, \Theta) \cdot d(I - f', B_{2,d}, \Theta).$$

we consider now the function

$$H_1(x, t) = x - tf'(x)$$

and we have that

$$H_1(x, t) = \Theta$$

if $x \in \partial B_{2,d}$ and $t \in [0, 1]$ iff f' has a characteristic value in $[0, 1]$ (or $Ix = 0$ for $x \in B_{2,d}$). Since in X_2 this is impossible we have that

$$d(I - f', B_{2,d}, \Theta) = d(I, B_{2,d}, \Theta) = 1.$$

Consider now the function

$$H_2(x, t) = ((2t - 1)I - tf')(x)$$

we have that $H_2(x, t) \neq \varnothing$ for $x \in \partial B_{1,d}$ which implies that

$$d(I - f', B_{1,d}, \Theta) = d(-I, B_{1,d}, \Theta)$$

Since we prove below that

$$d(-I, B_{1,d}, \Theta) = (-1)^\beta$$

the assertion of the proposition is proved.

PROPOSITION 12.5.28. *Let Θ be as above. Then the degree satisfies the relation*

$$d(-I, B_{1,d}, \Theta) = (-1)^\beta.$$

Proof. Using the additivity property of the degree (Definition 12.4.1, condition 3) we reduce the above formula to prove a formula for $d(-I, C_1, \Theta)$ where

$$C_1 = ((x_1, \ldots, x_\beta), |x_i| < 1, i = 1, 2, 3, \ldots.)$$

Further we use now the product property of the degree (Definition 12.5.1, condition 4) we reduce the problem to the computation of the number $d(-I, (-1, 1), \Theta)$, which must be equal to -1.

Indeed, for this we use the function

$$h(x) = \begin{cases} -x & x \in (-1, 1) \\ x - 2 & x \in (1, 4) \end{cases}$$

and from the additivity property of the degree, we obtain,

$$d(g, (-1, 4), \Theta) = d(-I, (-1, 1), \Theta) + d(x - 2, (1, 4), \Theta)$$

and since the function h is homotopic with

$$x \to (x + 6)/5$$

on $(-1, 4)$ we get further

$$d(g, (-1, 4), \Theta) = d((x + 6)/5, (-1, 4), \Theta).$$

From the additivity property we have the following property and, since it is very important, this is given as

PROPOSITION 12.5.29. *If f is a k-set contraction and $x - f(x) = y$ has no solutions in G then $d(I - f, G, y) = 0$*

We now use this property to prove that

$$d((x + 6)/5, (-1, 4), \Theta) = 0$$

and this is indeed so, because $x - (x + 6)/5 = 0$ is impossible on $(-1, 4)$. Once again using the additivity property we obtain,

$$d(x - 2, (-1, 4), \Theta) = d(x - 2, (-1, 1), \Theta) + d(x - 2, (1, 4), \Theta)$$

and

$$d(x - 2, (-1, 1), \Theta) = 0.$$

Now we remark that the function $f(x) = x - 2$ is homotopic on $[-1, 4]$ to the function $I(x) = x$ and since it has no solutions on the boundary, we have

$$d(x - 2, (-1, 4), \Theta) = d(I(x), (-1, 4), \Theta) = 1$$

since the degree satisfies the normalization property (Definition 12.4.2. condition 3).

Now since

$$d(g, (-1, 4), \Theta) = 0 = d(-I, (-1, 1), \Theta) + d(x - 2, (1, 4), \Theta)$$
$$= d(-I, (-1, 1), \Theta) + 1$$

which implies that

$$d(-I, (-1, 1), \Theta) = -1$$

and Proposition 12.5.29 is proved.

Now we note that for differentiable α-set contractions we can give a formula for the degree.

THEOREM 12.5.30. *Let G be an open set in a Banach space X and let* $f : G \to X$ *be any α-set contraction which is differentiable (i.e. Fréchet differentiable). Let y be a point of X. Then there exists a point y' which is a regular point of f such that*

$$d(I - f, G, y) = \sum_{x \in (I - f)^{-1}(y')} (-1)^{m(df_x)}$$

where $m(df_x)$ means the sum of the multiplicities of the characteristic values of df_x in $(0, 1)$.

Proof. According to the Sard–Smale version of the Brown–Sard theorem we may choose y' sufficiently near to y such that we have

$$d(I - f, G, y) = d(I - f, G, y')$$

and mapping the additivity and the result in 12.5.27 the assertion follows.

12.6. THE COMPUTATION OF THE TOPOLOGICAL DEGREE

In recent years there has been considerable interest in the computation of the fixed points (or roots) of continuous functions. The first important step in the computation of the topological degree is the paper of F. Stenger (1974). For the computation of the topological degree, Frank Stenger proves a basic formula which, in a simpler context, was anticipated by Jacques Hadamard in 1904. In what follows we prove this basic formula following a proof given by M. Stynes in his Ph.D. dissertation (University of Oregon, 1977). Next we present some algorithms for the computation of the topological degree which were obtained by R. Kearfoot and M. Stynes. For more information we refer to P. Erdelski (1973), R. Kearfoot (1977, 1978, 1979, 1979). F. Stenger (1974), M. Stynes (1977) and T. O'Neil and J. Thomas (1975).

First we note that if (a, b) is an open bounded interval and f is a continuous function on the closure of the interval with values in $\mathbf{R}$ then

$$\deg(f, (a, b), 0) = \tfrac{1}{2}(\operatorname{sgn} f(b) - \operatorname{sgn} f(a))$$

where

$$\operatorname{sgn} t = \begin{cases} 1 & t > 0 \\ 0 & t = 0 \\ -1 & t < 0. \end{cases}$$

$$\operatorname{sgn} f^{\sim} = (\operatorname{sgn} f_1, \ldots, \operatorname{sgn} f_n), \text{ if } f^{\sim} = (f_1, \ldots, f_n),$$
$$f^{\sim} : D \to \mathbf{R}^n.$$

In what follows we use also the following notation

$$\Delta^n(B_1, B_2, \ldots, B_n) = \begin{vmatrix} b_{11} & b_{12} & \cdots & b_{1n} \\ b_{21} & b_{22} & \cdots & b_{2n} \\ \cdot & \cdot & \cdots & \cdot \\ \cdot & \cdot & \cdots & \cdot \\ \cdot & \cdot & \cdots & \cdot \\ b_{n1} & b_{n2} & \cdots & b_{nn} \end{vmatrix}$$

and where

$$B_i = (b_{i1}, b_{i2}, \ldots, b_{in})$$

for $i = 1, 2, 3, \ldots, n$.

Suppose now that we have an m-simplex with vertices $(x_0, x_1, \ldots, x_m)$ in $\mathbf{R}^n$. We say that two orderings of the vertices are equivalent if one is obtained from the other by an even permutation of symbols. It is easy to see that this is in fact an equivalence relation and the set of all orderings of vertices of the simplex are partitioned into two equivalence classes. The one is called positive and the other, of course, negative. If $m = 0$ we have obviously only one class and this is by definition positive.

Let $(x_0, x_1, \ldots, x_n)$ be a set in $\mathbf{R}^n$; the orientation of $(x_0, x_1, \ldots, x_n)$ is by definition

$$\text{sgn} \begin{vmatrix} 1 & x_{01} & x_{02} & \cdots & x_{0n} \\ 1 & x_{11} & x_{12} & \cdots & x_{1n} \\ \cdot & \cdot & \cdot & \cdots & \cdot \\ \cdot & \cdot & \cdot & \cdots & \cdot \\ 1 & x_{n1} & x_{n2} & \cdots & x_{nn} \end{vmatrix}$$

where

$$x_i = (x_{i1}, x_{i2}, \ldots, x_{in}).$$

Now if $(e_0, e_1, \ldots, e_n)$ is the Cartesian basis for $\mathbf{R}^n$ this gives rise to an ordering of $\mathbf{R}^n$: $\mathbf{R}^n$ with this order is called the 'oriented $\mathbf{R}^n$'. It is not difficult to see that the orientation with respect to another basis is the same iff the determinant of the affine mapping taking one basis to another is positive. This implies that given an orientation in $\mathbf{R}^n$ we have an induced orientation on all subspaces. See also Section 12.1.

If we write $\langle x_0, x_1, \ldots, x_n \rangle$ then this means the oriented n-simplex with the vertices $x_0, x_1, \ldots, x_n$.

DEFINITION 12.6.1.   An m-chain, c^m, is a finite algebraic sum of oriented m-simplexes with integer coefficients.

Example 12.6.2.   Let $m = 3$ and then

$$\langle x_0, x_1, x_2, x_3 \rangle - \langle a_1, x_0, x_2, x_3 \rangle$$

is a 3-chain.

If c^m is a chain then we suppose that all possible cancellations have been made. This means that if c^m has the form

$$c^m = \sum_k p_k S_k^m$$

then $S_k^m \neq S_l^m$ for $k \neq l$ and $p_k \neq 0$.

DEFINITION 12.6.3.   If S^m is the simplex $\langle x_0, x_1, \ldots, x_m \rangle$ then the boundary is the $(m-1)$-chain defined as follows:

$$\partial S^m = \sum_{i=0} (-1)^i \langle x_0, \ldots, x_{i-1}, \hat{x}_i, x_{i+1}, \ldots x_n \rangle$$

(and if $m = 0$, $\partial S^m = 0$)

where $\hat{}$ means the omission of the corresponding point (or terms).

If c^m is an m-chain

$$c^m = \sum_k p_k S_k^m$$

then the boundary is defined by the formula

$$\partial c^m = \sum_k p_k \partial S_k^m$$

and thus it is an $(m-1)$-chain (for $m > 0$).

From the definition it is not difficult to see that the following proposition holds.

PROPOSITION 12.6.4.   *For any m-chain*

$$\partial(\partial c^m) = 0$$

i.e. $\partial^2 = 0$.

In what follows we use the notation S_j^m to denote the collection of all m-simplexes indexed by j and

$$\bigcup_j S_j^m = (p, p \in S_j^m \quad \text{for some } j)$$

$$\bigcap_j S_j^m = (p, p \in S_j^m \quad \text{for all } j).$$

If c^m is an m-chain, $c^m = \sum_j p_j S_j^m$ then $p \in c^m$ means that $p \in S_j^m$ for some j.

DEFINITION 12.6.5.   Let S^n be an oriented n-simplex in $\mathbf{R}^n$ (not necessarily the standard simplex) and consider $p \in \mathbf{R}^n \backslash (\partial S^n)$. We say that S^n and p are in general position.

The intersection number $i(S^n p)$ is defined as follows:

$$i(S^n, p) = \begin{cases} 0 & \text{if } p \notin S^n \\ \text{orientation of } S^n & \text{if } p \in S^n. \end{cases}$$

Consider now a hyperplane in the oriented space $\mathbf{R}^n$, say H^{n-1} and further we suppose that $H^{n-1} \cap G^1 = (p)$ where G^1 is a line in $\mathbf{R}^n$. Choose $\langle pa_1, \ldots, a_{n-1} \rangle$ and $\langle pb \rangle$ positively oriented in H^{n-1} and G^1 respectively.

Then it is not difficult to see that

$$i(H^{n-1}, G^1) = \text{orientation of } \langle pa_1, \ldots, a_{n-1}b \rangle$$

and that $i(H^{n-1}, G^1)$ is in fact independent of the choice of points $a_1, \ldots, a_{n-1}, b$.

DEFINITION 12.6.6. Two oriented simplexes S^{n-1} and T^1 are in general position in $\mathbf{R}^n$ if $S^{n-1} \cap T^1$ is a point or the empty set.

Then we define

$$i(S^{n-1}, T^1) = \begin{cases} 0 \text{ if } S^{n-1} \cap T^1 = \varnothing \\ i(H^{n-1}, G^1) \text{ if } S^{n-1} \cap T^1 \neq \varnothing \end{cases}$$

since $S^{n-1} \cap T^1$ determine a hyperplane H^{n-1} and a line G^1 such that

$$H^{n-1} \cap G^1 = S^{n-1} \cap T^1.$$

We indicate now how we can compute the number $i(S^{n-1}, T^1)$ in terms of orientation.

Suppose that $S^{n-1} = \langle a_1 a_2, \ldots, a_n \rangle$ and let $p \in S^{n-1} \backslash \partial S^{n-1}$. Then the orientation of S^{n-1} is the same as the orientation of $pa_2, \ldots, a_n$. Indeed, we have

$$p_k = \sum_i a_i a_{ik}, \qquad a_i > 0, \qquad \sum_1^n a_i = 1$$

and since a multiple of one row of a determinant can be substracted from another row without affecting the value of the determinant, we obtain that if in $\mathbf{R}^n$, $S^{n-1} \cap T^1$ is a point not in $\partial S^{n-1} \cup \partial T_1$ where $T^1 = \langle cb \rangle$ then $i(S^{n-1}, T^1) = \text{orientation of } \langle pa_2, \ldots, a_n b \rangle$.

DEFINITION 12.6.7. Let $a^p = \sum_i u_i X_i^p$, $b^q = \sum_j v_j Y_j^q$ be two oriented chains in $\mathbf{R}^n$ such that either $p = n-1$ and $q = 1$ or $p = n$ and $q = 0$. We say that a^p and b^q are in general position if for each pair (i, j), X_i^p and Y_0^q are in general position and then we define

$$i(a^p, b^q) = \sum_{i,j} u_i v_j i(X_i^p, Y^q).$$

For a detailed proof of the following result we refer to Theorem 2.2 of J. Cronin (1964)

THEOREM 12.6.8.   *If c^n and ∂b^1, also ∂c^n and d^1 are in general position then*

$$i(c^n, \partial d^1) = (-1)^n i(\partial c^n, d^1).$$

We define now the order of a point relative to $(n-1)$-boundaries.

DEFINITION 12.6.9.   Let a^{n-1} be an $(n-1)$-boundary, i.e., there exists a chain c^n such that $a^{n-1} = \partial c^n$ (may be many!) and suppose that c^n and p are in general position. Then the order of p relative to a^{n-1} is the number defined by

$$r(a^{n-1}, p) = i(c^n, p).$$

It is easy to see that $r(\,,\,)$ is independent of the chain c^n satisfying the relation $a^{n-1} = \partial c^n$.

DEFINITION 12.6.10.   An n-dimensional polyhedron K^n is a set with the following properties:

1. $$K^n = \bigcup_{i=1}^{m} S_i^n,$$

2. S_i^n are oriented n-simplices such that $S_i^n \cap S_j^n$ is empty or it is a common face whose vertices are vertices of both S_i^n and S_j^n.

Sometimes we shall use the following notations

$$K^n = \bigcup_{i=1}^{m} S_i^n = \sum_{i=1}^{m} S_i^n$$

depending upon the context.

DEFINITION 12.6.11.   An n-region is any connected n-dimensional polyhedron.

Now we define the topological degree for mappings defined on n-regions.

DEFINITION 12.6.12.   Let K^n be an n-region,

$$K^n = \sum_{i=1}^{m} S_i^n$$

and $f : K^n \to \mathbf{R}^n$ be a continuous mapping. Let $p \in \mathbf{R}^n \setminus f(\partial K^n)$ and suppose that a^{n-1} is an $(n-1)$-boundary which can be continuously deformed into $f(\partial K^n)$ without passing through the point p. Then the degree of f on K^n

relative to p is the number $d(f, K^n, p)$ defined by the formula

$$d(f, K^n, p) = r(a^{n-1}, p).$$

In Cronin (1964) it is proved that such an a^{n-1} exists and this may be constructed by breaking up each S_i^n into a finite sum of n-simplexes.

Also in Cronin (1964) it is proved that there exists a function $f^* : K^n \to \mathbf{R}^n$ which maps n-chains formed by the above sums into n-chains in $\mathbf{R}^n$ in such a way that $f^*(\partial K^n)$ is an $(n-1)$-boundary as is required in Definition 12.6.12.

The function f^* is called the 'chain approximation of f' on K^n with respect to the point p and has also the following important properties:

1. f^* is arbitrarily close to f on K^n,

2. $f^* (\partial K^n) = \partial(f^*(K^n))$.

In the case $n = 1$, $\partial K^2 = \langle x_m \rangle - \langle x_0 \rangle$ and we take $a^\circ = \partial(\langle f(x_0)f(x_m) \rangle)$ which gives

$$d(f, K^1, p) = \tfrac{1}{2}(\mathrm{sgn}\,((f(x_m) - p) - \mathrm{sgn}\,(f(x_0) - p)).$$

The properties of $d(f, K^n, p)$ are given in the following proposition; for a proof we refer to the book of J. Cronin (1964) quoted above.

PROPOSITION 12.6.13.   *The degree* $d(f, K^n, p)$ *defined above has the following properties*:

1.   *if $p \notin f(K^n)$ then $d(f, K^n, p) = 0$*,

2. $d(f, K^n, p)$ *is an integer*,

3. $d(f, K^n, p)$ *is homotopic invariant*.

Note 12.6.14.   According to the uniqueness of the degree satisfying certain properties we see that this $d(f, K^n, p)$ coincides with the degree defined in the beginning of this chapter.

Let K^n be an n-region which is written as a sum of oriented n-simplexes in such a way that K^n coincides with the topological boundary of K^n as a subset of $\mathbf{R}^n$.

Example 12.6.15.   Let us consider $n = 1$ and $K^1 = \sum_0^3 \langle x_i x_{i+1} \rangle$ and thus

$$\partial K^1 = \sum_0^3 \partial \langle x_i x_{i+1} \rangle = \sum_0^3 (\langle x_{i+1} \rangle - \langle x_i \rangle) = x_4 - x_0.$$

We note that K^1 may be represented as the set in the following diagram:

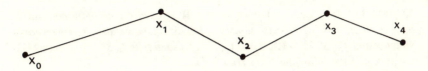

Let $f = (f_1, \ldots, f_n) : K^n \to \mathbf{R}^n$ be a continuous function and suppose that it has the property $f(x) \neq 0$ for $x \in \partial K^n$.

DEFINITION 12.6.16. We say that a boundary K^n is sufficiently refined relative to sgn f if:

1. if $n = 1$, $\partial K^1 = \langle x_m \rangle - \langle x_0 \rangle$ then $f(x_m) f(x_0) \neq 0$,

2. $n > 1$, if ∂K^n is $\bigcup_{i=1}^m \beta_i^{n-1}$ are $(n-1)$-regions then the following assertions hold:

(a) the $(n-1)$-dimensional interiors of β_i^{n-1} are pairwise disjoint,

(b) there exists at least one function f_{r_i} which is nonzero on each region β_i^{n-1},

(c) if $f_{r_i} \neq 0$ on β_i^{n-1} then β_i^{n-1} is sufficiently refined relative to sgn f_i^{n-1}, where $f_i^{n-1} = (f_1, \ldots, f_{i-1}, \hat{f}_i, f_{i+1}, \ldots, f_n)$ and $\hat{}$ signifies the omission.

Example 12.6.17. Take $n = 2$ and consider that

$$K^2 = \sum_{i=0}^{p-1} x_i x_{i+1}$$

where $x_p = x_0$. Let $f = (f_1, f_2)$ and this is sufficiently refined relative to sgn f if at least one of f_1, f_2 is nonzero on each segment (x_i, x_{i+1}) and neither is zero at any x_i's. Of course here

$$\beta_i^1 = \langle x_i x_{i+1} \rangle, \qquad 0 \leq i \leq 1.$$

DEFINITION 12.6.18. Let $n > 0$ and we say that a connected set in K^n is a Q^n-set if for $q \in Q^n$,

$$f_1(q) = 0 = \cdots = f_{i-1}(q) = \hat{f}_i(q) = f_{i+1}(q) = \cdots = f_n(q)$$

and sgn $f_i(q) = \Delta$, where $\Delta = +1$ or -1 (but fixed).

If two sets are associated with different i's in $(1, 2, 3, \ldots, n)$ or with different Δ, then we say that these are of different type. It is not difficult to see that since Q^n-sets are connected they must lie in the interiors of β_i^{n-1}.

LEMMA 12.6.19.   *Let $n > 1$ and $a \neq \Theta^n$ in $\mathbf{R}^n$ situated on one of the coordinate axes with the rth coordinate nonzero. Let S^{n-1} be an $(n-1)$-dimensional simplex and $f_0 : S^{n-1} \to \mathbf{R}^n$, $f_0 = (f_1, \ldots, f_n)$ such that $f_0(S^{n-1})$ is an $(n-1)$-simplex and the following properties hold:*

(a) $f_r \neq 0$ on S^{n-1} with $\operatorname{sgn} f_r / S^{n-1} = \operatorname{sgn} (rth$ *coordinate of a),*

(b) $i(f_0(S^{n-1}), \langle \Theta^n a \rangle$ and $i(f_r^{n-1}(S^{n-1}), \Theta^{n-1})$ *are defined and $i(,)$ is the intersection number and $f_r^{n-1} = (f_1, \ldots, f_{r-1}, \hat{f}_r, f_{r+1}, \ldots, f_n)$ and $\hat{\ }$ means the omission.*

Then we have

$$i(f(S^{n-1}), \langle \Theta^n a \rangle) = (-1)^{r+n} i(f_r^{n-1}(S^{n-1}), \Theta^{n-1}) \operatorname{sgn} f_r.$$

Proof. If

$$i(f(S^{n-1}), \langle \Theta^n a \rangle) = 0$$

then

$$f(S^{n-1}) \cap \langle \Theta^n a \rangle = \varnothing$$

and thus $f_r^{n-1}(S^{n-1}) \cap \langle \Theta^{n-1} \rangle = \varnothing$ by (a) and

$$i(f_r^{n-1}(S^{n-1}), \Theta^{n-1}) = 0.$$

Thus, in this case the assertion of the lemma is verified.
   Suppose now that $f(S^{n-1}) \cap \langle \Theta^n a \rangle = (\tilde{p})$

where

$$\tilde{p} = \left(\underset{1}{0}, \underset{2}{0}, \ldots, 0, \underset{r}{p}, 0, \ldots, \underset{n}{0}\right), \qquad p \neq 0.$$

Now according to (a) $\operatorname{sgn} p = \operatorname{sgn} f_r$ and

$$p \in f(S^{n-1}) \cap (f(S^{n-1})).$$

Since $i(f_r^{n-1}(S^{n-1}), \Theta^{n-1})$ is defined.
   Let us consider the point $\tilde{p}'$ defined as follows:

$$\tilde{p}' = (0, 0, \ldots, 0, 2p, 0, \ldots, 0)$$

and if $f(S^{n-1}) = \langle y_1 y_2 \cdots y_n \rangle$ then

$$i(f(S^{n-1}), \Theta^n a) = \text{orientation of } \langle py_2y_3 \cdots y_n p' \rangle$$

$$= \text{sgn}\,(-1)^{n+1}((1, p), (1, y_2), \ldots, (1, y_n), (1, p'))$$

$$= \text{sgn} \begin{vmatrix} 1 & 0 & 0 & \cdots & p & 0 & \cdots 0 \\ 1 & y_{12} & y_{13} \cdots y_{1r} & y_{1r+1} \cdots y_{1n} \\ \cdots\cdots\cdots\cdots\cdots\cdots\cdots\cdots\cdots \\ 1 & y_{n1} & y_{n3} \cdots y_{nr} & \cdots y_{nn} \\ 1 & 0 & 0 \cdots 2p & \cdots 0 \end{vmatrix}$$

where $y_m = (y_{m1}, \ldots, y_{mn})$ for $1 \leq m \leq n$.

To compute this we substract the first row from the last and expand the determinant in terms of the last row,

$$i(f(S^{n-1}), \Theta^n a)$$

$$= (-1)^{n+r} \text{sgn}\,p\,\text{sgn} \begin{vmatrix} 1 & 0 & 0 & \cdots & p & . & \cdots 0 \\ 1 & y_{21} & y_{22} \cdots y_{2r} & \cdots & y_{2n} \\ \cdots\cdots\cdots\cdots\cdots\cdots\cdots\cdots \\ 1 & y_{n1} & y_{n2} & \cdots & & y_{nn} \end{vmatrix}$$

$$= (-1)^{n+r} \text{sgn}\,f_r \cdot i(f_r^{n-1}(S^{n-1}), \Theta^{n-1})$$

and the lemma is proved.

LEMMA 12.6.20.   *Let K^n be an n-region and $f : K^n \to \mathbf{R}^n$ continuous with $f(x) \neq \Theta^n$ for $x \in \partial K^n$. Suppose further that K^n has been subdivided into a finite number of $(n-1)$-regions, say β_i^{n-1}, so that this is sufficiently refined relative to sgn f. If f^* is a chain approximation of f on K^n with respect to Θ^n with $(S_k^{n-1} : k \in I)$, the associated simplicial subdivision of $\partial K^n = \sum_j \beta_j^{n-1}$. If a is a point lying on one of the coordinate axes with the rth coordinate nonzero then we may arrange these so that the following Properties hold:*

(a) *the sufficient refinement of K^n relative to $\text{sgn} f$ is also a sufficient refinement of ∂K^n relative to sgn f^* with the property that $f_{r_j} \neq 0$ on β_j^{n-1} implies that $f_{r_j}^* \neq 0$ on β_j^{n-1},*

(b) $\|a\| > \max_{x \in K^n} \|f^*(x)\|$

(c) *for every S_k^{n-1} both $i(f^*(S_k^{n-1}), \langle \Theta^n a \rangle)$ and $i(f_r^{*n-1}(S_k^{n-1}), \Theta^{n-1})$ are defined.*

Proof. From the properties of f^* we know that we may suppose that it is very near to f and an induction argument on n implies that (a) is true for f^*. Now since f^* may be suppose very near to f this clearly gives that (b) holds for f^*. Now for (c) suppose that $i(f^*(S_k^{n-1}), \Theta^n a)$ is not defined and thus $f^*(S_k^{n-1}) \cap \langle \Theta^n a \rangle$ have a nontrivial segment. Thus $f_r^{*n-1}(S_k^{n-1})$ is on $(n-2)$-simplexes containing Θ^{n-1}. Now if $i(f_r^{*n-1}(S_k^{n-1}), \Theta^{n-1})$ is not defined, then $\Theta^{n-1} \in \partial(f_r^{*n-1}(S_k^{n-1}))$ which gives that Θ^{n-1} lies in an $(n-2)$-chain. Thus, to have that both $i(f^*(S_k^{n-1}), \Theta^n a)$ and $i(f_r^{*n-1}(S_k^{n-1}), \Theta^{n-1})$ are defined, it is sufficient to prove that Θ^{n-1} does not lie in a certain family of $(n-2)$-simplexes in $\mathbf{R}^{n-1}$. But, if necessary, we perturb $\langle \Theta^n a \rangle$ slightly in a direction perpendicular to itself so that the above intersection numbers are defined for all k. Since this is equivalent to leaving $\langle \Theta^n a \rangle$ as it was originally, but perturbing f^* slightly in the opposite direction, and this may be done without affecting the useful properties of f^*, the lemma is proved.

THEOREM 12.6.21. *Let $n > 1$ and K^n be a region and $f : K^n \to \mathbf{R}^n$ be a continuous function with the property that $f(x) \neq \Theta^n$ for $x \in \partial K^n$ and suppose further that $d(f, K^n, p)$ is defined. If ∂K^n is subdivided into $(n-1)$-regions β_j^{n-1} (a finite number, i.e. $j \in I$, I a finite set) sufficiently refined relative to $\operatorname{sgn} f$ then if $f = (f_1, \ldots, f_n)$ with $f_{r_i} \neq 0$ on β_i^{n-1} for each i, then*

$$d(f, K^n, p) = \tfrac{1}{2} n \sum_i (-1)^{r_i + 1} d(f_{r_i}^{n-1}, \beta_i^{n-1}, \Theta^{n-1}) \times$$

$$\times \operatorname{sgn} f_{r_i}/\beta_i^{n-1}.$$

Proof. From the fact that K^n is subdivided sufficiently relative to $\operatorname{sgn} f$, $d(f_{r_i}^{n-1}, \beta_i^{n-1}, \Theta^{n-1})$ is well defined and we have that $f_{r_i}^{n-1} \neq \Theta^{n-1}$ on β_i^{n-1}. We have also that, if f_s and f_t are nonzero on some β_i^{n-1} then

$$d(f_s^{n-1}, \beta_i^{n-1}, \Theta^{n-1}) = d(f_t^{n-1}, \beta_i^{n-1}, \Theta^{n-1}) = 0$$

and this gives that it is immaterial whether we associate f_s or f_t with β_i^{n-1}. Now we consider a to be a point on one of the coordinate axes of $\mathbf{R}^n$ such that

$$\|a\| > \sup_{x \in K^n} \|f(x)\|$$

and we let f^* be a chain approximation of f on K^n satisfying all the properties stated in Lemma 12.6.20. Then we have, by definition,

$$d(f, K^n, \Theta^n) = r(f^*(\partial K^n), \Theta^n) = i(f^*(K^n), \Theta^n)$$
$$= i(f^*(K^n), \Theta^n) - i(f^*(K^n), a)$$
$$= -i(f^*(K^n), a - \Theta^n) = -i(f^*(K^n), \partial(\langle \Theta^n a \rangle))$$
$$= (-1)^{n+1} i(\partial f^*(K^n), \langle \Theta^n a \rangle)$$
$$= (-1)^{n+1} i(f^*(\partial K^n), \Theta^n a)$$
$$= (-1)^{n+1} \sum_i i(f^*(\beta_i^{n-1}), \langle \Theta^n a \rangle)$$

since according to Lemma 12.6.20 we may assume that a is not in $f(K^n)$. Further, from the definition of the intersection numbers, we have

$$(+) \qquad d(f, K^n, \Theta^n) = (-1)^{n+1} \sum_{M_a} i(f^*(\beta_i^{n-1}), \langle \Theta^{n-1} a \rangle)$$

with

$$M_a = (i, f^*(\beta_i^{n-1}) \cap (\Theta^n a) \neq \varnothing).$$

Let i in M_a be fixed. Then $r_i = r$ with $\operatorname{sgn} f_{r_i} = \operatorname{sgn}$ (rth coordinate of a) and if

$$\beta_i^{n-1} = \sum_{k \in K_i} S_k^{n-1}$$

then by Lemma 12.6.19. and Lemma 12.6.20,

$$i(f^*(\beta_i^{n-1}), \langle \Theta^n a \rangle) = \sum_{k \in K_i} i(f^*(S_k^{n-1}), \langle \Theta^n a \rangle)$$
$$= \sum_{k \in k_i} (-1)^{r_i + n} i(f_{r_i}^{*n-1}(S_k^{n-1}), \Theta^{n-1}) \times$$
$$\times \operatorname{sgn} f_{r_i}^* / \beta_i^{n-1}.$$

From the formula $(+)$ given above we get further,

$$d(f, K^n, \Theta^n) = (-1)^{n+1} \sum_{i \in M_a} \sum_{k \in K_i} (-1)^{r_i + n} \times$$
$$\times i(f_{r_i}^{*n-1}(S_k^{n-1}), \Theta^{n-1}) \operatorname{sgn} f_{r_i}^* / \beta_i^{n-1}$$
$$= \sum_{i \in M_a} (-1)^{r_i + n} \times$$
$$\times i(f_{r_i}^{*n-1}(\beta_i^{n-1}), \Theta^{n-1}) \operatorname{sgn} f_{r_i} / \beta_i^{n-1}.$$

We remark that there are $2n$ sets M_a depending on the half of the coordinate axis on which a lies. We add all the formulas obtained for all M_a

and we get that

$$2nd(f, K^n, \Theta^n) = \sum_i (-1)^{r_i + n} \times$$

$$\times i(f_{r_i}^{*n-1}(\beta_i^{n-1}), \Theta^{n-1}) \operatorname{sgn} f_{r_i}^* / \beta_i^{n-1},$$

since we may argue as follows to show that the summation is over all the i which are connected with the subdivision of K^n into $(n\text{-}1)$-regions β_i^{n-1}.

Indeed, if some of these i are not in any M_a then we have $f^*(\beta_i^{n-1}) \cap \langle \Theta^n a \rangle = \varnothing$ for all a. Thus $f_{r_i}^{*n-1} \neq \Theta^{n-1}$ on β_i^{n-1} which implies that $i(f_{r_i}^{*n-1}(\beta_i^{n-1}), \Theta^{n-1})$ is zero and thus this i may be included in the sum without affecting its value. But it is impossible for this i to lie in a different M_a since this implies that no f_k^*, $1 \leq k \leq n$ was not zero on β_i^{n-1} and by (a) of Lemma 12.6.20 this is not the case.

Now if f^* is supposed to be a chain approximation of f on K^n with respect to Θ^n, this gives that $f_{r_i}^{*n-1}$ is a chain approximation of $f_{r_i}^{n-1}$ on β_i^{n-1} with respect to Θ^{n-1}, and we get the formula of the theorem.

THEOREM 12.6.22.   *Let $K^n = \sum_{j=1}^l \langle x_0^{(j)} x_1^{(j)}, x_n^{(j)} \rangle$ be an n-region with*

$$K^n = \sum_{j=1}^{l'} t_j \langle y_1^{(j)} y_2^{(j)}, \ldots, y_n^{(j)} \rangle, \qquad t_j = \pm 1.$$

If $g : K^n \to \mathbf{R}^n$ is a function then we have the relation

$$\sum_{j=1}^l \Delta^{n+1}((1, g(x_0^{(j)})), (1, g(x_1^{(j)})), \ldots, (1, g(x_n^{(j)})))$$

$$= \sum_{j=1}^{l'} t_j \Delta^n(g(y_1^{(j)}), \ldots, g(y_n^{(j)})).$$

Proof. We have, expanding Δ^{n+1} along its first column,

$$\sum_{j=1}^l \Delta^n((1, g(x_0^{(j)})), \ldots, (1, g(x_n^{(j)})))$$

$$= \sum_{j=1}^l \sum_{i=0}^q (-1)^{in}(g(x_0^{(j)}), \ldots, g(x_i^{(j)}), \ldots, g(x_n^{(j)}))$$

and since

$$\partial K^n = \sum_{j=1}^l \sum_{i=0}^q (-1)^i \langle x_0^{(j)}, \ldots, \hat{x}_i^{(j)} \cdots x_n^{(j)} \rangle$$

$$= \sum_{j=1}^{l'} t_j \langle y_1^{(j)}, \ldots, y_n^{(j)} \rangle$$

we obtain that

$$\sum_{j=1}^{l} \Delta^n((1, g(x_0^{(j)})), \ldots, (1, g(x_n^{(j)}))) = \sum_{j=1}^{l'} t_j \Delta^n(g(y_1^{(j)}), \ldots, g(y_n^{(j)})).$$

We are now ready to prove Stenger's formula.

THEOREM 12.6.23 (F. Stenger, 1974). *Let K^n be an n-region and $f : K^n \to \mathbf{R}^n$ be a continuous mapping satisfying the property that $f(x) \neq \Theta^n$ for $x \in \partial K^n$. Suppose that ∂K^n has been subdivided into $(n-1)$-regions β_k^{n-1}, $1 \leq k \leq m$ so that this division is sufficiently refined relative to $\operatorname{sgn} f$. If $\partial K^n = \sum_{j=1}^{l} t_j \langle y_1^{(j)}, \ldots, y_n^{(j)} \rangle$, $t_j = \pm 1$ then the following relationship holds:*

$$d(f, K^n, \Theta^n) = \frac{1}{2^n} n! \sum_{j=1}^{l} t_j \Delta^n(\operatorname{sgn} f(y_1^{(j)}), \ldots, \operatorname{sgn}(g(y_n^{(j)})).$$

Proof. We prove this formula by induction on n. For $n = 1$ this reduces to a formula mentioned at the beginning of this subsection and it is obviously true.

Suppose now that the assertion holds for $2, 3, \ldots, n-1$ and we prove it for n. According to Theorem 12.6.21 we have

$$d(f, K^n, \Theta^n) = \frac{1}{2^n} \sum_{k=1}^{m} (-1)^{r_k + 1} d(f_{r_k}^{n-1}, \beta_k^{n-1}, \Theta^{n-1}) \times$$
$$\times \operatorname{sgn} f_{rk}/\beta_k^{n-1}.$$

Since we suppose the subdivision of ∂K^n is sufficiently refined relative to $\operatorname{sgn} f_{r_k}^{n-1}$ we thus have,

$$\partial \beta_k^{n-1} = \sum_{i \in I_k} s_i \langle z_1^{(i)}, \ldots, z_{n-1}^{(i)} \rangle, \qquad s_i = \pm 1$$

and using the induction hypothesis, we have further,

$$d(f_{r_k}^{n-1}, \beta_k^{n-1}, \Theta^{n-1}) = \frac{1}{2^{n-1}} (n-1)! \sum_{i \in Ik} s_i \Delta^{n-1} \times$$
$$\times (\operatorname{sgn} f_{r_k}^{n-1}(z_1^{(i)}), \ldots, \operatorname{sgn} f_{r_k}^{n-1}(z_{n-1}^{(i)}))$$
$$= \frac{1}{2^{n-1}} (n-1)! \sum_{j \in J_k} s_j \Delta^n((1, \operatorname{sgn} f_{r_k}^{n-1}(y_1^{(j)})),$$
$$\ldots, (1, \operatorname{sgn} f_{r_k}^{n-1}(y_n^{(j)})))$$

Further we have that

$$d(f, K^n, \Theta^n) = \frac{1}{2^n} n! \sum_{k=1}^{m} \sum_{j \in J_k} (-1)^{r_k+1} t_j \cdot$$

$$\cdot \Delta^n((1, \operatorname{sgn} f_{r_k}^{n-1}(y_1^{(j)})), \ldots,$$

$$(1, \operatorname{sgn} f_{r_k}^{n-1}(y_n^{(j)}))) \cdot \operatorname{sgn} f_{r_k}/\beta_k^{n-1}$$

$$= \frac{1}{2^n} n! \sum_{k=1}^{m} \sum_{j \in J_k} t_j \Delta^n(\operatorname{sgn} f(y_1^{(j)}), \ldots, \operatorname{sgn} f(y_n^{(j)}))$$

and then moving the first column of $\Delta^n r_k$ columns to the right then multiplying it by $\operatorname{sgn} f_{r_k}/\beta_k^{n-1}$, we get

$$d(f, K^n, \Theta^n) = \frac{1}{2^n} n! \sum_{j=1}^{l} t_j \Delta^n(\operatorname{sgn} f(y_1^{(j)}), \ldots, \operatorname{sgn} f(y_n^{(j)}))$$

since

$$\beta_k^{n-1} = \sum_{j \in J_k} t_j \langle y_1^{(j)}, \ldots, y_n^{(j)} \rangle$$

with $t_j = \pm 1$. Thus Stenger's formula is proved.

Example 12.6.24.   Let K^2 be the unit square in $\mathbf{R}^2$ and $f(x, y) = (x, 2x^2 y)$. It is obvious that $(0, 0)$ is not in $f(\partial K^2)$ and clearly we have

$$\partial K^2 = \langle y_1 y_2 \rangle + \langle y_2 y_3 \rangle + \langle y_3 y_4 \rangle + \langle y_4 y_1 \rangle$$

where $y_1 = (1, 1)$, $y_2 = (-1, 1)$, $y_3 = (-1, -1)$ and $y_4 = (1, -1)$.

We note that this is a sufficiently refined relative to $\operatorname{sgn} f$ since x or $2x^2 y$ is nonzero on each line segment $\langle y_i y_{i+1} \rangle$ and $\langle y_4 y_1 \rangle$, $i = 1, 2$. Then according to Stenger's formula we have

$$d(f, K^2, (0, 0)) = \frac{1}{8}\left(\begin{vmatrix} 1 & 1 \\ -1 & 1 \end{vmatrix} + \begin{vmatrix} -1 & 1 \\ -1 & -1 \end{vmatrix} + \right.$$

$$\left. + \begin{vmatrix} -1 & -1 \\ 1 & -1 \end{vmatrix} + \begin{vmatrix} 1 & -1 \\ 1 & 1 \end{vmatrix}\right) = 1.$$

In his paper, F. Stenger (1974) has proposed an algorithm for the computation of the topological degree based on his formula just proved. Such an algorithm involves some conditions for the function f and Stenger's proof that a sufficiently refined subdivision is reached is by contradiction, and thus no indication about the number of steps can be given to assure that we have a sufficiently refined subdivision. In his

Ph.D. dissertation, M. Stynes gives an example to show that, in general, it is impossible to say, from the computation involved in that algorithm, when a sufficient refinement has been obtained. In order to overcome this difficulty in Stenger's algorithm, Stynes introduces the so-called 'impartial refinements' of ∂K^n which in many cases can be easy to verify.

DEFINITION 12.6.25. Let K^n be an n-dimensional polyhedron and $f : K^n \to \mathbf{R}^n$ be a continuous function. We say that ∂K^n is impartially refined relative to sgn f if the following assertions hold:

1. if $n = 1$, ∂K^1 is sufficiently refined relative to sgn f,

2. if $n > 1$, ∂K^n is impartially refined relative to sgn f if K^n has been subdivided in a finite number of $(n-1)$-regions $(\beta_k^{n-1})_{k=1}^m$ such that

(a) for each pair (k_1, k_2), the interiors of $\beta_{k_1}^{n-1}$ and $\beta_{k_2}^{n-1}$ have $\varnothing$ intersection,

(b) at least one of the functions f_i, $f = (f_1, \ldots, f_n)$, say f_{p_i}, is nonzero on each region β_i^{n-1},

(c) each region β_i^{n-1} is maximal insofar as if for $j \neq i$, $\beta_i^{n-1} \cap \beta_j^{n-1} = \varnothing$ then $p_i \neq p_j$,

(d) if S_i^{n-1} is an $(n-1)$-dimensional simplex in $\beta_{i'}^{n-1}$ such that S_i^{n-1} has a face of dimension $n-2$ lying on the boundary of $\beta_{i'}^{n-1}$ then this face is also a face of at least one $(n-1)$-simplex S_j^{n-1} lying in $S_{j'}^{n-1}$, for some j'.

Our aim here is to show that any impartial refinement is in fact a sufficiently refined subdivision relative to sgn f.

First we give the following result concerning the boundary of a polyhedron.

THEOREM 12.6.26.   *If K^n is an n-dimensional polyhedron then K^n is an $(n-1)$-dimensional polyhedron.*
    Proof. Let S^{n-1} and T^{n-1} be two $(n-1)$-simplexes in the $(n-1)$-chain of ∂K^n and to prove the theorem we must show that $S^{n-1} \cap T^{n-1}$ is either a common face or the empty set.
    Suppose that $S^{n-1} \cap T^{n-1} \neq \varnothing$ and since S^{n-1} and T^{n-1} are parts in the boundaries of the simplexes S^n and T^n of K^n we may suppose without loss of generality that

$$S^n = \langle s_0\, s_1, \ldots, s_n \rangle, \qquad T^n = \langle t_0\, t_1, \ldots, t_n \rangle$$

and

$$S^{n-1} = \langle s_1 \, s_2, \ldots, s_n \rangle, \qquad T^{n-1} = \langle t_0 \, t_1, \ldots, t_{n-1} \rangle.$$

We have then several cases:

Case 1. $S^n = T^n$, the assertion is obvious.

Case 2. $S^n \neq T^n$. Then since K^n is a polyhedron, $S^n \cap T^n$ is a common r-dimensional face, i.e., we have

$$S^n \cap T^n = (p, p = \sum_{i=0}^{r} a_i s_i, a_i \geq 0, \sum_0^r a_i = 1)$$

$$= (q, q = \sum_{i=0}^{r} b_i t_i, b_i \geq 0, \sum_0^r b_i = 1).$$

Since the vectors in the simplexes are supposed to be linearly independent, and the extreme points in $S^n \cap T^n$ are uniquely determined, we obtain that

$$S^n \cap T^n = (s_i, 0 \leq i \leq r) = (t_i, 0 \leq i \leq r)$$

and (changing the notation as necessary) we may suppose that $s_i = t_i$, $0 \leq i \leq r$. Now we have further that

$$S^{n-1} \cap (S^n \cap T^n) = (p, p = \sum_{i=0}^{r} a_i s_i, a_i \geq 0, \sum_0^r a_i = 1)$$

and

$$\varnothing \neq S^{n-1} \cap T^{n-1} \subseteq S^{n-1} \cap (S^{n-1} \cap T^{n-1}) \subseteq S^{n-1} \cap (S^n \cap T^n)$$

which implies that $r \geq 1$.

Similarly, we have for all T's,

$$T^{n-1} \cap S^{n-1} \subseteq T^{n-1} \cap (S^{n-1} \cap T^{n-1}) \subseteq T^{n-1} \cap (S^n \cap T^n)$$

which implies that $r \leq n - 1$.

But we have that $s_i = t_i$ for $0 \leq i \leq r$ and then

$$S^{n-1} \cap T^{n-1} = (S^{n-1} \cap T^{n-1}) \cap (S^n \cap T^n)$$

$$= (S^{n-1} \cap (S^n \cap T^n))(T^{n-1} \cap (S^n \cap T^n))$$

$$= (x, x = \sum p_i s_i, p_i \geq 0, \sum_{i=1}^{r} p_i = 1)$$

and this proves the theorem.

THEOREM 12.6.27.   *Suppose $n > 1$ and let K^n be an n-dimensional polyhedron. If $f : K^n \to \mathbf{R}^n$ is continuous and ∂K^n is impartially refined relative to $\operatorname{sgn} f$ then for each i' in $(1, 2, \ldots, m)$, $\beta_{i'}^{n-1}$ is impartially refined relative to $\operatorname{sgn} f_{r_{i'}}^{n-1}$.*

Proof. We may assume that $r_{i'} = 1 = i$ and $m > 1$ since, if $m = 1$, there is nothing to prove. Let $(S_i^{n-1}, i \in I)$ be a finite set of simplexes in β_1^{n-1} having an $(n$-2$)$-dimensional face lying in β_1^{n-1}. Let $(S_{j(i)}^{n-1} : j \in J)$ be the corresponding set of simplexes associated with (d) in Definition 12.6.25. Let $j \in J$ and take S_j^{n-1} in $\beta_{j'}^{n-1}$ and let S_{ij}^{n-2} be the simplex $S_i^{n-1} \cap S_{j(i)}^{n-1}$. Further we set

$$S_2 = (S_{ij}^{n-2} : i \in I, j \in J, S_j^{n-1} \subseteq \beta_{j'}^{n-1}, r_{j'} = 2)$$

and then, according to Theorem 12.6.26 the intersection of any two distinct $(n$-2$)$-simplexes S_{ij}^{n-2} is either empty or a common face. This gives that S_2 can be considered as a disjoint union of $(n$-2$)$-regions and taking connected components we may suppose that this decomposition in $(n$-2$)$-regions is unique. Let k be in $(3, 4, \ldots, n)$ and set

$$S_k = (S_{ij}^{n-2}, i \in I, j \in J, S_{j'}^{n-1} \subseteq \beta_{j'}^{n-1}, r_{j'} = k).$$

We decompose the set $S_k \backslash (S_2 \cup S_3 \cup \cdots \cup S_{k-1})$ into $(n$-2$)$-regions, say $\beta_1^{n-2}, \ldots, \beta_{m_1}^{n-2}$. In this way we obtain a decomposition of β_1^{n-1} according to (d) and we prove now that decomposition $(\beta_i^{n-2})_1^{m_1}$ is an impartial refinement of $\partial \beta_1^{n-1}$ relative to $\operatorname{sgn} f_1^{n-1}$.

First we note that (a) and (c) of Definition 12.6.25 are obviously satisfied.

Since β_1^{n-1} is maximal, $r_{j'} \neq 1$ and then a component of f_1^{n-1} is nonzero on each β_i^{n-2}.

Thus (b) is satisfied.

For (d) we consider, of course, that $n > 2$ and let $S_{l'}^{n-2}$ be any $(n-2)$-simplex of $\beta_{l'}^{n-2}$ such that an $(n-3)$-dimensional face of $S_{l'}^{n-2}$ lies in $\partial \beta_{l'}^{n-2}$. Since β_1^{n-1} is a region and $\partial^2 = 0$ we get that any $(n$-3$)$-simplex in $\partial \beta_1^{n-2}$ is an $(n$-3$)$-simplex in $\partial \beta_{t'}^{n-2}$ for some $t' \neq l'$ since $\beta_{l'}^{n-2} \subseteq \partial \beta_1^{n-1}$. This implies that it is a face of an $(n$-2$)$-simplex S_t^{n-2} in $\beta_{t'}^{n-2}$ and condition (d) is satisfied and thus the theorem is proved.

We are now ready to prove the theorem which forms the basis for the algorithm.

THEOREM 12.6.28.   *If K^n is a region and $f : K^n \to \mathbf{R}^n$ is continuous then, if ∂K^n is impartially refined relative to $\operatorname{sgn} f$, it is also sufficiently refined relative to $\operatorname{sgn} f$.*

Proof. We prove the assertion of the theorem by induction on n. First we note that if $n = 1$ the assertion is obviously true by the definitions of both refinements.

Suppose now that $n > 1$ and that the assertion of the theorem is true for $2, 3, \ldots, (n-1)$. Suppose now that ∂K^n is impartially refined relative to sgn f using $(n\text{-}1)$-regions, $\beta_1^{n-1}, \ldots, \beta_{m_1}^{n-1}$. For the proof of the theorem we remark that the conditions (a) and (b) of Definition 12.6.16. are obviously satisfied and thus we prove only (c). Let i be in $(1, 2, 3, \ldots, n)$, f_{r_i} nonzero on β_i^{n-1} and, by Theorem 12.6.27, $\partial \beta_i^{n-1}$ is impartially refined relative to sgn $f_{r_1}^{n-1}$ which gives us that $\partial \beta_i^{n-1}$ is sufficiently refined relative to sgn $f_{r_i}^{n-1}$ by induction. This clearly implies that (c) is satisfied and the theorem is proved.

We give now the alogrithm for the computation of the topological degree using 'impartial refinements'.

For this purpose we recall the notion of modulus of continuity associated with a function $F : (X, d_X) \to (Y, d_Y)$, where X, Y are metric spaces with the metrics d_X and d_Y respectively.

The modulus of continuity of F is the function of real variable $s \geq 0$ defined as follows:

$$w_F(s) = \sup_{d(x, y) \leq s} d_Y(f(x), f(y))$$

Let us now consider K^n as being an n-region and $f : K^n \to \mathbf{R}^n$ be continuous and for each i, w_i denotes the modulus of continuity of f_i, $f = (f_1, \ldots, f_n)$ of ∂K^n. We suppose further that there exist known functions Ω_i such that $w_i(s) \leq \Omega_i(s)$, $i = 1, 2, \ldots, n$.

Let i be in $(1, 2, \ldots, n)$ and consider x in ∂K^n with $f_i(x) \neq 0$ and take $s_i > 0$ such that $\Omega_i(s) < |f_i(x)|$.

If $\| , \|$ denotes the Euclidean norm on $\mathbf{R}^n$ we then have, for any $y \in \partial K^n$ such that $\|x - y\| \leq s_i$,

$$|f_i(y)| \geq |f_i(x)| - |f_i(x) - f_i(y)| \geq |f_i(x)| - w_i(s_i)$$
$$\geq |f_i(x)| - \Omega_i(s_i) > 0.$$

Consider now p in ∂K^n and set $\delta(p) = \max s_i$ where s_i are real numbers such that

$$\Omega_i(s_i) < |f_i(p)|$$

if $f_i(p) \neq 0$, if $f_i(p) = 0$ then we take $s_i = 0$.

Thus from the above we get that for at least one i,

$$|f_i(p)| \geq m = \min_{x \in \partial K_n} = \|f(x)\|$$

and thus

$$\delta(p) = \min s_i'$$

where $s_1' > 0$ are chosen such that $\Omega_i(s_1') < m$. This gives as that $\delta(p) \geq c$ where c is a constant independent of p.

Further we get that if $y \in \partial K^n$ has the property that $\|p - y\| \leq \delta(p)$ then $f_i(y) > 0$ for some i, independent of y. This implies that if $p \in \partial K^n$ is given we can surround it by a ball of radius at least c such that some component of f is nonzero on $B \cap \partial K^n$.

DEFINITION 12.6.29.   A simplex is acceptable if at least one component of f is known to be nonzero thereon.

It is obvious that any subdivision of an acceptable simplex is again an acceptable simplex.

DEFINITION 12.6.30.   An edge of a simplex is any of the one-dimensional faces of the simplex.

Suppose now that $S_1^{n-1}, \ldots, S_1^{n-1}$ are the oriented simplexes in ∂K^n such that

$$K^n = \sum_{i=1}^{l} S_i^{n-1}$$

and $p_1, \ldots, p_m$ is the set of vertices of S_j^{n-1}. The algorithm for the computation of the topological degree is as follows.

ALGORITHM 12.6.31.

1.   Set $i = 1$ and go to step 3,

2.   if $i = m$ terminate. Otherwise replace i by $i + 1$ and continue,

3.   compute $\delta(p_i)$,

4.   set $j = 1$ and go to step 6,

5.   if $j = 1$ go to step 2. Otherwise replace j by $j + 1$ and continue,

6.   if S_j^{n-1} is acceptable or if p_i is not a vertex of S_j^{n-1} go to step 5. Otherwise continue,

7.   Assume (without loss of generality) that

$$S_j^{n-1} = \langle p_i \, y_2 \cdots \, y_n \rangle.$$

Set $k = n$ and go to step 9,

8.   if $k = n$ list S_j^{n-1} as an acceptable answer and go to step 5. Otherwise replace k by $k + 1$ and continue,

9.   compute l_k the length of the edge $p_i p_k$ and

(a)   if $l_k \leq \delta(p_i)$ go to step 8,
(b)   if $l_k > \delta(p_i)$, let p_{m+1} be the point lying on $\langle p_i p_k \rangle$ at distance

$$\min \left(\delta(p_i), \tfrac{1}{2} l_k \right)$$

from p_i. In step 2 replace m by $m + 1$. Replace the oriented simplex S_j^{n-1} by two oriented simplexes:

$$S_j^{n-1} = \langle p_1 y_2 \cdots y_{k-1} p_{m+1} y_{k+1} \cdots y_n \rangle$$

and

$$S_{l+1}^{n-1} = \langle p_{m+1} y_2 \cdots y_n \rangle.$$

In step 5 replace l by $l + 1$ and replace every other oriented simplex having $\langle p_i p_k \rangle$ as an edge by two new oriented simplexes whose sum is the original simplex, increasing l in step 5 to $l + 1$ each time a new simplex is created and go to step 8.

Thus if we consider the vertices $p_1, \ldots, p_m$ in turn, then for each p_i we consider in turn all the simplexes having p_i as a vertex and are at the time unacceptable. Further, for each simplex S_j^{n-1} we consider all the edges emanating from p_i and subdivide all the edges whose length is greater than $\delta(p_i)$ in such a way that the length of the edge of S_j^{n-1} having p_i as an end point is at most $\delta(p_i)$. Then the simplex S_j^{n-1} is an acceptable simplex and similarly we proceed with all the other simplexes until we have the property that the length of the edges (of all new ones simplexes) is at most $\delta(p_i)$.

Concerning the above algorithm, it appears natural to ask whether it gives what we need. The answer is given in the following theorem.

THEOREM 12.6.32.   *Only a finite number of iterations are necessary in order to obtain all the acceptable simplexes of the subdivision.*

For the proof we need some results given as lemmas.

LEMMA 12.6.33. *Let ABC be a triangle as in the following figure*

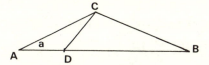

and suppose that the following properties hold:

(1)   *the diameter of the triangle is less or equal to k,*

(2) $AB > c$ *and D lies between A and C,*

(3)   *if* $AB \leq 2c$ *then* $AD = \frac{1}{2}AB$,

(4)   *if* $AB > 2c$ *then* $c \leq AD \leq \frac{1}{2}AB$

for some constants k and c.
    Then
$$CD^2 \leq k^2 - c^2/4.$$

Proof. Here and in what follows, we use the conventions in elementary geometry, i.e. for any triangle ABC, AB, CD, AC, ..., etc, denote the length of the above sets of the triangle. For example AB denotes the set of points lying on AB as well as the length of the set AB.
    For the proof we must consider several cases.
    Case 1. $AB = 2c$ and $AB = BC$.
    Since
$$BC^2 = AC^2 + AB^2 - 2AC\,AB\cos a$$
we get that
$$BC^2 - AB^2 = AC^2 - 2AC\,AB\cos a$$
and thus
$$BC \leq AB \leftrightarrow AC^2 - 2AC\,AB\cos a$$
$$\leq 0 \leftrightarrow AC - 2AB\cos a$$
$$\leq 0 \leftrightarrow \cos a \geq AC/2AB$$

Since we have $AB = 2c$, from the point (3) of the hypothesis we get that $AD = \frac{1}{2}AB$ and thus
$$CD^2 = AC^2 + AD^2 - AC\,AD\cos a$$
$$= AC^2 + \frac{1}{4}AB^2 - AC\,AB\cos a.$$

Since $BC \leq AB$ we have that $\cos a \geq AC/2AB$ and thus the last relations imply

$$CD^2 \leq AC^2 + \tfrac{1}{4}AB^2 - AC\,AB\,AC/2AB$$
$$= \tfrac{1}{4}AC^2 + \tfrac{1}{4}AB^2 = \tfrac{3}{4}k^2$$

and since $c < k$ the assertion of the lemma is true in this case.

Case 2. $AB \leq 2c$ and $AB < BC$. If $AB > AC$ then by symmetry we can obtain the same result as in Case 1. Thus we consider the case when $AB \leq AC$. We have (supposing without loss of generality, that $BC \geq AC$) from

$$BC^2 = AC^2 + AB^2 - 2AC\,AB\cos a$$

that

$$AB^2 \geq 2AC\,AB\cos a$$

or that

$$\cos a \leq AB/2AC.$$

Further we obtain the following relations,

$$CD^2 = BC^2 - \tfrac{3}{4}AB^2 + AC\,AB\cos a$$
$$\leq BC^2 - \tfrac{3}{4}AB^2 + AC\,AB\,AB/2AC$$
$$\leq BC^2 - \tfrac{1}{4}AB^2 \leq k^2 - c^2/4$$

and this implies that the lemma is true also in this case.

To complete the proof we must analyse the remaining case, namely,
Case 3. $AB > 2c$. In this case we have,

$$c \leq AC \leq \tfrac{1}{2}AB$$

and

$$CD^2 = AC^2 + AD^2 - 2AC\,AD\cos a.$$

Since CD depends only on AD, differentiating we get that

$$\mathrm{d}(CD)/\mathrm{d}(AD) = 2AD - 2AC\cos a$$

and this is zero iff $\cos a = AD/AC$ i.e., when $AD \perp AB$. This gives that the minimum value of CD is when $AD = c$ or when $AD = \tfrac{1}{2}AB$. If $AD = \tfrac{1}{2}AB$ occurs, then Cases 1 and 2 implies the assertion of the lemma. Thus we suppose that $AD = c$. Then we have that,

$$AC^2 + c^2 - 2c\,AC\cos a \geq AC^2 + \tfrac{1}{2}AB^2 - AC\,AB\cos a$$

or

$$c^2 - \tfrac{1}{4}AB^2 \geq (AC\cos a)(c - \tfrac{1}{2}AB).$$

Thus we have, since $c - \tfrac{1}{2}AB$ is strictly negative,

$$c + \tfrac{1}{2}AB \leq 2\,AC\cos a$$

or that

$$\cos a = (c + \tfrac{1}{2}AB)/2AC$$

and this implies the assertion of the lemma.

For the algorithm described above under 12.6.31 we note that by a new edge we mean an edge formed by subdivision which is not a part of any edge previously presented. Thus, in the case of the above triangle, only CD is a new edge (and not AD or DA which are parts of one of the old edges). We note also the following remarkable property of the algorithm, namely that it does not increase the number of simplices which have p_i as a vertex. We define d_0 as the maximum of the diameters of the simplices S_i^{n-1} in the decomposition of ∂K^n,

$$\partial K^n = \sum_{i=1}^{l} S_i^{n-1}$$

and let d_r be the largest of the diameters of the acceptable simplexes which are present after we have surrounded $p_1, \ldots, p_r$ by acceptable simplexes (we set $d_r = 0$ if all the simplexs are acceptable).

THEOREM 12.6.34.   *Let p_i be any vertex. Then we are dealing with this in the algorithm, and then any new edge constructed in an unacceptable simplex lies in a triangle with the following property: the edge and the triangle satisfy the properties of CD and ABC of the lemma 12.6.33 with $k = d_{i-1}$ and c.*

Proof. Suppose that $(p_i p_{i'})$ is the edge which we divide at p_j (different from p_i and $p_{i'}$) then the new edge has the following form: $(p_k p_j)$ with $k = i, i'$. The triangle $p_i p_{i'} p_k$ contains any such an edge and the triangle is a face of an unacceptable simplex. Thus we can apply the above result taking into account the following

$$p_i \to A, \qquad p_{i'} \to B, \qquad p_k \to C.$$

Since we must have (the length of) $p_i p_{i'}$ greater than c (since in the contrary case this edge would not have divided) the diameter of the triangle is at most d_{i-1} by definition, and since $\delta(s_i) \geq c$, step 9 of the algorithm implies all the restrictions on the length of the edge $p_i p_j$ and the theorem is proved.

We are now ready to prove Theorem 12.6.32. Obviously we have the relations

$$0 \leq d_{r+1} \leq d_r$$

for $r = 0, 1, 2, \ldots$ and we show that for some $r_0, d_{r_0} \leq c$.

It is clear that the algorithm will return then to step 8.

Consider now the vertices of S_i^{n-1} be $p_1, \ldots, p_m$ and after we are dealing with these vertices, any part of the original edges in an unacceptable simplex will have length at most $\max(\frac{1}{2}d_0, d_0 - c)$ since it is a result of a bisection or from a subdivision which removed a length of at least c. Other edges of the unacceptable simplexes are as CD in Theorem 12.6.34, and this gives us that their length is at most $(d_0 - \frac{1}{4}c^2)^{1/2}$.

Thus we have

$$d_m^2 = \max(\tfrac{1}{4}d_0^2, (d_0 - c)^2, d_0^2 - \tfrac{1}{4}c^2)$$

and considering the list of vertices

$$p_1 p_2 \cdot \cdot \cdot p_m p_{m+1} p_{m+2} \cdot \cdot \cdot p_{m'}$$

as well as

$$p_{m+1} p_{m+2} \cdot \cdot \cdot p_{m'}$$

we obtain that

$$d_{m'}^2 = \max(\tfrac{1}{4}d_m^2, (d_m - c)^2, d_m^2 - \tfrac{1}{4}c^2).$$

We can continue and we eventually obtain that this sequence has terms less or equal to c. From steps 6, 5 and 2 of the algorithm we see that after a number of iterations all the simplexes are acceptable, since surrounding p_i by acceptable simplexes, every simplex having p_i as a vertex is acceptable.

Now for the computation of the topological degree we may proceed as follows: first we note that since K^n is supposed to be an n-region ∂K^n is a polyhedron by Theorem 12.6.26 and thus Algorithm 12.6.31 preserves this property. Thus we have a decomposition of ∂K^n into acceptable simplexes which give an impartial refinement of ∂K^n relative to sgn f. We can take the maximal connected collection of simplexes to be $(n-1)$-regions avoiding overlapping, and using the fact that the boundary of the boundary of K^n

obtained in the way just described is zero, we find that we can apply Stenger's formula for the calculation of the topological degree.

For an excellent summary of recent experiments in the computation of the topological degree we refer to B. Kearfoot (February, 1978), Kearfoot has also presented very interesting and useful approaches to the computation of the topological degree. A detailed account of these is given in his Ph.D. dissertation (Univeristy of Utah, 1977). Concerning the bisection method and the topological degree see B. Kearfoot (1978, 1979, 1980) and the references quoted there.

12.7. SOME APPLICATIONS OF THE TOPOLOGICAL DEGREE

There are numerous applications of the topological degree scattered throughout the literature. In what follows we present some specific applications related to fixed point theory. The first application we note is the new proof of the Schauder fixed point theorem.

THEOREM 12.7.1. (Schauder fixed point theorem). *Let* $f : C \to C$ *be a continuous compact mapping defined on a closed, bounded and convex subset of a Banach space* X. *Then there exists a fixed point of* f *in* C.

Proof. We consider the family of mappings

$$g_t(x) = x - t f(x) \qquad 0 \le t \le 1$$

defined on C with values in X and clearly this is a homotopy. But if B is a ball containing C we have

$$\deg(0, g_0, B) = 1$$

and thus

$$\deg(0, g_1, B) = 1$$

which implies that there exists $x_0 \in C$ such that $f(x_0) = x_0$, i.e. f has a fixed point.

We now present a proof of Altman's theorem

THEOREM 12.7.2. (Altman's theorem). *Let* C *be as in Theorem* 12.7.1 *and let* $f : C \to X$ *be a continuous mapping which is compact. Suppose that* $0 \in C$ *and that* f *has the property that for any* $x \in \partial C$,

$$\|x - f(x)\|^2 \ge \|f(x)\|^2 - \|x\|^2.$$

Then f *has a fixed point in* C.

Proof. We consider the mappings

$$g_t(x) = tf(x) \qquad 0 \leq t \leq 1$$

and we remark that we have

$$\deg(0, I - g_0, C) = 1$$

and thus

$$\deg(0, I - g_1, C) = 1$$

since, using the above property of f, we can prove that g_t avoids the point $0, 0 \leq t < 1$. The above relation implies that f has a fixed point in C.

The following result involves the so-called 'Leray–Schauder condition'.

THEOREM 12.7.3. (Leray–Schauder). *Let O be a bounded open neighbourhood of zero in a Banach space and suppose that $f : \bar{O} \to X$ satisfies the following property, called the 'Leray–Schauder condition':*

'if for some x_0 in O and $t > 0$, $f(x_0) = tx_0$ then necessarily $t = 1$'

then there exist fixed points for f in $\bar{O}$, when f is compact.

Proof. We consider the following functions

$$g_t(x) = x - tf(x)$$

and clearly we may assume that 0 is not in $g_1(O)$. Since f satisfies the Leray–Schauder condition we get that for every t in $[0, 1]$ and every $x \in O$,

$$O \neq (l - t)g_0(x) + tg_1(x) = x - tf(x)$$

and then

$$\deg(0, I, O) = 1 = \deg(0, g_1, O)$$

and thus there exist fixed points for f.

Let Ω be a symmetric set in a Banach space X. A function $f : \Omega \to X$ is said to be odd (respectively even) if $f(-x) = -f(x) (f(x) = f(-x))$.

THEOREM 12.7.4 (The Borsuk antipodal theorem). *Suppose now that Ω is a bounded open and symmetric set in a Banach space X and $f : \Omega \to X$ is continuous, compact and odd, $\theta \notin (I - f)(\partial\Omega)$. Then*

$$d(I - f, \Omega, \theta) \equiv 1 \pmod 2$$

and f has a fixed point in Ω.

Proof. We have proved this theorem in Chapter 5 for the case of finite dimensional spaces and, of course, without using the Brouwer degree. We give now a proof for this case using the topological Brouwer degree and we note that the proof extends immediately to infinite dimensional spaces using an approximation, a step which was used in extending the Brouwer degree as well as in other cases.

Now, if X is finite-dimensional we can suppose without loss of generality that $X = \mathbf{R}^n$ for some n.

Consider $\varepsilon > 0$ such that $\varepsilon \bar{B} = \varepsilon \bar{B}(\theta, 1)$ is in Ω. Consider further that the mapping $g : \bar{\Omega} \to \mathbf{R}^n$ is continuous and $g/\partial\Omega = (I - f)/\partial\Omega, g/\varepsilon\bar{B} = I/\varepsilon\bar{B}$ which is possible by the Urysohn–Tietze theorem. In this case we have

$$d(I - f, \Omega, \theta^n) = d(g, \Omega, \theta^n) = d(I, \varepsilon B, \theta^n) + d(g, \Omega \backslash \varepsilon \bar{B}, \theta^n)$$
$$= 1 + d(g, \Omega \backslash \varepsilon \bar{B}, \theta^n)$$

and to prove the assertion of the theorem in this case we must show that

$$d(g, \Omega \backslash \varepsilon \bar{B}, \theta^n) = 0 \quad (\mathrm{mod}\ 2).$$

Let g_1 be the restriction of the map g to the set $(\partial\Omega \cup (\varepsilon\partial B(\theta^n, 1)) \cap \mathbf{R}^{n-1}$ where $\mathbf{R}^{n-1}$ is considered canonically embedded in $\mathbf{R}^n$. The function g_1 is then continuous, odd and has no zeros. Then it can be extended to a function $h : (\bar{\Omega} \backslash \varepsilon B) \cap \mathbf{R}^{n-1} \to \mathbf{R}^n$, again without zeros. The function

$$h_1(x) = \begin{cases} g(x) & x \in \partial\Omega \bigcup (\varepsilon\partial B(\theta^n, 1)) \\ h(x) & x \in (\bar{\Omega} \backslash \varepsilon B(\theta^n, 1)) \cap \mathbf{R}^{n-1} \end{cases}$$

is continuous, odd and without zeros. If $D = \Omega \backslash \varepsilon B(\theta^n, 1)$ then we find, by the continuity of the Brouwer degree, a function $h_1 : \bar{D} \to \mathbf{R}^n$ and $\theta^n \notin h_1(\partial D \cup (\bar{D} \cap \mathbf{R}^{n-1}))$ with the property that

$$d(g, D, \theta^n) = d(h_1, D, \theta^n)$$

and since we may suppose h_1 is odd (if not then we take the odd part) and set $D^+ = D \cap (\mathbf{R}^{n-1} \times (x, x \in \mathbf{R}^n, x > 0)), D^- = D \cap (\mathbf{R}^{n-1} \times (x, x \in \mathbf{R}^n, x < 0))$.

Then we get, from the additivity property of the degree,

$$d(h_1, D^+, \theta^n) + d(h_1, D^-, \theta^n) = d(h_1, D, \theta^n).$$

If y is a regular value of h_1 such that

$$d(h_1, D, \theta^n) = d(h_1, D, y)$$

then from the fact that h_1 is odd we obtain that $-y$ is a regular value of h_1 and thus

$$(x, x \in D^+, h_1(x) = y) = -(x, x \in D^-, h_1(x) = -y).$$

Now we know (and can readily prove) that the derivative of h_1 is even and then we obtain

$$d(h_1, D^+, \theta^n) = d(h_1 - y, D^+, \theta^n)$$

$$= \sum_{x \in h_1^{-1}(y) \cap D^+} \text{sgn det } h_1'(x)$$

$$= \sum_{x \in h_1^{-1}(-y) \cap D^-} \text{sgn det } h_1'(-x)$$

$$= d(h_1 + y, D^-, \theta^n)$$

where we have applied the translation invariance of the degree. From these we obtain that $d(h_1, D, \theta^n) = 2d(h_1, D^+, \theta^n) = 0 \pmod 2$ and Borsuk's antipodal theorem is proved in this case.

From the definition of the degree it is clear that we can assume that $I - f$ is odd and finite-dimensional and the result now follows from the definition of the Leray–Schauder degree.

THEOREM 12.7.5.   (The Borsuk–Ulam theorem). *Let Ω be an open and bounded symmetric neighbourhood of zero in some finite-dimensional Banach space X and $f : \partial\Omega \to X$ is continuous such that $f(\partial\Omega)$ is contained in a proper subspace of X. Then there exists at least one point x such that $f(x) = f(-x)$.*

For the proof we need the following

LEMMA 12.7.6.   *Let X be a finite-dimensional Banach space and Ω be a bounded, open, symmetric neighbourhood of zero. If $f : \partial\Omega \to X$ is continuous and odd, $f(\partial\Omega)$ is contained in a proper subspace of X then there exists at least one point x in $\partial\Omega$ such that $f(x) = \theta$.*

Proof. If 0 is not in $f(\partial\Omega)$ then we can extend f to $f_{\text{ext}}/\bar{\Omega}$ such that the values of the extension are contained again in a proper subspace of X and then $d(f_{\text{ext}}, \Omega, O) \neq 0$ which implies that f_{ext} (the extension of f) has the property that $\bar{f}_{\text{ext}}(\Omega)$ is a neighbourhood of zero and this contradicts the fact that the values of f_{ext} are in a proper subspace.

Proof of Borsuk–Ulam's theorem. We apply Lemma 12.7.6 to the odd part of f and this implies just the assertion of the theorem.

In Chapter 2, Theorem 2.6.3 on computing the measure of noncompactness for the unit ball in Banach spaces, it was mentioned that the result follows from the Liusternik–Schnirelman–Borsuk theorem. We present now this theorem.

THEOREM 12.7.7 (Liusternik–Schnirelmann–Borsuk). *Let E be a finite-dimensional Banach space and let $n = \dim E$. If Ω is an open, bounded, symmetric neighbourhood of zero and if $\partial\Omega$ is covered by n closed sets, then at least one of these sets must contain a pair of antipodal points.*

Proof. Let $F_1, \ldots, F_n$ be the closed sets which cover $\partial\Omega$. Suppose that $F_i \cap (-F_i) = \varnothing$ for $i = 1, 2, 3, \ldots, n-1$. We define $n-1$ functions $f_i : \partial\Omega \to [0, 1]$ such that $f_i(-F_i) = (1)$ and $f_i(F_i) = (0)$ and f_i continuous. These functions exist by the Urysohn–Tietze theorem. Consider now $(a_1, \ldots, a_{n-1})$, which are $n-1$ linearly independent elements in E and define

$$f_i : \partial\Omega \to (e_1, \ldots, e_{n-1}) = \text{the linear space generated by}$$
$$\text{the elements } e_1, \ldots, e_{n-1}.$$

by the formula

$$f(x) = \sum_1^{n-1} f_i(x) e_i$$

and then by the Borsuk–Ulam theorem there exists at least one x_0 in Ω such that $f(x_0) = f(-x_0)$ which implies that for each i, $f_i(-x_0) = f_i(-x_0)$ and x_0 is not in $\bigcup_{i=1}^{n-1} (F_i \cup (-F_i))$. But we know that the sets F_i, $i = 1, 2, \ldots, n$ cover $\partial\Omega$ and thus $-F_i$, $i = 1, 2, \ldots, n$ cover $\partial\Omega$. This gives us that x_0 is in $F_n \cap (-F_n)$ and the theorem is proved.

For the formulation of the Invariance of Domain Theorem we need the following notion:

DEFINITION 12.7.8. *If $f : S \to S_1$, S, S_1 are two topological spaces, then f is said to be locally injective if for each $s \in S$, there exists a neighbourhood of s, V_s such that $f : V_s \to S_1$ is an injective mapping.*

THEOREM 12.7.9. *Let X be a Banach space and let Ω be an open subset of X. If $f : \Omega \to X$ is such that $I - f$ is locally injective and f is a compact mapping then $I - f$ is open.*

Proof. Consider Ω_1 to be an open neighbourhood of Ω. Then we must show that $(I - f)(\Omega_1)$ is an open set in X. Consider $y \in (I - f)(\Omega_1)$, and show that there exists an open subset V which is a neighbourhood of y in X such that $V \subset (I - f)(\Omega_1)$.

Let x_0 be in Ω_1 such that $x_0 - f(x_0) = y$ and let $\varepsilon > 0$, such that $(I - f)$ is injective on $x_0 + \varepsilon \bar{B}$, which is possible since $(I - f)$ is locally injective. We now consider the mapping

$$x \to (I - f)(x + x_0) - y$$

on the set $\Omega_1 - x_0$, which is an open set. Thus we may assume that the mapping $I - f$ defined on $\varepsilon \bar{B}$ with values in X is injective and that $\theta = \theta - f(\theta) = f(\theta)$. We consider the compact mapping defined by

$$h(x, t) = f(x/(1 + t)) - f(-tx/(1 + t))$$

on $\varepsilon \bar{B} \times [0, 1]$ with values in X.

It is obvious that $h(., t)$ is odd on $\varepsilon \bar{B}$ and, since $I - f$ is injective, we get that for any $t \in [0, 1]$ and any $x \neq \theta$, $h_1(x, t) \neq \theta$ where $h_1(x, t)$ is the function $x - h(x, t)$. The homotopy invariance of the degree implies that

$$d(I - f, \varepsilon B, \theta) = d(I - h(., 1), \Omega, \theta)$$

and since the latter number is not zero, by Borsuk's antipodal theorem $d(I - f, \Omega, \theta)$ is not zero and this implies that $(I - f)(\varepsilon B)$ is a neighbourhood of zero in X. This proves the theorem.

For the version of the Jordan theorem for Banach spaces we refer the reader to the book of J. Schwartz (Chapter 3, Theorem 3.57).

Bibliography

R. Abraham, A. Kelly and J. Robbin, *Transversal Mappings and Flows*, Benjamin, New York, 1967. *MR* **39** ≠ ≠ 2181.

B. L. McAllister, 'Hyperspaces and Multifunctions. The First Half-Century (1900–1950)', *Nieuw Archief voor Wisk.* **26** (1978) 309–329.

B. L. McAllister, 'Convergence of Sequences of Sets', (to appear).

B. L. McAllister, 'Cyclic elements in Topology, a History'. *Amer. Math. Monthly* **73** (1966) 337–350.

M. Altman, 'A Fixed Point Theorem in Hilbert Space', *Bull. Acad. Polon. Sci. Cl. III* **5** (1957) 89–92.

M. Altman, 'A fixed Point Theorem in Banach Space', *Bull. Acad. Polon. Sci. Cl. III* **5** (1957) 17–22. *MR* **19** ≠ ≠ 297.

M. Altman, 'An Extension to Locally Convex Spaces of Borsuk's Theorem on Antipodes', *Bull. Acad. Polon. Sci.* **6** (1958) 293–295. *MR* **20** ≠ ≠ 3531.

M. Altman, 'Continuous Transformations of Open Sets in Locally Convex Spaces', *Bull. Acad. Polon. Sci.* **8** (1958) 297–301.

M. Altman, 'Contractor Directions, Directional Contractors and Directional Contractions', *Pacif. J. Math.* **62** (1976) 1–18.

M. Altman, 'An Integral Test and Generalized Contractions', *Amer. Math. Monthly* **8** (1975) 817–829.

M. Altman, 'Series of Iterates', *Colloq. Math.* **38** (1978) 305–317.

J. W. Alexander, 'On Transformations with Invariant Points', *Trans. Amer. Math. Soc.* **23** (1922) 89–95.

P. Alexandrov and H. Hopf, *Topologie*. Reprint, Chelsea, New York, (1965). *MR* **32** ≠ ≠ 3023.

Z. Altshuler, E. Odell and Y. Sternfeld, 'A Fixed Point in c_0', *Notices Amer. Math. Soc.* **25** (1978) A–294.

A. Ambrosetti, 'Un teorema di esistenza per le equazioni differenziali negli spazi di Banach', *Rend. Sem. Mat. Univ. Padova* **39** (1967) 349–361. *MR* **36** ≠ ≠ 5478.

A. Ambrosetti, 'Proprieta spetrali di certi operatori lineari non compatti', *Rend. Sem. Mat. Univ. Padova* **42** (1969) 189–200.

N. Huu Anh, 'Note on a Fixed Point of F. E. Browder', *J. Math. Anal. Appl.* **18** (1967) 391–392.

G. F. Andrus and T. Nishiura, 'Fixed Points of Random Set-Valued Maps', *Notices Amer. Math. Soc.* **25** (1978) A–154.

K. Anzai and I. Ishikawa, 'On Common Fixed Points of Several Continuous Affine Mappings', *Pacif. J. Math.* **72** (1977) 1–4.

N. Aronszajn, 'La correspondant topologique de l'unicite dans la theorie des equations differentielles', *Ann. Math.* (1942) 730–742. *MR* **4** ≠ ≠ 100.

N. Aronszajn and P. Panitchpakdi, 'Extension of Uniformly Continuous Transformations', *Pacif. J. Math.* **6** (1956) 405–439.

K. J. Arrow and F. H. Hahn, *General Competitive Analysis*, Holden Day, San Francisco, 1971.

B. H. Arnold, 'A Topological Proof of the Fundamental Theorem of Algebra', *Amer. Math. Monthly* **56** (1954) 465–466.

N. Assad and W. A. Kirk, 'Fixed Point Theorems for Set-Valued Mappings of Contractive Type', *Pacif. J. Math.* **43** (1972) 553–562.

E. Asplund, 'Positivity of Duality Mappings', *Bull. Amer. Math. Soc.* **73** (1967) 200–203. *M R* **34** ≠ ≠ 6481.

E. Asplund, 'Fréchet Differentiability of Convex Functions', *Acta Math.* **121** (1968) 31–47. *MR* **37** ≠ ≠ 6754.

E. Asplund and R. T. Rockafellar, 'Gradients of Convex Functions', *Trans. Amer. Math. Soc.* **139** (1969) 443–467. *MR* **39** ≠ ≠ 1968.

E. Asplünd and I. Namioka, *See* I. Namioka.

M. F. Atiyah, *Lectures on K-theory*. Mimeographed Notes, Harvard University, 1964.

M. F. Atiyah, 'Algebraic Topology and Elliptic Operators', *Comm. Pure Appl. Math.* **20** (1967) 237–249.

M. F. Atiyah and R. Bott, 'On the Periodicity Theorem for Complex Vector Bundles', *Acta Math.* **112** (1964) 229–247.

J. P. Aubin, *Applied Abstract Analysis*, John Wiley, 1977.

R. J. Aumann, 'Existence of Competitive Equilibria in Markets with a Continuum of Traders', *Econometrika* **34** (1966).

L. M. Averina and V. N. Sadovskii, 'On the Local Solubility of Functional-Differential Equations of Neutral Type', *Trudy Mat. Fac. No. 3. Voronezh Gos. Univ.* (1971) 1–12.

W. L. Ayres, 'Some Generalizations of the Scherer Fixed Point Theorem', *Fund. Math.* **16** (1930) 332–336.

A. L. Badoev and V. N. Sadovskii, 'Examples of Condensing Operators in the Theory of Differential Equations with Deviating Argument of Neutral Type', *Doklady Akad. Nauk SSSR* **186** (1969) 1239–1242 = *Soviet Math. Doklady* **10** (1969) 724.

I. A. Bahtin, 'On the Existence of Common Fixed Points for Commutative Sets of Nonlinear Operators', *Funct. Analiz. i ego Priloj.* **4** (1970) 86–87.

I. A. Bahtin, 'Existence of Common Fixed Points for Abelian Families of Discontinuous Operators', *Siberian Math. J.* **13** (1972) 167–172.

D. F. Bailey, 'Some Theorems on Contractive Mappings', *J. London Math. Soc.* **41** (1966) 101–106.

D. F. Bailey, 'Krasnoselskii's Theorem on the Real Line', *Amer. Math. Monthly*, **81** (1974) 506–507.

D. F. Baily and B. E. Rhoades, 'A Counterexample for some Fixed Point Iteration Procedures', (to appear).

J. A. Bakker, 'Isometries in Normed Spaces', *Amer. Math. Monthly*, **78** (1971) 655–658.

W. C. Bakker, 'A method of Iteration for Solving Functional Differential Equations with Advanced and Retarded Argument', Ph.D. Thesis, Emory Univ. 1978.

S. Banach, 'Sur les operations dans les ensembles abstraites et leurs applications', *Fund. Math.* **3** (1922) pp. 133–181.

S. Banach, *Theorie des operations lineaires,* Monograph, PWN, Warszawa, 1932.

S. Banach and S. Mazur, 'Über mehrdeutige Stetige Abbildungen', *Studia Math.* **5** (1934) 174–178.

H. T. Banks and M. O. Jacob, 'A Differential Calculus for Multifunctions', *J. Math. Anal. Appl.* **29** (1970) 246–272.

R. G. Bartle, 'Implicit Functions and Solutions of Equations in Groups', *Math. Zeitschr.* **62** (1955) 335–346. *MR* **17** ≠ ≠ 62.

R. G. Bartle, 'On the Openess and Inversion of Differentiable Mappings', *Ann. Fenn. Ser. A. No.* 257 (1958). *MR* **28** ≠ ≠ 2932.

R. G. Bartle and L. M. Graves, 'Mappings Between Function Spaces', *Trans. Amer. Math. Soc.* **72** (1952) 400–413. *MR* **13** ≠ ≠ 951.

V. Barbu and A. Cellina, 'On the Surjectivity of Multi-Valued Dissipative Mappings', *Boll. Union. Mat. Ital.* **29** (1970) pp. 817–826.

V. Barbu, *Nonlinear Semigroups*, Noordhoff, Groningen, 1974.

F. L. Bauer, 'An Elementary Proof of the Hopf Inequality for Positive Operators', *Numer. Math.* **7** (1965) 331–337.

F. L. Bauer, E. Deutsch and J. Stier, 'Abscatzungen für die Eigenverte positiver linearer Operatoren', *Linear Algebra and Its Applications*, **2** (1960) pp. 275–301.

Glen Baxter, 'On Fixed Points of the Composite of Commuting Functions', *Proc. Amer. Math. Soc.* **15** (1964) 851–856.

B. Beuzamy, 'Un cas de convergence des iterées d'une contraction dans un éspace uniformement convexe', *C. R. Acad. Paris* (to appear).

B. Beuzamy and B. Maurey, 'Points minimaux et ensembles optimaux dans les espaces de Banach', *J. Funct. Anal.* **24** (1977) 107–139.

L. P. Belluce and W. A. Kirk, 'Some Fixed Point Theorems in a Metric and Banach Spaces', *Canad. Math. Bull.* **12** (1969) pp. 481–489.

L. P. Belluce and W. A. Kirk, 'Fixed Point Theorems for Certain Classes of Nonexpansive Mappings', *Proc. Amer. Math. Soc.* **20** (1969) 141–146. *MR* **38** ≠ ≠ 1663.

L. P. Belluce and W. A. Kirk, 'Fixed Point Theorems for Families of Contraction Mappings', *Pacif. J. Math.* **18** (1966) 213–217. *MR* **33** ≠ ≠ 7846.

L. P. Belluce and W. A. Kirk, 'Nonexpansive Mappings and Fixed Points in Banach Spaces', *Ill. J. Math.* **11** (1967) 471–479. *MR* **35** ≠ ≠ 5988.

L. P. Belluce, W. A. Kirk and E. Steiner, 'Normal Structure in Banach Spaces', *Pacif. J. Math.* **26** (1968) 433–440. *MR* **38** ≠ ≠ 1501.

C. Berge, *Topological spaces*, Olyver and Boyd, Edinburgh, 1970.

L. Bers, *Topology*, Courant Institute, New York, 1954–1955.

C. Bessaga, 'On the Converse of the Banach "Fixed Point Principle"', *Colloq. Math.* **7** (1969) 41–43.

A. Beurling and A. E. Livingston, 'A Theorem on Duality Mappings in Banach Spaces', *Ark. för Math.* **4** (1962) 405–411. *MR* **26** ≠ ≠ 2851.

R. H. Bing, 'Extending a Metric', *Duke Math. J.* **14** (1947) 511–519.

R. H. Bing, 'Challenging Conjectures', *Amer. Math. Monthly* **74** (1967) 56–64.

R. H. Bing, 'The Elusive Fixed Point Property', *Amer. Math. Monthly*, **76** (1969) 119–132.

G. Bocşan and G. Constantin, 'The Kuratowski Function and Some Applications to Probabilistic Metric Spaces', *Atti Accad. Nazionale dei Lincei* **8** (1973) 236–240.

G. Birkhoff and O. D. Kellogg, 'Invariant Points in Function Spaces', *Trans. Amer. Math. Soc.* **23** (1922) 96–115.

G. Birkhoff, 'Extensions of Jentzsch's Theorem', *Trans. Amer. Math. Soc.* **85** (1957) 219–227.

G. D. Birkhoff, 'Proof of Poincaré's Geometric Theorem', *Trans. Amer. Math. Soc.* **14** (1913) 14–22.

G. D. Birkhoff, *Dynamical Systems*, Amer. Math. Colloq. Publ. R. I. 1927.

R. Blair, 'Proofs of Urysohn', Lemma and Related Theorems by Means of Zorn's Lemma', *Math. Mag.* **47** (1974) 71–78.

F. S. de Blasi, 'On Differentiability of Multifunctions', *Pacif. J. Math.* **66** (1976) 67–81.

L. M. Blumenthal and G. E. Wahlin, 'On the Spherical Surface of Smallest Radius Enclosed a Bounded Set of n-Dimensional Euclidean Space', *Bull. Amer. Math. Soc.* **47** (1941) 771–777.

G. Bocsan and G. Constantin, 'The Kuratowski Function and Some Applications to Probabilistic Metric Spaces', *Rend. Accad. Naz. dei Lincei* **55** (1973) 236–240.

P. Bohl, 'Über die Bewegung eines mechanisches Systems in der Nähe einer Gleichgewishstlage', *J. Reine Angew. Math.* **127** (1904) 279–286.

G. Bojarski, 'On the Index Problem for Systems of Integral Equations, III', *Bull. Acad. Polon.Sci.* **13** (1963) 633–637.

V. A. Bondarenko, 'On the Existence of a Universal Measure of Non-Compactness', *Probl. Mat. Analiza Slozjn. System.* No. 2, Voronezh (1968) 18–21.

R. Bonic, 'Four Brief Examples Concerning Polynomials', *J. Diff. Geometry* **2** (1968) 391–392.

F. F. Bonsall, *Lectures on some Fixed Point Theorems of Functional Analysis, Tata Institute,* Bombay, 1962.

F. F. Bonsall and D. G. Stirling, 'Square Roots in $*$ — Algebras', *Glasgow Math. J.* **13** (1972) 74.

Ju. G. Borisovich, 'Rotation of Weakly Continuous Fields', *Doklady Akad. Nauk SSSR* **131** (1960) 230–233. = *Soviet Math. Doklady* **1** (1960) 214–217.

Ju. G. Borisovich and Ju. I. Sapronov, 'A Contribution to the Topological Theory of Condensing Operators', *DAN SSSR* **183** (1968) 18–20. = *Soviet Math. Doklady* **9** (1969) 1304–1307.

Ju. G. Borisovich änd Ju. I. Sapronov, 'On the Topological Theory of Compactly Contracting Mappings', *Trudy Sem. po Funktionalnomy Analiz.* No. 12, Voronezh (1969) 43–68.

Ju. G. Borisovich, 'An Application of the Concept of the Rotation of a Vector Field', *Doklady Akad. Nauk SSSR* **153** (1963) 12–15. = *Soviet. Math. Doklady* **4** (1963) 1584.

K. Borsuk, 'Drei Sätze über die n-dimensionale Euklidische Sphäre', *Fund. Math.* **21** (1933) 177–190.

K. Borsuk, 'Sur un continuu acyclique qui se laisse transformer en lui-même sans points invariants', *Fund. Math.* **24** (1935) 51–58.

K. Borsuk, 'A Theorem on Fixed Points', *Bull. Acad. Polon. Sci. Cl. III* **2** (1954) 17–20.

K. Borsuk, *Theory of Retracts*, Monograph. Math. PWN, Warszawa, 1967. *MR* **35** ≠ ≠7306.

K. Borsuk, 'Concerning a Problem due to Sam Nadler', *Colloq. Math.* **35** (1973) 63–65.

S. C. Bose, 'Common Fixed Points of Mappings in a Uniformly Convex Spaces', *J. London Math. Soc.* **18** (1978) 151–156.

S. C. Bose, 'Weak Convergence to the Fixed Points of Asymptotically Nonexpansive Map', *Proc. Amer. Math. Soc.* **68** (1978) 304–308.

R. Bott, Stable Homotopy of the Classical Groups', *Arr. Math.* **70** (1959) 313–337.

N. Bourbaki, '*Espaces vectoriels topologiques*, Herman, Paris, 1966.

N. Bourbaki, *General Topology*, Hermann, Paris, 1962.

D. G. Bourgin, 'Fixed Points on Neighborhood Retracts', *Rev. Roum. Math. Pure Appl.* **2** (1957) 371–374. *MR* **20** $\neq$ $\neq$ 1297.

D. G. Bourgin, 'Un indice dei punti uniti, I'. *Rend. Accad. Naz. dei Lincei* **19** (1955) 435–440. *MR* **19** $\neq$ 571.

D. G. Bourgin, 'Un indice dei punti uniti, II'. *Rend. Accad. Naz. dei Lince* **20** (1956) 43–48. *MR* **19** $\neq$ $\neq$ 571.

D. G. Bourgin, 'Un indice dei punti uniti', *Rend. Accad. Naz. dei Lincei* **21** (1956) 395–400. *MR* **19** $\neq$ $\neq$ 1971.

D. G. Bourgin, *Modern Algebraic Topology*, Macmillan, New York, 1963. *MR* **28** $\neq$ $\neq$ 3415.

C. L. Bowman, 'Error Bounds for Labelings used in Computing Fixed Points', Ph.D. Thesis, Univ. of California, Irvine, Calif. 1977.

W. M. Boyce, 'Commuting Functions with no Common Fixed Points', *Trans. Amer. Math. Soc.* **137** (1969) 77–92.

W. M. Boyce, 'Γ-Compact Mappings on an Interval and Fixed points', *Trans. Amer. Mat. Soc.* **160** (1971).

D. Boyd and J. S. W. Wong, 'On Nonlinear Contractions', *Proc. Amer. Math. Soc.* **20** (1969) 458–464.

L. de Branges, 'The Stone-Weierstrass Theorem'. *Proc. Amer. Math. Soc.* **10** (1959) 822–824.

H. Brezis, 'Equations et inequations nonlineaires dans les espaces en dualité', *Ann. Inst. Fourier Grenoble* **18** (1968) 115–175.

H. Brezis, *Operateurs maximaux monotones*. North Holland, Notas de Matematica No. 50 (1973).

H. Brezis and F. E. Browder, 'A General Principle on Ordered Sets in Nonlinear Functional Analysis', *Advances in Math.* **21** (1976) 355–364.

M. S. Brodski and D. P. Milman, 'On the Center of a Convex Set', *Dokl. Akad. Nauk SSSR* **59** (1948) 837–840. *MR* **9** $\neq$ $\neq$ 448.

R. B. S. Brooks, 'Coincidence, Roots and Fixed Points', Ph.D. Thesis, Univ. California Los Angeles, 1967.

A. Brønsted and R. T. Rockafeller, 'On the Subdifferentiability of Convex Functions', *Proc. Amer. Math. Soc.* **16** (1965) 605–611. *MR* **31** $\neq$ $\neq$ 2361.

A. Brønsted, 'Fixed Points and Partial Orders', *Proc. Amer. Math. Soc.* **60** (1976) 365–366.

L. E. J. Brouwer, 'Über Abbildungen von Mannigfaltigkeiten', *Math. Ann.* **71** (1912) 97–115.

L. E. J. Brouwer, 'An Intuitionist Correction of the Fixed Point Theorem on the Sphere', *Proc. Royal Soc. London* (A) **213** (1952) 1–2.

F. E. Browder, 'The Topological Fixed Point Theory and its Applications to Functional Equations', Ph.D. Thesis, Princeton Univ. Princeton, 1948.

F. E. Browder, 'Non linear Functional Equations in Locally Convex Spaces', *Duke Math. J.* **24** (1957) 579–589. *MR* **19** $\neq$ $\neq$ 1184.

F. E. Browder, 'On a Generalization of the Schauder Fixed Point Theorem', *Duke Math. J.* **26** (1959) 291–303. *MR* **21** $\neq$ $\neq$ 4368.

F. E. Browder, 'On the Fixed Point Index for Continuous Mappings of Locally Connected Spaces', *Summa Brasi. Math.* **4** (1960) 253–293. *MR* **26** $\neq$ $\neq$ 4354.

F. E. Browder, 'On the Continuity of Fixed Points under Deformations of Continuous Mappings', *Summa Brasil. Math.* **4** (1960) 183–191. *MR* **24** $\neq$ $\neq$ A543.

F. E. Browder, 'The Solvability of Nonlinear Functional Equations', *Duke Math. J.* **30** (1963) 557–566. *MR* **27** $\neq$ $\neq$ 6133.

F. E. Browder, 'Nonlinear Equation of Evolution', *Ann. Math.* **80** (1964) 485–523 *MR* **30** $\neq$ $\neq$ 4167.

F. E. Browder, 'Continuity Properties of Monotone Nonlinear Operators in Banach Spaces', *Bull. Amer. Math. Soc.* **70** (1964) 551–553. *MR* **29** ≠ ≠ 502.

F. E. Browder, 'On a Theorem of Beurling and Livingston', *Canad. J. Math.* **17** (1965) 617–622. *MR* **31** ≠ ≠ 595.

F. E. Browder, 'Multivalued Monotone Non-Linear Mappings and Duality Mappings in Banach Spaces', *Trans. Amer. Math. Soc.* **118** (1965) 338–351. *MR* **31** ≠ ≠ 5114.

F. E. Browder, 'Another Generalization of the Schauder Fixed Point Theorem', *Duke Math. J.* **32** (1965) 399–406. *MR* **34** ≠ ≠ 3567.

F. E. Browder, 'A Further Generalization of the Schauder Fixed Point Theorem', *Duke Math. J.* **32** (1965) 575–578. *MR* **34** ≠ ≠ 3287.

F. E. Browder, 'Fixed Point Theorems on Infinite Dimensional Manifolds', *Trans. Amer. Math. Soc.* **119** (1965) 179–194. *MR* **33** ≠ ≠ 3287.

F. E. Browder, 'Non-Expansive Nonlinear Operators in a Banach Space', *Proc. Nat. Acad. Sci. U.S.A.* **54** (1965) 1041–1044. *MR* **32** ≠ ≠ 4574.

F. E. Browder, 'Nonlinear Operators in Banach Spaces', *Math. Ann.* **162** (1965/66) 280–283. *MR* **32** ≠ ≠ 8187.

F. E. Browder, 'Fixed Point Theorems for Non-Linear Semi-Contractive Mappings in Banach Spaces', *Arch. Rat. Mech. Anal.* **21** (1965/66) 259–269. *MR* **34** ≠ ≠ 641.

F. E. Browder, 'Problemes nonlineaires', *Sem. Math. Montreal*, No. 15 (1966). *MR* **40** ≠ 3380.

F. E. Browder, 'Convergence of Approximants to Fixed Points of Non-Expansive Nonlinear Mappings in Banach Spaces', *Arch. Rat. Mech. Anal.* **24** (1967a) 82–90. *MR* **34** ≠ ≠ 6582.

F. E. Browder, 'Convergence Theorems for Sequences of Nonlinear Operators in Banach Spaces', *Math. Zeitschr.* **100** (1967b) 201–225. *MR* **35** ≠ ≠ 5984.

F. E. Browder, 'A new Generalization of the Schauder Fixed Point Theorem', *Math Ann.* **174** (1967c) 285–290. *MR* **36** ≠ ≠ 6991.

F. E. Browder, 'Nonlinear Maximal Monotone Operators on Banach Spaces', *Math. Ann.* **175** (1968) 89–113. *MR* **36** ≠ ≠ 6989.

F. E. Browder, 'The Fixed Point Theory of Multi-Valued Mappings in Topological Vector Spaces', *Math. Ann.* **177** (1968) 283–301. *MR* **37** ≠ ≠ 4679.

F. E. Browder, 'Asymptotic Fixed Point Theorems', *Math. Ann.* **185** (1970) 38–60. *MR* **43** ≠ ≠ 1165.

F. E. Browder, 'Nonlinear Mappings of Analytic Type in Banach Spaces', *Math. Ann.* **185** (1970) 259–278. *MR* **41** ≠ ≠ 4318.

F. E. Browder, 'Topology and Nonlinear Functional Equations', *Studia Math.* **31** (1968) 189–204. *MR* **38** ≠ ≠ 6410.

F. E. Browder, 'Nonlinear Eigenvalue Problems and Group Invariance', *Functional Analysis and Related Fields*, Proc. Sympos. in Honor of M. Stone, Chicago, Iu. 1968. Springer Verlag, New York, 1970. pp. 1–58. *MR* **42** ≠ ≠ 6662.

F. E. Browder, *Proc. Sympos. Pure Math.* Vol. 18. Part. 2. Amer. Math. Soc., Providence R. I. 1976.

F. E. Browder and D. Figueiredo, 'J-Monotone Nonlinear Operators in Banach Spaces', *Indag. Math.* **28** (1966) 412–420. *MR* **34** ≠ ≠ 4957.

F. E. Browder and R. D. Nussbaum, 'The Topological Degree for Non-Compact Nonlinear Mappings in Banach Spaces', *Bull. Amer. Math. Soc.* **74** (1968) 671–676. *MR* **38** ≠ ≠ 583.

F. E. Browder and W. V. Petryshyn, 'The Solution by Iteration of Nonlinear Functional Equations in Banach Spaces', *Bull. Amer. Math. Soc.* **72** (1966) 571–576. *MR* **32** ≠ ≠ 8155.

F. E. Browder and W. V. Petryshyn, 'Construction of Fixed Points of Nonlinear Mappings in Hilbert Spaces', *J. Math. Anal. Appl.* **20** (1967) 197–228. *MR* **36** ≠ 747.

F. E. Browder and H. Brezis, *See* H. Brezis and F. E. Browder.

R. Brown, 'On the Nielsen Fixed Point Theorem for Compact Maps', *Duke Math. J.* **36** (1969) 699–708.

R. Brown, 'An Elementary Proof of the Uniqueness of the Fixed Point Index', *Pacif. J. Math.* **35** (1970) 548–558.

R. Brown, *The Lefschetz Fixed Point Theorem*, Glenview, Scott & Foreman, 1971.

R. Brown, 'Elementary consequences of Noncontractibility of the Circle', *Amer. Math. Monthly* **81** (1974) 247–252.

R. E. Bruck, 'Properties of Nonexpensive Mappings in Banach Spaces', *Trans. Amer. Math. Soc.* **179** (1973) 251–262.

P. J. Bushell, 'On the Projective Contraction Ratio for Positive Linear Mappings', *J. London Math. Soc.* **5**(1972) 235–239.

P. J. Bushell, 'On the Solutions of the Matrix Equation $T'AT = A^2$,' *Linear Algebra* (to appear).

P. J. Bushell, 'Hilbert's Metric and Positive Contraction Mappings', *Arch. Rat. Mech. Math.* **52** (1973) 330–338.

W. L. Bynum, 'Normal Structure and the Banach–Mazur Distance Coefficient', *Notices Amer. Math. Soc.* **25** (1978) A-116.

V. W. Bryant, 'A remark on a Fixed Point Theorem for Iterates Mapping', *Amer. Math. Monthly* **75** (1968) 399–400.

R. Cacciopoli, 'Una teorem generale sull'existenza di elementi uniti in una transformazione funzionale', *Rend. Accad. Nazionale deli Lincei*, **11** (1930) 794–799.

R. Cacciopoli, 'Problemi nonlineari in analisi funzionale', *Rend. Sem. Mat. Roma* **1** (1931) 13–22.

R. Cacciopoli, 'Sugli elementi uniti delle transformazione funzionale; un osservazione sui problemi di valori ai limiti', *Rend. Accad. Nazionale dei Lincei* **13** (1931) 498–502.

R. Cacciopoli, 'Sugli elementi uniti delle transformazioni funzionale; un teorema di esistenza e di unicita ed alcune applicazioni', *Rend. Mat. Sem. Univ. di Padova* **3** (1932) 1–15.

R. Cacciopoli, 'Un principio di inversione per le correspondencze funzionali e sue applicazioni alle equazioni a derivate parziali', *Rend. Accad. Naz. dei Lincei* **16** (1932) 484–489.

G. L. Cain, 'Fixed Points and Stability of Condensing Multifunctions on Random Normed Spaces', *Notices Amer. Math. Soc.* **26** (1978) A–109.

A. Calderon, 'Singular Integrals', *Bull. Amer. Math. Soc.* **72** (1966) 427–465.

M. R. Calvin, 'Multi-Valued Mappings of a Closed and Bounded Set of a Normed Linear Space–A Mapping Degree', *Notices Amer. Math. Soc.* **25** (1978) A–128.

B. Calvert and J. R. L. Webb, 'An Existence Theorem for Quasi–Monotone Operators', *Rend. Accad. Nazionale dei Lincei* **8** (1971) 362–368.

C. Caratheodory and H. Rademacher, 'Über eindeutigkeit in kleinen und in grossen Stetiger–Abbildungen von Gebieten', *Arch. Math. Phys.* **26** (1917) 1–9.

J. Caristi, 'Fixed Point Theorems for Mappings Satisfying the Inwardness Condition', *Trans. Amer. Math. Soc.* **215** (1976) 241–251.

J. Caristi and W. A. Kirk, 'Geometric Fixed Point Theory and Inwardness Condition', *Proc. Conf. on Geometry of Metric and Linear Spaces* (Michigan, 1974), Lecture Notes in Math. No. 490 (1975) pp. 75–83.

P. H. Carter 'An Improvement of the Poincaré–Birkhoff Fixed Point Theorem', Ph.D. Thesis, The University of Florida, Florida, 1978.

M. L. Cartwright and J. L. Littlewood, 'Some Fixed Point Theorems', *Ann. Math.* **54** (1951) 1–37.

A. L. Cauchy, *Leçons sur les calculs différentiel et integral*, Vols. 1 and 2, Paris, 1884.

L. Cesari, 'Functional Analysis and Periodic Solutions of Nonlinear Differential Equations', *Contrib. to Diff. Eq.* **1** (1963) 149–187.

L. Cesari, 'Functional Analysis and the Galerkin Method', *Mich. Math. J.* **11** (1964) 385–414. *MR* **30** ≠ ≠ 1622.

L. Cesari and R. Kannan, 'Solutions in the large of Lienard Systems with Forcing Terms', *Annali di Mat. IV*, **61** (1976) 101–124.

E. W. Cheney, 'Applications of Fixed Point Theorems to Approximation Theory', *Proc. Sympos. Theory of Approximation*, Academic Press, 1976.

S. K. Chaterjea, 'Fixed Point Theorems', *C. R. Acad. Bulgar Sci.* **25** (1972) 727–730.

S. K. Chaterjea, 'Fixed Point Theorems for a Sequence of Mappings with a Contractive Iterate', *Publ. Math. Inst. Beograd*, **14** (1972) 15–18.

G. Chichilsky and P. J. Kalman, 'Rothe's Fixed Point Theorem and the Existence of Equilibria in Monetary Economics', *J. Math. Anal. Appl.* **65** (1978) 56–65.

W. G. Chinn and N. Steenrod, *First Concepts of Topology*, Random House, New Mathematical Library, 1966, Yale University.

S. C. Chu and R. D. Meyer, 'On Continuous Functions, Commuting Functions and Fixed Points', *Fund. Math.* **59** (1966) 91–95.

S. C. Chu and J. B. Diaz, 'Remarks on a Generalization of Banach's Principle of Contraction Mappings', *J. Math. Anal. Appl.* **11** (1965) 440–446.

L. B. Čirič, 'Generalized Contractions and Fixed Point Theorems', *Publ. Math. Beograd* **12** (1971) 19–26.

L. B. Čirič, 'A Generalization of Banach's Contraction Principle', *Proc. Amer. Math. Soc.* **45** (1974) 267–273.

F. Clarke, 'Pointwise Contraction Criteria for the Existence of Fixed Points' (to appear).

D. Cohen, 'On the Sperner Lemma', *J. Comb. Theory* **2** (1967) 585–587.

J. E. Conjura, 'Solvable Extensions for Linear and Nonlinear Operators, the Nature of their Domains and Error Analysis for Approximate Solutions and Residuals', Ph.D. Rutgers State Univ. New Jersey, New Brunchwick, 1977.

G. Constantin, Some Spectral Properties for the Local a-Contraction Operators', *Boll. Union. Mat. Ital.* **4** 1972.

C. Corduneanu, 'Equazioni differenziali negli spazi di Banach', *Rend. Accad. Nazionale dei Lincei* **23** (1957).

C. Corduneanu, 'Some Applications of a Fixed Point Theorem to the Theory of Differential Equations', *Annalele Stiint. Univ. Al. I. Cuza din Iasi, T.* **IV** (1958) 43–47. (in Roumanian).

C. Corduneanu, 'Sur certains systems differentielles non-linéeaires', *Annalele Univ. Al. I. Cuza, Iasi, T.* **IV** (1960) 257–260.

C. Corduneanu, 'Problemes globaux dans la theorie des equations integrales de Volterra', *Annali di Mat.* **67** (1965) 349–363.

C. Corduneanu, 'Sur certaines Equations Fonctionneles de Volterra', *Funkcialaj Ekvacioj* **9** (1966) 119–127.

C. Corduneanu, *Integral Equations and Stability of Feedback systems*, Academic Press, 1973.

C. Corduneanu, *Almost Periodic Functions*, John Wiley, 1968.

M. Crandall, 'Differential Equations in Convex Sets', *J. Math. Soc.* Japan, **22** (1970) 443–455. *MR* 30 ≠ ≠4047.

M. Crandall, 'On Accreative Sets in Banach Spaces', *J. Funct. Anal.* **5** (1970) 204–217. *MR* **41** ≠ ≠4201.

M. Crandall, and A. Pazy, 'Semigroups of Nonlinear Contractions and Dissipative Sets', *J. Func. Anal.* **3** (1969) 376–418. *MR* **39** ≠ ≠4705.

M. Crandall, 'An Introduction to Constructive Aspects of Bifurcation and the Implicit Function Theorem', *Appl. of Bifurcation Theory.* Proc. Adv. Sem. Univ. of Wisconsin, 1976. Publ. Res. Center No. 38.

J. Cronin, 'Branch Points of Solutions of Elliptic Differential Equations', *Trans. Amer. Math. Soc.* **69** (1950) 208–231. *MR* 12 ≠ ≠716.

J. Cronin, 'Branch Points of Solutions of Elliptic Differential Equations, II'. *Trans. Amer. Math. Soc.* **76** (1950) 207–222. *MR* **16** ≠ ≠97.

J. Cronin, 'Analytic Functional Mappings', *Ann. Math.* **58** (1953) 175–181.

J. Cronin, 'Topological Degree of Some Mappings', *Proc. Amer. Math. Soc.* **5** (1954) 175–178. *MR* **16** ≠ ≠47.

J. Cronin, 'The Dirichlet Problem for Nonlinear Elliptic Equations', *Pacif. J. Math.* **5** (1955) 335–344. *MR* **17** ≠ ≠494.

J. Cronin, 'Fixed Points and Topological Degree in Nonlinear Analysis', *Amer. Math. Soc. Surveys,* 1964.

J. Cronin, 'Using Leray–Schauder Degree', *J. Math. Anal. Appl.* **25** (1969) 175–178.

J. Cronin, 'Upper and Lower Bound for the Number of Solutions of Nonlinear Equations', *Proc. Sympos. Pure Math.* Vol. 18. Part 1. Amer. Math. Soc. 1970.

J. Cronin, 'Eigenvalues of some Nonlinear Operators', *J. Math. Anal. Appl.* **38** (1972) 659–667.

J. Cronin, 'Equations with Bounded Nonlinearities', *J. Diff. Eq.* **14** (1973) 581–596.

J. Cronin, 'A Stability Criterion', *Applicable Analysis,* **1** (1971) 23–30.

J. Cronin, 'A Mathematical Model of Aneurysm of the Circle Q: II.A Qualitative Analysis of the Equation of Austin'. *Math. Biosci.* **22** (1974) 237–275.

J. Cronin, 'Some Qualitative Analysis of Singularly Perturberd equations', *7th Intern. Conf. über nichtlineare Schwingungen,* Akademie Verlag, 1977, Berlin. 161–199.

D. Cudia, 'The Geometry of Banach Spaces. Smoothness'. *Trans. Amer. Math. Soc.* **110** (1964) 284–314.

D. Cudia, 'Rotundity', *Proc. Sympos. Pure Math. Vol.* 7. Amer. Math. Soc. R. I. 1963.

G. Darbo, 'Punti uniti in transformazioni a codominio non compatto', *Rend. Sem. Mat. Univ. Padova* **24** (1955) 84–92. *MR* **16** ≠ ≠1140.

J. Daneš, 'Some Fixed Point Theorems', *Comment. Mat. Univ. Carolinae* **9** (1968) 223–235. *MR* 38 ≠ ≠3744.

J. Daneš. 'Generalized Contractive Mappings', *Summer School on Fixed Points,* Krkonose, Czechoslovakia, Aug. 31–Sept. 6, 1968.

J. Daneš, 'Fixed Point Theorems, Nemitzkii and Urysohn Operators, and Continuity of Nonlinear Mappings', *Comment. Mat. Univ. Carolinae* **11** (1970) 481–500.

J. Daneš, 'Generalized Concentrative Mappings and their Fixed Points', *Comment. Mat. Univ. Carolinae* **11** (1970) 115–135. *MR* **43** ≠ ≠7668.

J. Daneš, 'A Geometrically Useful Theorem in Nonlinear Analysis', *Boll. Union. Mat. Ital.* **6** (1972) 369–375.

J. Daneš, 'On the Istrǎţescu Measure of Noncompactness', *Bull. Math. Soc. R. S. Roumanie* **16** (1972) 403–406.

J. Daneš, 'On Densifying and Related Mappings and their Applications in Nonlinear Analysis', *Theory of Nonlinear Operators*, Summer Schoos, 1972. Neuendorff, Berlin 1974.

J. Daneš, 'Two Fixed Point Theorems in Topological and Metric Spaces', *Bull. Austral. Math. Soc.* **14** (1979) 259–265.

M. M. Day, 'Polygons Circumscribed about Closed Convex Curves', *Trans. Amer. Math. Soc.* **62** (1947a) 315–319.

M. M. Day, 'Some Characterizations of Inner Product Spaces', *Trans. Amer. Math. Soc.* **62** (1947b) 320–337.

M. M. Day, 'Amenable semigroups', *Ill. J. Math.* **1** (1957) pp. 509–544.

M. M. Day, 'Semigroups and Amenability', *Semigroups*. Proc. Symp. at Wayne State Univ. (Ed. by K. Folley) Academic Press (1969) pp. 5–53.

M. M. Day, 'Fixed Point Theorems for Compact Convex Sets', *Ill. J. Math.* **5** (1961) 581–590. (Correction, ibid., **8** (1964) 713).

M. M. Day, 'Invariant renorming', *Fixed Points*. (Ed. by S. Swaminathan) Academic Press 1977. 51–63.

M. M. Day, *Normed Linear Spaces*, Springer Verlag, New York, 1958.

K. Deimling, 'Zeros of Accreative Operators', *Manuscripta Math.* **13** (1974) 365–374.

A. Deleanu, 'Théorie des points fixes sur les retractes de voisinage des éspaces convexoids', *Bull. Soc. Math. France* **89** (1959) 235–243. *M R* **23** ≠ ≠ 763.

A. Deleanu, 'Une generalization du théorem de point fixe de Schauder', *Bull. Math. Soc. France* **89** (1961) pp. 223–226. *M R* **26** ≠ ≠ 769.

J. B. Diaz and F. T. Metcalf, 'On the Structure of the Set of subsequential Limits of Successive Approximations', *Bull. Amer. Math. Soc.* **73** (1967) 516–519. *M R* **35** ≠ ≠ 2268.

J. B. Diaz and F. T. Metcalf, 'On the Set of Subsequential Limits of Successive Approximations', *Trans. Amer. Math. Soc.* **135** (1961) 459–485.

J. B. Diaz and B. Margolis, 'A Fixed Point Theorem of the Alternative for Contractions on a Generalized Complete Metric Space', *Bull. Amer. Math. Soc.* **74** (1968) 305–309.

C. L. Dolph, 'Nonlinear Integral Equations of Hammerstein Type', *Trans. Amer. Math. Soc.* **66** (1949) 289–307. *M R* **11** ≠ ≠ 367.

C. L. Dolph and G. Minty, 'On Nonlinear Integral Equations of the Hammerstein Type', *Proc. Symp. Univ. of Wisconsin*, 1964, 99–154.

W. G. Dotson, 'On the Mann Iterative Processes', *Trans. Amer. Math. Soc.* **149** (1970) 65–73.

W. G. Dotson, 'Fixed Point Theorems for Nonexpansive Mappings on Star-Shaped Subsets of a Banach Space', *J. London Math. Soc.* **4** (1972) 408–410.

W. G. Dotson and W. R. Mann, 'A Generalized Corollary of the Browder–Kirk Fixed Point Theorem', *Pacif. J. Math.* **26** (1968) 455–459.

D. Downing and W. A. Kirk, 'A Generalization of Caristi's Theorem and Applications to Nonlinear Mapping Theory', *Pacif. J. Math.* **69** (1977) 339–347.

D. Downing and W. A. Kirk, 'Fixed Point Theorems for Set-Valued Mappings in Metric and Banach Spaces', *Math. Jap.* **22** (1977) 99–112.

D. Downing, 'Some Coincidence Theorems in Metric and Banach Spaces', *Notices Amer. Math. Soc.* **25** (1978) A–110.

W. G. Dotson and W. R. Mann, 'The Schauder Fixed Point Theorem for Nonexpansive Mappings', *Amer. Math. Monthly* **84** (1977) 363–366.

J. Dugundji, 'An Extension of Tietze's Theorem', *Pacif. J. Math.* **1** (1951) 353–367. *MR* **13** ≠ ≠ 373.

J. Dugundji, 'Absolute Neigbourhood Retracts and Local Connectedness', *Com. Math.* **13** (1958) 229–241.

J. Dugundji, *Topology*, Allyn and Bacon, 1954.

N. Dunford and J. Schwartz, *Linear Operators*, I and II. Interscience, New York. 1958, 1962.

B. C. Eaves, 'An Old Theorem', *Proc. Amer. Math. Soc.* **26** (1970) 509–513.

B. C. Eaves, 'Computing Kakutani Fixed Points', *SIAM J. Appl. Math.* **21** (1971) 236–244.

B. C. Eaves, 'On the Basic Theorem of Complementarity', *Math. Progr.* **1** (1971) 68–75.

B. C. Eaves, 'Homotopies for Computation of Fixed Points', *Math. Progr.* **3** (1972) 1–22.

B. C. Eaves, 'Linear Complementarity Problem', *Working Paper No.* 275. Center for Research in Manag. Sciences, Univ. of Calif. 1968.

B. C. Eaves and R. Saigal, 'Homotopies for the Computation of Fixed Points on Unbounded Domains', *Math. Progr.* **3** (1972) pp. 225–237.

C J. Earle and R. Hamilton, 'A Fixed Point Theorem for Holomorphic Mappings', *Proc. Symp. Pure Math.* **16** (1970) 61–65. *MR* **42** ≠ ≠ 918.

J. Eells, 'A Setting for Global Analysis', *Bull. Amer. Math. Soc.* **72** (1966). *MR* **34** ≠ ≠ 3590.

J. Eells and J. Fournier, 'La théorie des points fixes des applications a itérée condensante', *Bull. Soc. Math. France*, Mem. 46, 91–120

M. Edelstein, 'An Extension of Banach's Contraction Principle', *Proc. Amer. Math. Soc.* **12** (1961) 7–10. *MR* **22** ≠ ≠ 11375.

M. Edelstein, 'On Fixed and Periodic Points Under Contractive Mappings', *J. London Math. Soc.* **37** (1962) 74–79. *MR* **24** ≠ ≠ A 2936.

M. Edelstein, 'On Predominantly Contractive Mappings', *J. London Math. Soc.* **38** (1963) 81–86. *MR* **26** ≠ ≠ 6726.

M. Edelstein, 'A Theorem on Fixed Points Under Isometries', *Amer. Math. Soc. Monthly* **70** (1963) 298–300.

M. Edelstein, 'On Nonexpansive Mappings', *Proc. Amer. Math. Soc.* **15** (1964) 689–695.

M. Edelstein, 'On Nonexpansive Mappings in Banach Spaces', *Proc. Camb. Phil. Soc.* **60** (1964) 437–447. *MR* **29** ≠ ≠ 1521.

M. Edelstein, 'A Remark on a Theorem of Krasnoselskii', *Amer. Math. Monthly*, **13** (1966) 509–510. *MR* **33** ≠ ≠ 3072.

M. Edelstein and A. C. Thompson, 'Contractions, Isometries and some Properties of Inner-Product Spaces', *Indag. Math.* **29** (1967) 326–331. *MR* **36** ≠ ≠ 5672.

M. Edelstein, 'The Construction of an Asymptotic Center with a Fixed Point Property', *Bull. Amer. Math. Soc.* **78** (1972) 206–208.

M. Edelstein, 'Fixed Point Theorems in Uniformly Convex Spaces', *Proc. Amer. Math. Soc.* **44** (1974) 369–374.

M. Edelstein, 'On Some Aspects of Fixed Point Theory in Banach Spaces', *The Geometry of Metric and Linear Spaces*, Lectures Notes in Math. No. 490 (1975).

M. Edelstein, 'Remarks and Questions Concerning Nonexpansive Mappings', *Fixed Point Theory and Its Applications*, Academic Press, 1976.

D. E. Edmunds and J. R. L. Webb, 'A Leray–Schauder Theorem', *Math. Ann.* **182** (1969) 207–212.

D. E. Edmunds, C. A. Stuart and A. J. Potter, 'Noncompact Positive Operators', *Proc. Roy. Soc. Ser. A.*, **328** (1972) 67–81.

D. E. Edmunds and J. R. L. Webb, 'Nonlinear Operator Equations in Hilbert Spaces', *J. Math. Anal. Appl.* **34** (1971) 471–478. *MR* **43** ≠ ≠ 6778.

D. E. Edmunds, 'Remarks on Nonlinear Functional Equations', *Math. Ann.* **174** (1967) 233–239. *MR* **36** ≠ ≠ 3180.

H. H. Ehrmann, 'On Implict Function Theorems and the Existance of Solutions of Nonlinear Equations', *Enseig. Math.* **9** (1963) 129–176.

S. Eilenberg and D. Montgomery, 'Fixed Points Theorems for Multi-Valued Transformations', *Amer. J. Math.* **68** (1946) 214–222.

D. Eisenbud and H. Levine, 'An Algebraic Formula for the Degree of a C^∞ Map Germ', *Ann. Math.* **106** (1977) 19–44.

D. Eisenbud, 'An Algebraic Approach to the Topological Degree of a Smooth Map', *Bull. Amer. Math. Soc.* **84** (1978) 751–764.

I. Ekeland, 'Sur les problèmes variationnelles', *C. R. Paris, Ser. A. B.* **275** (1972) 1057–1059.

W. Engel, 'Random Fixed Point Theorems for Multi-Valued Mappings', *Pacif. J. Math.* **76** (1978) 351–360.

P. J. Erdelski, 'Computing the Brouwer Degree in $\mathbf{R}^2$', *Math. Comp.* **27** (1963) 121.

Ky Fan, 'Fixed Points and Minimax Theorems in Locally Convex Spaces', *Proc. Nat. Acad. Sci. U.S.A.* **38** (1952) 121–126.

Ky Fan, 'A Generalization of Tucker's Combinatorial Lemma with Topological Applications', *Ann. Math.* **56** (1952) 431–437. *MR* **14** ≠ ≠ 490.

Ky Fan, 'Existence Theorems and Extreme Solutions For Inequalities Concerning Convex Functions or Linear Transformations', *Math. Zeitschr.* **68** (1957) 205–217, *MR* **19** ≠ ≠ 1183.

Ky Fan, and I. Glicksberg, 'Some Geometric Properties of the Sphere in a Normed Linear Space', *Duke Math. J.* **25** (1958) 553–568. *MR* **20** ≠ ≠ 542.

Ky Fan, 'Invariant Cross Sections and Invariant Linear Spaces', *Israel J. Math.* **2** (1964) 19–26. *MR* **30** ≠ ≠ 1382

Ky Fan, 'A Generalization of Tichonoff's Fixed Point Theorem', *Math. Ann.* **142** (1961) 305–310.

Ky Fan, 'Sur une theorem de minimax', *C. R. Acad. Paris* **259** (1964) 3925–3928.

Key Fan, 'Applications of a Theorem Concerning Sets with Convex Sections', *Math. Ann.* **163** (1966) 189–203. *MR* **32** ≠ ≠ 8101.

Ky Fan, 'Extensions of Two Fixed Point Theorems of F. E. Browder', *Math. Zeitschr.* **112** (1969) 234–240.

Ky Fan, 'A Minmax Inequality and Applications', *Inequalities III.* Academic Press, 1972.

Ky Fan, 'Orbits of Semi-Groups of Contractions and Groups of Isometries', *Abhand. Math. Sem. Univ. Hamburg,* **45** (1976) 245–250.

Cheng Martin Fan, 'On Continuity of Fixed Points of Collectively Condensing Maps', *Proc. Amer. Math. Soc.* **63** (1977) 74–76.

E. Fadell, 'On a Coincidence Theorem of F. B. Fuller', *Pacif. J. Math.* **15** (1965) 825–834.

E. Fadell, 'Some Examples in Fixed Point Theory', *Pacif. J. Math.* **33** (1970) 89—101.

E. Fadell, 'Recent Results in the Fixed Point Theory of Continuous Maps', *Bull. Amer. Math. Soc.* **76** (1970) 10–29.

F. A. Ficken, 'The Continuation Method for Functional Equations', *Comm. Pure Appl. Math.* **4** (1951) 435–456. *MR* **13** ≠ ≠ 562.

G. Fichera, 'Problemi eletrostatici con vincoli unilaterali: il problema di Signorini con ambique condizioni al contorno', *Atii Acad. Naz. dei Lincei*, I(8) **VII** (1963–1964) 91–140. *MR* **31** ≠ ≠ 2888.

G. Fichera, 'Electrostatic Problem with Unilateral Constraints', *Sem. sur les equations aux derivées partielles*, College de France, 1966/1967.

D. Figueiredo, *Topics in Nonlinear Analysis*, Lectures Notes, Inst. for Fluid Dynamics and Appl. Math. Univ. of Maryland, College Park Md. 1967.

D. Figueiredo, and L. A. Karlowitz, 'On Radial Projection in Normed Spaces', *Bull. Amer. Math. Soc.* **73** (1967a) 364–368. *MR* **35** ≠ ≠ 2130.

D. Figueiredo and L. A. Karlowitz, 'On the Projection onto Convex Sets and the Extension of Contractions', *Proc. Conf. on Projections and Related Topics*, (Preliminary Edition) Clemson University, Clemson (1967b). 1967.

D. Figueiredo and L. A. Karlowitz, 'The Extension of Contractions on Normed Spaces', *Proc. Symp. Pure Math.* Vol. 18 Part 1. (1970) pp. 95–104. Amer. Math. Soc. Providence R. I., 1970. *MR* **43** ≠ ≠ 877.

D. Figueiredo and L. A. Karlowitz, 'The Extension of Contractions and the Intersection of Ball in Banach Spaces', *J. Funct. Anal.* **11** (1972) 168–172.

W. M. Fleischman (Ed.) *Set-Valued Mappings, Selections and Topological Properties of '2^X'*, Lectures Notes in Math. 171, 1970.

B. Fisher, 'A Fixed Point Theorem', *Math. Mag.* **48** (1975) 223–225.

M. L. Fisher, F. J. Gould and J. W. Tolle, 'A Simplicial Approximation Algorithm for Solving Systems of Nonlinear Equations', *Symposia Math.* **19** (1976) Instituto Nazional di Alta Matematica, Roma.

B. Fisher, 'Results on Fixed Points', *Bull. Acad. Sci. Polon.* **25** (1977) 1253–1256.

R. B. Fraser and S. B. Nadler, 'Sequences of Contractive Maps and Fixed Points', *Pacif. J. Math.* **31** (1969) 659–667.

M. Fréchet, 'Sur quelques points du calcul fonctionnel', *Rendiconti di Circolo Mat. di Palermo*, **22** (1906) 1–74.

M. Freeman, 'The Inverse as a Fixed Point in a Function Space', *Amer. Math. Monthly* **83** (1976) 344–349.

W. Forster, 'Fixed Point Algorithms, Background and Estimates for Implementation on Array Processors', Univ. of Southampton Preprint Ser. No. 9 (1978).

W. Forster, 'On Numerical Approximation of Fixed Points in $C_{[0,1]}$', Univ. of Southampton Preprint Ser. No. 13. (1978).

W. Forster, 'Die Struktur numerischer Methoden: Konstruktive Fixpunktsätze', *ZAMM* **57** (1977) 277–279.

M. Fort, 'Some Properties of Continuous Functions', *Amer. Math. Monthly*, **59** (1952) 372–375.

R. L. Frum-Ketkov, 'Mapping into a Banach Sphere', *D. A. N.* **175** (1967) 1229–1231. = *Sov. Math. Doklady* **8** (1967) 1004–1006.

M. Furi and M. Martelli, 'Succesioni di transformazioni in uno spazio metrico e punti fissi', *Rend. Acad. Naz. dei Lincei*, **47** (1969) 27–31.

M. Furi and A. Vignoli, 'A Fixed Point Theorem in Complete Metric Spaces', *Boll. Union. Mat. Ital.* **IV** No. 4–5 (1969) 505–506.

M. Furi and A. Vignoli, 'On a-Nonexpansive Mappings and Fixed Points', *Rend. Acad. Naz. dei Lincei* **48** (1970) 195–198.

M. Furi and M. Martelli, 'A Characterization of Compact Filter Bases in Complete Metric Spaces', *Rend. Inst. Mat. Univ. di Trieste* **2** (1970) 109–113.

M. Furi, M. Martelli and A. Vignoli, 'On Minimum Problems for Families of Functionals', *Annali di Mat. Pura ed Appl.* **86** (1970) 181–187.

M. Furi and A. Vignoli, 'About Well-Posed Optimization Problems for Functionals in Metric Spaces', *J. Optimization Theory and Appl.* **5** (1970) 225–229.

M. Furi and A. Vignoli, 'A Characterization of Well-Posed Minimum Problems in a Complete Metric Space', *Optimization Theory and Appl.* **5** (1970).

M. Furi, M. Martelli and A. Vignoli, 'Contributions to the Spectral Theory of Nonlinear Operators in Banach Spaces', *Annali di Mat. Pura ed Appl.* (to appear).

S. Fučik, J. Nečas et al., *Spectral Analysis of Nonlinear Operators,* Lectures Notes in Math. No. 343. Springer Verlag, New York.

J. Gatica and W. A. Kirk, 'Fixed Point Theorems for Lipschitzian Pseudocontractive Mappings', *Proc. Amer. Math. Soc.* **36** (1972) 111–115

J. Gatica and W. A. Kirk, 'A Fixed Point Theorem for k-Set-Contractions Defined on a Cone', *Pacif. J. Math.* **53** (1974) 131–136.

E. Gagliardo and Clifford Kottman, 'Fixed Points for Orientation Preserving Homomorphisms of the Plane which Interchange Two Points', *Pacif. J. Math.* **59** (1975) 27–32.

A. L. Garkavi, 'On the Chebyshev Center of a Set in a Normed Space', *Investigations of Comtemporary Problems in the Constructive Theory of Functions*, Moscow, 1961, pp. 328–331,

A. L. Garkavi, 'The Best Possible Net and the Best Possible Cross-Section of a Set in a Normed Space', *Izv. Akad. Nauk SSSR., Ser.* 2, **39** (1964) 111–132.

A. Genel and J. Lindenstrauss, 'An example Concerning Fixed Points', *Israel J. Math.* **22** (1975) 81–86.

D. Gilbarg and N. Trudinger, 'Elliptic Differential Equations of Second Order', *Grund. Math. Wiss.* **224** (1977). Springer Verlag, New York.

I. Glicksberg, 'A Further Generalization of the Kakutani Fixed Point Theorem with Application to Nash Equilibrium Points', *Proc. Amer. Math. Soc.* **3** (1952) 170–174. *MR* **13** ≠ ≠ 764.

I. T. Gochberg, L. S. Goldenstein and A. S. Markus, 'Investigations of some Bounded Linear Operators in Connection with their q-Norms', *Uchen. Zap. Kishinevskogo Inst.* **29** (1957) 29–36.

K. Goebel, 'An Elementary Proof of the Fixed Point Theorem of Browder and Kirk', *Mich. Math. J.* **16** (1969) 381–383.

K. Goebel, 'On Fixed Point Theorems for Multi-Valued Nonexpansive Mappings', *Annales Univ. Maria Curie-Sklodowska* (to appear).

K. Goebel, W. A. Kirk and T. N. Shimi, 'A Fixed Point Theorem in Uniformly Convex Spaces', *Boll. Union. Mat. Ital.* **7** (1973) 67–75.

K. Goebel and W. A. Kirk, 'A Fixed Point Theorem for Transformations whose Iterates Have Uniform Lipschitz Constant', *Studia Math.* **67** (1973) 135–140.

K. Goebel and T. Kuczumov, 'A Contribution to the Theory of Nonexpansive Mappings' (to appear).

A. J. Goldman and P. R. Meyers, 'Simultaneous Contractifications', *J. Res. Bur. Standards* **74**B (1969) 301–305.

M. Golomb, 'Zur Theorie der nichtlinearen Integralgleichungen, Integralgleichungensysteme und allgemeinen Funktionalgleichungen', *Math. Zeitschr.* **39** (1935) 45–75.

D. B. Goodner, 'Projections in Normed Linear Spaces', *Trans. Amer. Math. Soc.* **69** (1950) 89–108.

G. M. Goncharov. 'On Some Existence Theorems for the Solution of a Class of Nonlinear Operator Equations', *Mat. Zametki* **7** (1970) 229–239.

D. Graffi, 'Forced Oscillations for Several Nonlinear Circuits', *Ann. Math.* **54** (1951) 262–271.

D. Graffi, 'Sulle oscillazioni nella mecanica nonlineare', *Riv. Mat. Univ. Parma* **2** (1952) 317–326.

D. Graffi, 'Oscillazioni nonlineari', *Conf. Sem. Mat. di Univ. Bari* No. 1 (1954). A. Granas, 'On Certain Classes of Nonlinear Mappings in Banach Spaces', *Bull. Acad. Sci. Polon. Cl. III* **9** (1957) 867–871.

A. Granas, 'On a Certain Geometrical Theorem in Banach Spaces', *Bull. Acad. Sci. Polon. Cl. III* **9** (1957) 873–877.

A. Granas, 'Über einen Satz von K. Borsuk', *Bull. Acad. Sci. Polon. Cl. III* **10** (1957) 959–961.

A. Granas, 'On Continuous Mappings of Open Sets in Banach Spaces', *Bull. Acad. Sci. Polon.* **1** (1958) 25–30.

A. Granas, 'Homotopy Extension in Banach Spaces and some of its Applications to the Theory of Nonlinear Equations', *Bull. Acad. Sci. Polon.* **7** (1959) 387–394 (in Russian).

A. Granas, 'Theorems on Antipodes and Theorems on Fixed Points for a Certain Class of Multivalued Mappings in Banach Spaces', *Bull. Acad. Sci. Polon.* **8** (1959) 271–275,

A. Granas, 'Sur la notion de degree topologique pour une certaine classe de transformations multivalentes dans les éspaces de Banach', *Bull. Acad. Sci. Polon.* **7** (1959) 191–194, *MR* **21** ≠ ≠ 7457.

A. Granas, 'On the Disconnection of Banach Spaces', *Fund. Math.* **48** (1960) 189–200.

A. Granas, *Introduction to Topology of Functional Spaces*, Lectures Notes, Univ. of Chicago, 1961.

A. Granas, *Introduction à la topologie des éspaces de Banach*, Lecture Notes, Ins. H. Poincare, Paris, 1966.

A. Granas, 'The Theory of Compact Fields and some of its Applications to the Topology of Functional Spaces', *Rozprawy Mat.* **30** (1962). *MR* **26** ≠ ≠ 6743.

A. Granas, 'An Extension to Functional Spaces of the Borsuk–Ulam Theorem on Antipodes', *Bull. Acad, Polon. Sci.* **10** (1962) 87–90.

A. Granas, 'A Note on Deformation in Functional Spaces', *Bull. Acad. Polon. Sci.* **10** (1962) 87–90.

A. Granas, 'Sur la multiplication cohomotopique dans les éspaces de Banach', *C. R. Acad. Paris* **254** (1962) 56–57.

A. Granas, 'A Note on Schauder's Theorem on Invariance of Domain', *Bull Acad. Polon. Sci.* **10** (1962) 233–238. (in Russian).

A. Granas, 'Generalizing the Hopf–Lefschetz Fixed Point Theorem for Non-Compact ANR's', *Sympos. on Infinite dim. Topology*, (R. D. Anderson, Ed.) *Ann. of Math. Studies*, Princeton (1972) 119–130.

A. Granas and L. Gorniewicz, 'Fixed Point Theorems for Multi-Valued Mappings of the Absolute Neighbourhood Retracts', *J. Math. Pure et Appl.* **49** (1970) 381–395.

W. J. Gray and L. Waugham, 'The Almost Fixed Point Property for Hereditary Unicoherent Continua', *Proc. Amer. Math. Soc.* **27** (1971) 381–386.

M. Greenberg, *Lectures on Algebraic Topology*, Benjamin, New York, 1967.

F. P. Greenleaf, *Invariant Means on Topological Groups and their Applications* Van Nostrand, 1969.

C. W. Groetsch, 'A Nonstationary Iterative Process for Nonexpansive Mappings', *Proc. Amer. Math. Soc.* **43** (1974) 155–158.

J. de Groot, H. de Vries and T. Van der Walt, 'Almost Fixed Point Theorems for the Euclidean Plane', *Proc. Niederl. Akad. Wettenshappen Ser. A.* **66** (1963) 606–612.

A. Grotheudieck, *Topological Linear Spaces*. São Paulo, 1954.

B. Grünbaum, 'A Generalization of a Theorem of Kirszbaum and Minty', *Proc. Amer. Math. Soc.* **13** (1962) 812–814. *MR* **27** ≠ ≠ 6110.

V. K. Gupta and P. Srinivasan, 'A Note on Common Fixed Points', *Yokohama Math. J.* **19** (1971) pp. 91–95.

V. K. Gupta and P. Srinivasan, 'On Common Fixed Points', *Rev. Roum. Math. Pure Appl.* **17** (1972) 531–538.

V. I. Gurarii, 'On the Differential Properties of the Modulus of Convexity in a Banach Space', *Mat. Issled.* **2** (1967) 141–148 (in Russian).

L. F. Guseman, Jr. 'Fixed Point Theorems for Mappings with a Contractive Iterate at a Point', *Proc. Amer. Math. Soc.* **26** (1970) 615–618. *MR* **42** ≠ ≠ 919.

J. Hadamard, 'Sur quelques applications de l'indice de Kronecker', in J. Tanery *Introduction à la théorie des fonctions d'une variable* Vol. 2 Paris (1910) 437–477.

Ch. Hagopian, 'Fixed Point Problems for Disk-Like Continua', *Amer. Math. Monthly* **83** (1976) 471–472.

Jack K. Hale, 'Continuous Dependence of Fixed Points of Condensing Maps', *J. Math. Anal. Appl.* **46** (1974) 388–393.

B. Halpern, 'A General Fixed Point Theorem', *Proc. Sympos. Pure Math.* **18** Part. 1. Amer. Math. Soc. R. I. (1970) 114–131.

B. Halpern and G. Bergman, 'A Fixed Point Theorem for Inward and Outward Maps', *Trans. Amer. Math. Soc.* **130** (1968) 353–358. *MR* **36** ≠ ≠ 4397.

O. Hamilton, 'Fixed Points under Transformations of Continua which are not Connected, im kleinem', *Trans. Amer. Math. Soc.* **44** (1938) 18–24.

O. Hamilton, 'A Fixed Point Theorem for Pseudo-Arcs and Certain Other Metric Continua', *Proc. Amer. Math. Soc.* **2** (1951) 173–174.

H. Hanani, E. Netanyahu and M. Reichaw-Reichbach, 'The Sphere in the Image', *Israel J. Math.* **1** (1963) 188–195. *MR* **29** ≠ ≠ 453.

G. E. Hardy and T. D. Rogers, 'A Generalization of a Fixed Point Theorem of Reich', *Canad. Math. Bull.* **16** (1973) 201–206.

F. Hausdorff, *Set Theory*, Second Edition, Chelsea Publ. Comp. N. Y. 1962.

M. Hazewinkel, 'Meteic topologic Fixed Point Theorems and Applications' *Lectures Notes Econometric Inst.* Erasmus Univ. Rotterdam, 1972.

M. Hazewinkel and M. van de Vel, 'On Almost Fixed Point Theory', *Canad. Math. J.* **30** (1978) 673–699.

E. Hewitt and K. A. Ross, *Abstract Real Analysis*, Springer Verlag, New York.

Troy L. Hicks, 'Fixed Point Theorems in Locally Convex Spaces', *Notices Amer. Math. Soc.* **25** (1978) A–126.

D. Hilbert, 'Über die gerade Linie als kürzeste Verbindung zweier Punkte, *Math. Ann.* **46** (1895) 91–96.

W. Hildebrand, *Core and Equilibria of a Large economy*, Princeton Univ. Press, Princeton, 1974.

B. P. Hilam, 'A Characterization of the Convergence of Successive Approximations', *Amer. Math. Monthly* **83** (1976) 273.

E. Hille and R. S. Phillips, 'Functional Analysis and Semi Groups', *Amer. Math. Soc. Colloq. Publ.* Vol. 31. *MR* **19** ≠ ≠ 664.

C. H. Himmelberg, 'Measurable Selections', *Fund. Math.* **87** (1975) 53–72.

C. H. Himmelberg, J. R. Porter and F. S. van Vleck, 'Fixed Point Theorems for Condensing Multifunctions', *Proc. Amer. Math. Soc.* **23** (1969) 635–641. *MR* **39** ≠ ≠ 7480.

M. W. Hirsch, 'A Proof of the Nonretractability of a Cell onto its Boundary', *Proc. Amer. Math. Soc.* **14** (1963) 364–365.

M. W. Hirsch and S. Smale, 'On Algorithms for Solving $f(x) = 0$', *Notices Amer. Math. Soc.* **25** (1978) A–544.

R. D. Holmes, 'Fixed Points for Local Radial Contractions', *Proc. Sympos. Fixed Point Theory and Its Applications*, Dalhousie Univ. 1975; Academic Press (1976) 79–89.

H. Hopf, 'Eine Verallgemeinerung der Euler–Poincaréschen Formel', *Götting. Nachr. Gs. Wiss.* (1928) 127–136.

H. Hopf, 'Über die algebraische Anzahl von Fixpunkten', *Math. Zeitschr.* **29** (1929) 493–524.

P. Holm and E. Spanier, Involutions and Fredholm maps', *Topology*, **10** (1971) 203–218.

C. W. Howe, 'An Alternate Proof of the Existence of general Equilibria in a von Neumann Model', *Econometrica*, **28** (1960) No. 3.

S. T. Hu, *Theory of Retracts*, Detroit, Univ. of Wayne Press, 1965. *MR* **31** ≠ ≠ 6202.

Thakyin Hu, 'An Extension of the Markov–Kakutani Theorem', *Notices Amer. Math. Soc.* **25** (1978) A–111.

Thakyin Hu and W. A. Kirk, 'Local Contractions in Metric Spaces', (to appear).

M. Hukuhara, 'Sur l'existence des points invariants d'une transformation dans l'espace fonctionnel', *Japan J. Math.* **20** (1950) 1–4.

R. E. Huff, 'Some Applications of a General Lemma on Invariant Means', *Ill. J. Math.* **14** (1970) 216–221.

R. E. Huff, 'Existence and Uniqueness of Fixed Points for Semigroups of Affine maps', *Trans. Amer. Math. Soc.* **152** (1970) 99–106.

T. L. Hayden and T. J. Suffridge, 'Biholomorphic Maps in Hilbert Space have a Fixed Point', *Pacif. J. Math.* **38** (1971) 419–422. *MR* **46** ≠ ≠ 4288.

T. L. Hayden and T. J. Suffridge, 'Fixed Points of Holomorphic Maps in Banach Spaces', *Proc. Amer. Math. Soc.* **60** (1976) 95–105.

R. D. Holmes and A. T. Lau, 'Nonexpansive Actions of Topological Semigroups and Fixed Points', *J. London Math. Soc.* **5** (1972) 330–336.

T. L. Hicks and Ed W. Huffman, 'Fixed Point Theorems of Generalized Hilbert Spaces', *J. Math. Anal. Appl.* **64** (1978)

Ed W. Huffman, 'Strict Convexity in Locally Convex Spaces and Fixed Point Theorems in Generalized Hilbert Spaces', Ph.D. thesis. Univ. of Missouri Rolla, Missouri, 1977.

Ed W. Huffman, 'Strict Convexity in Locally Convex Spaces and Fixed Point Theorems', *Math. J.* **22** (1977) 323–333.

W. Hurewicz and H. Wallman, *Dimension Theory*, Princeton, 1941.

J. P. Huneke, 'On Common Fixed Points of Commuting Continuous Functions on an Interval', *Trans. Amer. Math. Soc.* **139** (1969) 371–381.

J. R. Isbell, 'Commuting Mapping of Tree', *Bull. Amer. Math. Soc.* **63** (1957) 419.

K. Iseki, 'On Banach's Theorem on Contraction Principle', *Proc. Jap. Acad. Sci.* **41** (1965) 145–146.

K. Isteki, 'On Common Fixed Points of Mappings', *Bull. Austral. Math. Soc.* **10** (1974) 365–370.

K. Iseki, 'Some Fixed Point Theorems in Metric Spaces', *Math. Jap.* **20** (1975) 101–110.

K. Iseki, 'A Simple Application of Reich's Fixed Point Theorem', *Math. Sem. Notes,* **5** (1977) 135–136.

K. Iseki, 'An Approach to Fixed Point Theorems', *Math. Sem. Notes* **20** (1975) No. 30.

S. Ishikawa, 'Fixed Points and Iteration of a Nonexpansive Mapping in a Banach Space', *Proc. Amer. Math. Soc.* **59** (1976) 65–71.

Ana I. Istrătescu and Vasile I. Istrățescu, 'On The Theory of Fixed Points for Some Classes of Mappings.'

    I. *Bull. Math. Soc. Roumanie.* **14** (1970a),

    II. *Rev. Roum. Math. Pure Appl.* **16** (1971a)

    III. *Rend, Acad, Naz, dei Lincei.* **49** (1970b),

    IV. *Bull. Soc. Math. Roumanie.* **15** (1970),

    V. *Rend. Acad. Naz. dei Lincei.* **51** (1971b).

    VI. *Rev. Roum. Math. Pure Appl.* **17** (1972a),

    VII. *Rend. Acad. Naz. dei Lincei.* **52** (1972b).

Ana Istrățescu and Vasile Istrățescu, 'On the Existence of a Solution of $f(x) = kx$ for a Continuous not Necessarily Linear Operator', *Proc. Amer. Math. Soc.* **56** (1976).

Vasile I. Istrățescu and I. Săcuiu, 'Fixed Point Theorems for Contraction Mappings in Probabilistic Metric Spaces', *Rev. Roum. Math. Pure Apple.* **18** (1973) 1375–1380.

Vasile I. Istrățescu, 'A Weierstrass Theorem for Real Banach Spaces', *J. Approx. Theory* **19** (1977).

Vasile I. Istrățescu, 'A Weierstrass Theorem for Complex Banach Spaces'.

Vasile I. Istrățescu, *Probabilistic Metric Spaces. An Introduction.* Ed. Technica. Bucharest, 1974.

Vasile I. Istrățescu, 'A Generalization of Contraction Mappings' (to appear) (1979a),

Vasile I. Istrățescu, 'A Simple Example of a Nonexpansive Mapping on a Rotund Banach Space Without Fixed Points' (to appear) (1979b).

Vasile I. Istrățescu, 'A Fixed Point Theorem for Locally Power Diminishing Diameter Mappings' (to appear) (1979c)

Vasile I. Istrățescu, *A Bibliography for Fixed Point Theory* (in preparation, contains a list of 3,600 papers).

Vasile I. Istrățescu, *Topics in Linear Operator Theory*, Contributi del Centro Linceo Interdisciplinare di Scienze Matematiche e loro applicazioni, No. 42. Roma. 1978.

Ioana I. Istrățescu, 'On Some Fixed Point Theorems in Generalized Metric Spaces', *Boll. Union. Mat. Ital.* **13**–A (1976) 95–100.

Ioana I. Istrățescu, 'Fixed Point Theorems for Some Classes of Contraction Mappings on Nonarchimedean Probabilistic Metric Spaces', *Publ. Math. Debrecen* **25** (1978) 29–35.

S. Itoh, 'A Random Fixed Point Theorem for Multivalued Contraction Mapping', *Pacif. J. Math.* **68** (1977) 85–90.

G. L. Itzkovitz, 'On Nets of Contractive Maps in Uniform Spaces', *Pacif. J. Math.* **35** (1970) 417–425.

I. M. James, 'Inner Products in Normed Linear Spaces', *Bull. Amer. Math. Soc.* **53** (1947) 559–566.

L. Janos, 'A Converse of Banach's Contraction Principle', *Proc. Amer. Math. Soc.* **18** (1967) 287–289.

L. Janos, 'Topological Homotheties on Compact Metrizable Spaces', *Amer. J. Math.* **90** (1968) 877–880.

L. Janos, 'Linearization of a Contractive Homeomorphism', *Amer. J. Math.* **90** (1968) 881–884.

L. Janos, 'Topological Homotheties on Compact Hausdorff Spaces', *Proc. Amer. Math. Soc.* **21** (1969) 562–568.

L. Janos and J. L. Solomon, 'A Fixed Point Theorem and Attractors', *Notices Amer. Math. Soc.* **25** (1978) A–130.

G. G. Johnson, 'Fixed Points by Mean Value Iterations', *Proc. Amer. Math. Soc.* **34** (1972) 193–194.

John Jones Jr. 'Approximation of Zeros of Functions', *Notices Amer. Math. Soc.* **25** (1978) A–162.

G. Stephen Jones, 'Asymptotic Fixed Point Theorems and Periodic Systems of Functional Differential Equations', *Contrib. to Diff. Eq.* **2** (1963) 385–405. *MR* **28** $\neq$ $\neq$ 1361.

G. Stephen Jones, 'Periodic Motions in Banach Spaces and Applications to Functional Differential Equations', *Contrib. to Diff. Eq.* **3** (1964) 75–106. *MR* **29** $\neq$ $\neq$ 342.

G. Stephen Jones, 'Stability and Asymptotic Fixed Point Theory', *Proc. Nat. Acad. Sci. U.S.A.* **53** (1965) 1262–1264. *MR* **49** 59.

G. Stephen Jones, 'Asymptotic Fixed Point Theorems', *Proc. Sympos. on Infinite Dim. Topology.* Edited by R. D. Anderson, Ann. of Math. Studies 1972, pp. 177–184.

G. Jungck, 'Commuting Mappings and Fixed Points', *Amer. Math. Soc. Monthly* **83** (1976) 261–263.

R. I. Kachurovski, 'On Monotone Operators and Convex Functionals', *Uspechi Mat. Nauk* **15** (1960) 213–215.

R. I. Kachurovski, 'On Some Fixed Point Principles', *Uch. Zap. Moskov. obl. Ped. Inst.* (1960) 215–220.

R. I. Kachurovski, 'On Monotone Operators and Convex Functionals,' *Uch. Zap. Moskov. Ped. Inst.* (1962) 231–243.

R. I. Kachurovski, 'Non-Linear Monotone and Other Operators in Banach Spaces', Information Bull. no. 7 ICM. Moscow 1966.

R. I. Kachurovski, 'Non-Linear Operators of Bounded Variation, Monotone and Convex Operators', *Uspechi Math. Nauk* **21** (1966) 256–257. *MR* **34** $\neq$ $\neq$ 6579.

R. I. Kachurovski, 'Monotone Non-Linear Operators in Banach Spaces', *DAN SSSR* **163** (1965) 559–562. = *Soviet Math. Doklady* **6** (1965) 935–938.

R. I. Kachurovski, 'Non-Linear Equations with Monotone and Other Operators', *DAN SSSR* **8** (1967) 515–518. = *Soviet Math. Doklady* **8** (1967) 427–430.

R. I. Kachurovski, 'Non-Linear Monotone Operators in Banach Spaces', *Russian Math. Surveys* **23** (1968) 117–165.

M. I. Kadets, 'Spaces Isomorphic to a Locally Uniformly Convex Space', *Izv. Uceb. Zaved. Matematica* **6** (1959) 51–57.

M. I. Kadets, 'Topological Equivalence of all Separable Banach Spaces', *DAN SSSR* **167** (1966) 23–25. = *Soviet Math. Doklady* **7** (1966) 319–322. *MR* **34** $\neq$ $\neq$ 1828.

S. Kakutani, 'Two Fixed Point Theorems Concerning Compact Convex Sets', *Proc. Imper. Acad. Sci. Tokyo* **14** (1938) 242–245.

S. Kakutani and K. Yosida, 'Operator Theoretic Treatment of Markov's Processes and Mean Ergodic Theorems', *Ann. of Math.* **42** (1941) 188–228.

S. Kakutani, 'Topological Properties of the Unit Sphere of a Hilbert Space', *Proc. Imper. Acad. Sci. Tokyo*, **19** (1943) 269–273.

M. I. Kamenskii, 'On the Theory of the Degree of a Mapping', *Trudy Mat. Fac. Voronezh Gos. Univ. No.* **4** (1971) 54–60.

M. I. Kamenskii, 'On Peano's Theorem in Infinite Dimensional Spaces', *Mat. Zametki* (1972) = *Mathmematical Notes of the Academy of Science of the USSR* (1972).

S. Kaniel, 'Quasi-Compact Nonlinear Operators in Banach Spaces and Applications', *Arch. Rat. Mech. Anal.* **21** (1966) 259–278. *MR* **32** ≠ ≠ 4575.

S. Kaniel, 'Construction of a Fixed Point for Contractions in Banach Spaces', *Israel J. Math.* **9** (1971) 935–940.

R. Kannan, 'Some Results on Fixed Points', *Bull. Calcutta Math. Soc.* **60** (1968) 71–78.

R. Kannan, 'Some Results on Fixed Points, II'. *Amer. Math. Monthly* **76** (1969) 405–408.

R. Kannan, 'Some Results on Fixed Points, III'. *Fund. Math.* **70** (1971) 169–177.

L. V. Kantorovich and G. P. Akilov, *Functsionalnyi analiz v normirovannyh Prostranstvah*, Moscow (1959) = *Functional Analysis in Normed Spaces*, Academic Press, New York, 1964.

S. Karamardian, 'Generalized Complementarity Problem, *J. Optimization Theory and Appl.* **8** (1971) 161–167.

S. Karamardian, 'A Further Generalization of Kirszbraun's Theorem', *Inequalities III*, Los Angeles, 1969.

S. Karamardian (Editor) *Fixed Points. Alogrithms and applications*, Academic Press, 1977.

S. Karlin, *Mathematical Methods and Theory in Games, Programming and Economics*, Vols. I and II. J. Wiley, 1958.

S. Kasahara, 'A Remark on the Contraction Principle', *Proc. Japan Acad. Sci.* **44** (1968) 21–26.

S. Kasahara, 'Functional Equations and Some Fixed Point Theorems in Metric Spaces', *Mat. Sem. Notes* **2** (1974) No. 28.

S. Kasahara, 'Surjectivity and Fixed Points of Nonlinear Mappings', *Math. Jap.* **20** (1975) 57–64.

S. Kasahara, 'On Some Generalizations of the Banach Contraction Theorem', *Math. Sem. Notes* **3** (1975) No. 23.

S. Kasahara, 'On Fixed Points in Partially Ordered Sets and Kirk–Caristi Theorem', *Math. Sem. Notes* **3** (1975) No. 35.

S. Kasahara, 'Boundedness of Semicontinous Finite Real Functions', *Proc. Japan Acad. Sci.* **33** (1977) 183–186.

S. Kasahara, 'Some Fixed Point Coincidence Theorems', *Math. Sem. Notes* **3** (1975) No. 28.

S. Kasahara, 'Fixed Point Theorems and Some Abstract Equations in Metric Spaces', *Math. Jap.* **21** (1976) 165–178.

S. Kasahara, 'On Some Generalizations of the Banach Contraction Theorem', *Publ. RISM Kyoto Univ.* **12** (1976) 427–437.

S. Kasahara, 'Remarks on Some Fixed Points', *Math. Sem. Notes* **4** (1976) 43–50.

S. Kasahara, 'A Fixed Point Criterion in *L*-Spaces', *Math. Sem. Notes* **4** (1976) pp. 205–210.

S. Kasahara, 'Fixed Points of Some Pairwise Contractive Mappings on Compact *L*-Spaces', *Math. Sem. Notes* **4** (1976) 121–133.

S. Kasahara, 'A Remark on a Fixed Point Theorem in Ranked Spaces', *Math. Sem. Notes* **4** (1976) 233–236.

S. Kasahara and K. Iseki, 'On Pairwise Contractive Semigroups', *Math. Sem. Notes* **4** (1976) 91–100.

S. Kasahara, 'Common Fixed Point Theorems in Certain *L*-Spaces, *Math. Sem. Notes* **5** (1977) 173–178.

S. Kasahara, 'Fixed Point Theorems in Certain *L*-Spaces', *Math. Sem. Notes* **5** (1977) 29–35.

T. Kato, 'Demicontinuity, Hemicontinuity and Monotonicity', *Bull. Amer. Math. Soc.* **70** (1964) 548–550.

T. Kato, 'Demicontinuity, Hemicontinuity and Monotonicity', *Bull. Amer. Math. Soc.* **73** (1967) 886–890.

R. B. Kearfoot, 'Computing the Degree of Maps and a Generalized Method of Bisection', Ph.D. Thesis University of Utah, 1977.

R. B. Kearfoot, 'A Proof of Convergence and Error Bound for the Method of Bisection in $\mathbf{R}^n$', *Math. of Com.* **32** (1978) 1147–1153.

R. B. Kearfoot, 'An Efficient Degree Computation Method for a Generalized Method of Bisection', *Numer. Math.* **32** (1979), 109–127.

R. B. Kearfoot, 'A Summary of Recent Experiments to Compute the Topological Degree (February, 1978).

J. L. Kelley, 'Fixed Sets under Homomorphisms', *Duke Math. J.* **5** (1939) 535–537.

J. L. Kelley, 'Banach Spaces with the Extension Property', *Trans. Amer. Math. Soc.* **72** (1952) 323–326.

J. L.Kelley, *General Topology*, Van Nostrand, 1955.

R. B. Kellogg, 'Uniqueness in the Schauder Fixed Point Theorem', *Proc. Amer. Math. Soc.* **60** (1970) 207

R. R. Kellogg, T. Y. Li and J. Yorke, 'A Constructive Proof of the Brouwer Fixed Point Theorem and the Computational Results', *SIAM J.*, **13** (1976) 473–480.

J. B. Keller and S. Antman (Editors), *Bifurcation Theory and Nonlinear Eigenvalue Problems*, Benjamin, Reading, Mass. 1969.

B. von Kerekiarto, *Vorlesungen über Topologie*, Berlin, 1923.

B. von Kerekiarto, 'zur theorie der mehrdeutigen stetigen Abbildungen', *Math. Zeitschr.* **8** (1920) 310–319.

L. Khanzanchi, 'Results on Fixed Points in Complete Metric Spaces', *Math. Jap.* **19** (1974) 283–289.

Vo-Khac Khoah, 'Q-solutions d'un systeme differential', *C.R. Acad. Paris* **258** (1964) 3430–3433.

W. A. Kirk, 'A Fixed Point Theorem for Mappings which do not Increase Distance', *Amer. Math. Monthly* **72** (1965) 1004–1006. *MR* **32** ≠ ≠6436.

W. A. Kirk, 'On Mappings with Diminishing Diameters', *J. London Math. Soc.* **44** (1969) 107–111.

W. A. Kirk, 'A Fixed Point Theorem for Nonexpansive Mappings', *Proc. Sympos. Pure Math.* **18** American Math. Soc. R. I. 1970.

W. A. Kirk, 'Fixed Point Theorems for Nonlinear Nonexpansive Mappings and Generalized Contraction Mappings', *Pacif. J. Math.* **38** (1971a) 89–94. *MR* **46** ≠ ≠4290.

W. A. Kirk, 'On Successive Approximations for Nonexpansive Mappings in Banach Spaces', *Glasgow Math. J.* **12** (1971b) 6–9.

W. A. Kirk, 'Fixed Point Theorems for Nonexpansive Mappings Satisfying Certain Boundary Conditions', *Proc. Math. Soc.* **50** (1975) 143–149.

W. A. Kirk and J. Caristi, 'Mapping Theorems in Metric and Banach Spaces *Bull. Acad. Polon. Sci.* **23** (1975) pp. 891–894.

W. A. Kirk, 'Caristi's Fixed Point Theorem and the Theory of Normal Solvability', *Fixed Point Theory and Its Applications*, Academic Press, 1976.

W. A. Kirk, 'Caristi's Fixed Point Theorem and Metric Convexity', *Colloq. Math.* **36** (1976) 81–86.

W. A. Kirk and D. Downing, 'Fixed Point Theorems for Set-Valued Mappings in Metric and Banach Spaces', *Math. Jap.* **22** (1977) 99–112.

W. A. Kirk and R. Schlomberg, 'Some Results on Pseudo-Contractive Mappings', *Pacif. J. Math.* **71** (1977) 89–100.

W. A. Kirk and R. Schlomberg, 'Mapping Theorems for Local Expansions in Metric and Banach Spaces', *Notices Amer. Math. Soc.* **26** (1979) A–212.

W. A. Kirk and W. O. Ray, 'A Remark on Directional Contractions' (to appear).

W. A. Kirk and W. O. Ray, 'A Note on Lipschitzian Mappings in Convex Metric Spaces (to appear)

M. D. Kirszbraun, 'Über die Zussamenziehende und Lipschiftsche Transformationen', *Fund. Math.* **22** (1934) 77–108.

S. Kinoshita, 'On Some Contractible Continua Without the Fixed Point Property', *Fund. Math.* **40** (1953) 96–98.

V. Klee, 'Convex Bodies and Periodic Homeomorphisms in Hilbert Spaces', *Trans. Amer. Math. Soc.* **74** (1953) 10–43. *MR* **14** $\neq$ $\neq$ 989.

V. Klee, 'Some Topological Properties of Convex Sets', *Trans. Amer. Math. Soc.* **78** (1955) 30–45.

V. Klee, 'Leray–Schauder Theory Without Local Convexity', *Math. Ann.* **141** (1960) 286–296.

I. Kolodner, 'Fixed Points', *Amer. Math. Monthly* **71** (1964) 906.

I. Kolodner, 'Equations of Hammerstein Type in Hilbert Spaces', *J. Math. Mech.* **13** (1964) 701–750. *MR* **30** $\neq$ $\neq$ 1415.

I. Kolodner, 'On Completeness of Partially Ordered Sets and Fixed Points for Isotone Mappings', *Amer. Math. Monthly* **75** (1968) 48–49.

J. Kolomy, 'The Solvability of Nonlinear Integral Equations', *Comment. Math. Univ. Carolin.* **8** (1967) 273–289. *MR* **35** $\neq$ $\neq$ 5878.

J. Kolomy, 'Normal Solvability, Solvability and Fixed Points Theorems', *Colloq. Math.* **29** (1974) 253–266.

A. N. Kolmogorov and S. Fomin, *Elements of the Theory of Functions and Functional Analysis*, Vol. 1. Graylock Press, Rochester, 1957.

Y. Komura, 'Nonlinear Semigroups in Hilbert Space', *J. Math. Soc. Jap.* **19** (1967) 493–507. *MR* **35** $\neq$ $\neq$ 7176.

Y. Komura, 'Differentiability of Nonlinear Semigroups', *J. Math. Soc. Japan* **21** (1969) pp. 375–404. *MR* **40** $\neq$ $\neq$ 3358.

G. Köthe, *Topological Vector Spaces*. Vol I: Springer Verlag. Vol. 2. 1980.

M. A. Krasnoseskii, 'Two Observations about the Method of Successive Approximations', *Uspechi Mat. Nauk* **10** (1955) 123–127. *MR* **16** $\neq$ $\neq$ 833.

M. A. Krasnoselskii, 'Some Problems of Nonlinear Analysis', *Uspechi Math. Nauk*, (*N.S*) **9** (1961) 57–114. = *Amer. Math. Transl.* **10** (1958).

M. A. Krasnoselskii, *Topological Methods in the Theory of Nonlinear Integral Equations*, The MacMillan Co. New York 1964. *MR* **20** ≠ ≠ 3464; *MR* **28** ≠ ≠ 2414.

M. A. Krasnoselskii, *Positive Solutions of Operator Equations*, Noordhof, Groningen, 1964. *MR* **26** ≠ ≠ 2862. *MR* **31** ≠ ≠ 6107.

M. A. Krasnoselskii, 'Small Solutions of a Class of Nonlinear Operator Equations', *Sov. Math. Doklady* **9** (1968) 579–581.

M. G. Krein, 'A new Application of the Fixed Point Principles in the Theory of Operators on a Space with Indefinite Metric', *Doklady Akad. Nauk SSSR* **154** (1964) 1023–1026. = *Soviet Math. Doklady* **5** (1964) 224–228. *MR* **41** ≠ ≠ 6314.

B. Knaster, C. Kuratowski and S. Mazurkiewicz, 'Ein beweis des Eixpunktsatzes für n-dimensionale Simplexe', *Fund. Math.* **14** (1929) 132–137.

R. J. Knill, 'Cones, Products and Fixed Points', *Fund. Math.* **60** (1967) 35–46.

R. J. Knill, 'A Fixed Point Theorem Involving Angle Conditions', *Notices Amer. Math. Soc.* **25** (1978) A–141.

K. Kuga, 'Brouwer's Fixed Point Theorem: An Alternative Proof', *SIAM J. Math. Anal.* **5** (1974) 893–897.

P. Kuhfittig, 'Fixed Point Theorems for Mappings with Non-Convex Domain and Range', *Rocky Mount. J. Math.* **7** (1977)

H. W. Kuhn, 'Some Combinatorial Lemmas in Topology', *IBM Journal for Research Development* **4** (1960) 518–524.

H. W. Kuhn, 'Simplicial Approximation of Fixed Points', *Proc. Nat. Acad. Sci. U.S.A.* **61** (1968) 1238–1242.

C. Kuratowski, 'Sur les espaces complets', *Fund. Math.* **15** (1930).

C. Kuratowski, *Topologie, I.* Warzawa, 1962.

A. Lasota, 'Une generalization du premier théorème de Fredholm et ses applications dans la théorie des équations différentielles ordinaires', *Ann. Polon. Math.* **18** (1966) 65–77. *MR* **33** ≠ ≠ 2849.

A. Lasota and Z. Opial, 'On the Existence and Uniqueness of Solutions of Nonlinear Functional Equations', *Bull. Acad. Polon. Sci. Ser. Sci. Math. Astr. Phys.* **15** (1967) 797–800. *MR* **38** ≠ ≠ 5078.

A. Lasota and A. Cellina, 'A New Approach to the Definition of Topological Degree for Multivalued Mappings', *Rend. Acad. Acad. Naz. dei Lincei* **47** (1969) 434–440. *MR* **43** ≠ ≠ 2677.

C. Lascaret, 'Cas d' addition des applications monotones maximales dans espaces de Hilbert', *C. R. Acad. Paris* **261** (1965) 1160–1163. *MR* **34** ≠ ≠ 645.

A. C. Lazer, 'On Schauder's Fixed Point Theorem and Forced Second Order Nonlinear Oscillations', *J. Math. Anal. Appl.* **21** (1968) 421–425.

S. Lefschetz, 'Manifolds with Boundary and Their Transformations', *Trans. Amer. Math. Soc.* **29** (1927) 429–467.

S. Lefschetz, *Algebraic Topology*, Amer. Math. Soc. Colloq. Publ. Vol. 27. Providence R.I. (1942). *MR* **4** ≠ ≠ 84.

S. Lefschetz, *Topics in Topology*, Ann. of Math. Studies No. 10. Princeton N. J. 1942. *MR* **4** ≠ ≠ 86.

E. Lami Dozo, 'Multivalued Nonexpansive Mappings and Opial's Condition', *Proc. Amer. Math. Soc.* **38** (1973) 286–292.

E. Lami Dozo and J. P. Gossez, 'Some Geometric Properties Related to the Fixed Point

Theory for Nonexpansive Mappings', *Pacif. J. Math.* **40** (1972) 565–573. *MR* **46** ≠ ≠ 9815.

R. Legget, 'Remarks on α-Set Contractions and Condensing Maps', *Math. Zeitschr.* **132** (1973) 361–366.

R. Legget, 'A Note on Locally α-Set Contracting Linear Operators', *Boll. Union. Mat. Ital.* **12** (1975) 124–126.

S. Leader, 'A Topological Characterization of Banach Contractions', *Pacif. J. Math.* **69** (1977) 461–466.

J. Leray, 'Sur les équations et les transformations', *J. Math. Pure et Appl.* **24** (1946) 201–248. *MR* **7** ≠ ≠ 468.

J. Leray, 'Théorie des points fixes; indice total et nombres de Lefschetz', *Bull. Soc. Math. France* **87** (1959) 221–233. *MR* **26** ≠ ≠ 762.

J. Leray, 'Fixed Point Index and Lefschetz Number', *Proc. Sympos. on Infinite Dim. Topology* (Edited by R. D. Anderson) Ann. of Math. Studies, 1972. pp. 219–234.

J. Leray et J. Schauder, 'Topologie et equations fonctionnelles', *Ann. Ecole Norm. Sup.* **51** (1934) 45–78.

J. Leray, 'Les problèmes nonlinéaires', *Enseign. Math.* **30** (1936) 141.

J. Leray and J. L. Lions, 'Quelques resultats de Visic sur les problèmes elliptiques non-linéaires par les méthodes de Minty-Browder', *Bull. Math. Soc. France* **93** (1965) 97–107. *MR* **33** ≠ ≠ 2939.

E. A. Lifsič and V. N. Sadovskii, 'A Fixed Point Theorem for Generalized Condensing Operators', *Doklady Akad. Nauk SSSR* **183** (1968) 278–279 = *Soviet Math. Doklady* **9** (1968) 1370–1371. *MR* **38** ≠ ≠ 5083.

J. L. Lions, 'Sur certains équations paraboliques nonlinéaires', *Bull. Soc. Math. France* **93** (1965) 155–175. *MR* **33** ≠ ≠ 2966.

J. L. Lions, 'Sur certains systèmes hyperboliques nonlinèaires', *C. R. Acad. Paris* **257** (1963) 2057–2060. *MR* **28** ≠ ≠ 351.

J. L. Lions, *Problèmes aux limites dans les équations aux dérivées partielles*, Les Presses de 1' Universite de Montreal, 1965.

J. L. Lions and W. A. Strauss, 'Some Non-Linear Evolution Equations', *Bull. Soc. Math. France* **93** (1965) 43–96. *MR* **33** ≠ ≠ 7663.

J. L. Lions and W. A. Strauss, 'Sur certain problémes nonlinéaires', *C. R. Acad. Paris* **257** (1963) 3267–3270. *MR* **29** ≠ ≠ 3905.

T. C. Lim, 'A Characterization of Normal Structure', *Proc. Amer. Math. Soc.* **43** (1974) 313–319.

T. C. Lim, 'A Fixed Point Theorem for Multivalued Nonexpansive Mappings in a Uniformly Convex Space', *Bull. Amer. Math. Soc.* **80** (1974) 1123–1126.

T. C. Lim, 'Remarks on Some Fixed Point Theorems', Proc. Amer. Math. Soc. **60** (1976) 179–182.

T. C. Lim, 'On Asymptotic Centers and Fixed Points of Nonexpansive Mappings' (to appear).

T. C. Lim, 'Asymptotic Centers and Nonexpansive Mappings in Some Conjugate Banach Spaces (to appear).

T. C. Lim, 'A Fixed Point Theorem for Families of Nonexpansive Mappings', *Pacif. J. Math.* **53** (1974) 497–493. *MR* **51** ≠ ≠ 1503.

R. Lipschitz, *Lehrbuch der Analyse*, Bonn, 1877.

L. A. Liusternik and L. G. Schnirelman, *Méthodes topologiques dans les problèmes variation-nelles*, Actaualité Sci. Industr. 188, Herman, Paris, 1930.

L. A. Liusternik, 'The topology of Calculus of Variations in the Large', *Trudy Mat. Inst. Steklova* **19** (1947); English transl. *Math. Monograph. No.* 16 Amer. Math. Soc. R. I. *MR* **9** $\neq$ $\neq$ 596 and *MR* **36** $\neq$ $\neq$ 906.

A. R. Lovaglia, 'Locally Uniformly Convex Spaces', *Trans. Amer. Math. Soc.* **78** (1955) 225–238. *MR* **16** $\neq$ $\neq$ 596.

V. I. Lomonosov, 'On Invariant Subspaces of Families of Operators Commuting with a Completely Continuous Operator', *Function. Analiz i ego priloj.* **7** (1963) 55–56.

J. L. Liouville, 'Sur la théorie des équations différentielles linéaires et sur le dévelopement des funuctions en series', *J. Math. Pure et Appl.* **1** (1836) 561–614.

G. Lumer, 'Semi-Inner Product Spaces', *Trans. Amer. Math. Soc.* **100** (1961) 29–43.

G. Lumer and R. S. Phillips, 'Dissipative Operators on Banach Spaces', *Pacif. J. Math.* **11** (1961) 679–698. *MR* **74** $\neq$ $\neq$ A2248.

T. W. Ma, 'Topological Degree of Set-Valued Fields in Locally Convex Spaces', *Diss. Mat.* **92** (1972).

T. W. Ma, 'Non-Singular Set-Valued Compact Fields in Locally Convex Spaces', *Fund. Math.* **75** (1972) 249–259.

T. W. Ma, 'Degree Theory for Set-Valued Compact Perturbations', *Bull. Acad. Polon. Sci. Ser. Sci. Math. Astron. et. Phys.* **20** (1972) 169–175.

R. Machuca, 'A Coincidence Theorem', *Amer. Math. Monthly*, **74** (1967) 569.

R. Manka, 'Two Fixed Point Theorems on Contraction', *Bull. Polon. Acad. Sci.* **26** (1978) 41–48.

W. R. Mann, 'Mean Value Methods in Iteration', *Proc. Amer. Math. Soc.* **4** (1953) 506–510.

B. Margolis, 'On Some Fixed Point Theorems in Generalized Complete Metric Spaces', *Proc. Amer. Math. Soc.* **19** (1968) 275–282.

J. T. Martin, 'A Fixed Point Theorem for Set Valued Mappings', *Bull. Amer. Math. Soc.* **74** (1968) 639–640. *MR* **37** $\neq$ $\neq$ 3409.

J. T. Martin, 'Fixed Point Theorems for Set-Valued Contractions', *Notices Amer. Math. Soc.* **15** (1968) 373.

A. A. Markov, 'Quelques théorèmes sur les ensembles abeliens', *C. R. (Doklady) Acad. Sci. URSS*, **1** (1936) 311–313.

R. de Marr, 'Common Fixed Points for Commuting Contractions', *Pacif. J. Math.* **13** (1963) 1139–1141. *MR* **28** $\neq$ $\neq$ 2446'

J. Matkowski, 'Fixed Point Theorems for Mappings with a Contractive Iterate at a Point', *Proc. Amer. Math. Soc.* **62** (1975) 344–350.

J. Mawhin, *Topological Degree Methods in Nonlinear Boundary Value Problems*, CBMS Regional Conference Series in Math. No. 40. Amer. Math. Soc. R. I. (1978).

S. Mazur, 'Über die kleinste konvexe Menge, die eine gegebene kompakte Menge enthält', *Studia Math.* **2** (1930) 7–9.

K. Menger, 'Statistical Metrics', *Proc. Nat. Acad. Sci. U.S.A.* **28** (1942) 535–537.

Cl. Meyer, 'Points invariants dans les éspaces localement convexes', *Sem. Choquet, Initiation a la Analyse*, 4e Anée, 1964/65, No. 10.

P. R. Meyers, 'Some Extensions of Banach's Contraction Theorem', *J. Res. Bur. Standards* **69** B (1965) 179–184.

P. R. Meyers, 'A Converse to Banach's Contraction Theorem', *J. Res. Bur. Standards* **71**B (1967) 73–76.

P. R. Meyers, 'On Contractive Semigroups and Uniform Asymptotic Stability', *J. Res. Bur. Standards*, **74**B (1970a) 115–120.

P. R. Meyers, 'Contractifiable Semigroups', *J. Res. Bur. Standards*, **74**B (1970b) 315–322.

P. R. Meyers and A. J. Goldman, *See* A. J. Goldman, and P. R. Meyers.

M. D. Meyerson and A. L. Wright, 'A New Constructive Proof of the Borsuk–Ulam Theorem' (to appear). Proc. A. M. S. **73** (1979) pp. 134–136.

O. H. Merrill, 'Applications and Extensions of an Algorithm that Computes Fixed Points of Certain Non-Empty Convex Upper Semi-Continuous Point Set Mappings', Dept. of Industrial Engineering, Univ. of Mich. *Techn. Report* 71–7, Sept. 1971.

A. Meir and E. Keeler, 'A Theorem on Contraction Mappings', *J. Math. Anal. Appl.* **28** (1969) 326–329.

F. T. Metcalf and T. D. Rogers, 'The Cluster Set of Sequences of Successive Approximations', *J. Math. Anal. Appl.* **31** (1970) 206–212.

E. Michael, 'Topologies on Spaces of Subsets', *Trans. Amer. Math. Soc.* **71** (1951) 152–182.

E. Michael, 'Continuous Selections, I.' *Ann. Math.* **63** (1956).

E. Michael, 'Continuous Selections in Banach Spaces', *Studia Math.* **22** (1963) 75–76.

S. G. Mihlin, *The Problem of Minimum of a Quadratic Functional*, Holden Day, San Francisco, 1965.

J. Milnor, *Topology from the Differentiable Viewpoint*, Univ. of Virginia, Charlottesville, 1965. *MR* **37** $\neq$ $\neq$ 2239.

J. Milnor, 'Analytic Proof of the "Hairy Ball Theorem" and the Brouwer Fixed Point Theorem', *Amer. Math. Monthly* **85** (1978) 521–524.

G. J. Minty, 'On the Maximal Domain of a Monotone Function', *Mich. Math. J.* **8** (1961) 135–137. *MR* **24** $\neq$ $\neq$ A2224.

G. J. Minty, 'On the Simultaneous Solution of a Certain System of Linear Inequalities', *Proc. Amer. Math. Soc.* **13** (1962) 11–12. *MR* **26** $\neq$ $\neq$ 573.

G. J. Minty, 'Monotone (Nonlinear) Operators in a Hilbert Space'. *Duke Math. J.* **29** (1962) 341–346. *MR* **29** $\neq$ $\neq$ 6319.

G. J. Minty, 'On the Monotonicity of the Gradient of a Convex Function', *Pacif. J. Math.* **14** (1964) 243–247. *MR* **29** $\neq$ $\neq$ 5125a.

G. J. Minty, 'On the Solvability of Nonlinear Functional Equations of Monotonic Type', *Pacif. J. Math.* **14** (1964) 249–255. *MR* **29** $\neq$ $\neq$ 5125b.

G. J. Minty, 'On a "Monotonicity" Method for the Solution of a Nonlinear Equations in Banach Spaces', *Proc. Nat. Acad. Sci. U.S.A.* **50** (1963) 1038–1041. *MR* **28** $\neq$ $\neq$ 5358.

G. J. Minty, 'Two Theorems on Nonlinear Functional Equations in Hilbert Space', *Bull. Amer. Math. Soc.* **69** (1963) 691–692. *MR* **32** $\neq$ $\neq$ 8188.

G. J. Minty, 'A Theorem on Maximal Monotonic Sets in Hilbert Spaces', *J. Math. Anal. Appl.* **11** (1965) 434–439. *MR* **33** $\neq$ $\neq$ 6462.

G. J. Minty, 'Monotone Operators and Certain Systems of Nonlinear Ordinary Differential Equations', *Proc. Symp. System Theory*, Brooklyn Polytechnic Institute, 1965, 39–55.

G. J. Minty, 'On a Generalization of the Direct Method of the Calculus of Variations', *Bull. Amer. Math. Soc.* **73** (1967) 315–321. *MR* **35** $\neq$ $\neq$ 3501.

G. J. Minty, 'On the Extension of Lipschitz–Hölder Continuous and Monotone Functions', *Bull. Amer. Math. Soc.* **76** (1970) 334–339.

G. J. Minty, 'A Finite-Dimensional Tool-Theorem in Monotone Operator Theory', *Adv. in Math.* **12** (1974) 1–7.

T. Mitchell, 'Function Algebras, Means and Fixed points', *Trans. Amer. Math. Soc.* **130**(1968) 117–126.

T. Mitchell, 'Topological Semigroups and Fixed Point Theorems', *Ill. J. Math.* **14** (1970) 630–641.

T. Mitchell, 'Fixed Points of Reversible Semigroups of Nonexpansive Mappings', *Kodai Math. Sem. Rep.* **21** (1970a) 322–323.

C. Miranda, 'Teoremi di esistenza in Analisi funzionale', *Quaderni Matematica No.* 3, Litografia Tacchi, Pisa, 1950. *MR* **12** ≠ ≠ 265.

C. Miranda, 'Partial Differential Equations of Elliptic Type', *Ergeb. der Math. Grenzgeb.* No. 2. 1970 Springer Verlag, New York. *MR* **19** ≠ ≠ 421.

C. Miranda, 'Un'osservazione su un teorema di Brouwer', *Boll. Union. Mat. Ital.* **2** (1940) 5–7.

F. Monna, 'Surure théorème de M. Luxemburg concernant les points fixes d'un classe d'applications d'un éspace metric dans lui même', *Indag. Math.* **23** (1961) 89–96.

F. Monna, 'Analyse nonarchimedienne', *Ergebn. der Math. Grenzgeb.* No. 56, 1970.

J. Moreau, 'Proximité et dualité dans un éspace Hilbertien', *Bull. Soc. Math. France* **93** (1965) 273–299. *MR* **34** ≠ ≠ 1829.

J. Moreau, 'Un cos des convergence des iterées d'une contraction d'un éspace Hilbertien', *C. R. Acad. Paris* **286** (1978) 143–144.

J. Moreau, *Fonctionnelles convexes.* Lectures Notes, Collège de France, 1967.

U. Mosco, 'A Remark on a Theorem of F. Browder', *J. Math. Anal. Appl.* **20** (1967) 90–92. *MR* **36** ≠ ≠ 3178.

U. Mosco, 'Approximation of the Solutions of some Variational Inequalities', *Ann. Scuola Norm. Sup. Pisa* **21** (1967) 373–394. *MR* **37** ≠ ≠ 1966.

U. Mosco, 'Convergence of Sets and of Solutions of Variational Inequalities', *Adv. in Math.* **3** (1969) 510–585.

U. Mosco, 'Perturbations of Variational Inequalities', *Proc. Symp. Pure Math.* **18** Amer. Math. Soc. 1970 182–194.

U. Mosco, 'Implicit Variational Problems and Quasi-Variational Inequalities', in *Lecture Notes in Mathematics,* No. 543. 1976.

K. K. Mukerjea, 'Coincidence Theory for Infinite Dimensional Manifolds', *Bull. Amer. Math. Soc.* **74** (1968) 493–496. *MR* **36** ≠ ≠ 5965.

R. N. Mukerjee, 'On some Fixed Point Theorems in a Banach Space', *Comment. Math. Univ. St. Pauli* **23** (1974) 15–21.

A. Mukerjea and K. Pothoven, *Real and Functional Analysis,* Plenum Press, New York and London, 1978.

S. B. Nadler, 'Sequences of Contractions and Fixed Points', *Pacif. J. Math.* **27** (1968) 579–585.

S. B. Nadler, 'Multi-Valued Contraction Mappings', *Pacif. J. Math.* **30** (1969) 475–487.

S. B. Nadler and R. B. Fraser, *See* R. B. Fraser.

S. B. Nadler and H. Covitz, *See* H. Covitz.

S. B. Nadler, 'Some Problems Concerning the Stability of Fixed Points', *Colloq. Math.* **27** (1973) 263–268.

M. Nagumo, 'Degree of Mappings in Convex Linear Topological Spaces', *Amer. J. Math.* **73** (1951a) 497–511. *MR* **13** ≠ ≠ 150.

M. Nagumo, 'A Theory of Degree of Mapping Based on Infinitesimal Analysis', *Amer. J. Math.* **73** (1951b) 485–496. *MR* **13** ≠ ≠ 150.

M. Nagumo, 'A Note on the Theory of Degree of Mappings in Euclidean Spaces', *Osaka J. Math.* **4** (1952) 1–9.

S. A. Naimpally, 'Contractive Mappings in Uniform Spaces', *Indag. Math.* **27** (1965) 477–481.

I. Namioka and E. Asplund, 'A Geometric Proof of Ryll-Nardzewski's Fixed Point Theorem', *Bull. Amer. Math. Soc.* **73** (1967) 443–445.

E. Natanyahu and M. Reichauw, 'A Theorem on Infinite Positive Matrices', *Proc. Amer. Math. Soc.* **16** (1965) 361–363.

V. Nemitzki, 'The Method of Fixed Points in Analysis', *Uspechi Math. Nauk* **1** (1936) 141–174 = *Amer. Math. Soc. Transl.* **34** (1963) 1–37.

T. Negishi, 'The Stability of a Competitive Economy', *Econometrica* **30** (1962). Nr. 4.

John von Neumann, 'Zür allgemeinen Theorie des Masses', *Fund. Math.* **13** (1929) 73–116.

John von Neumann, 'Über ein Ökonomisches Gleichumgssystem und eine Verallgemeinerung des Brouwerschen Fixpunktsatz', *Ergebnisse eines Math. Colloq.* **8** (1937) 73–83.

C. Ryll-Nardzewski, 'On Fixed Points of Semi-Groups of Endomorphisms of Linear Spaces,', *Proc. Fifth Berkeley Sympos.* Vol. 2 Part. 1 (1967) 55–61.

H. Nikaido, 'On von-Neumann's Mini-Max Theorem', *Pacif. J. Math.* **4** (1953)

H. Nikaido, 'A Proof of the Invariant Mean-Value Theorem on Almost Periodic Functions', *Proc. Amer. Math. Soc.* **16** (1965) 361–363.

H. Nikaido, *Convex Structures and Economic Theory*, Academic Press, 1968.

R. Nevanlinna, 'Über die Methode der Sukzessive Approximationen', *Ann. Acad. Fenn. Ser. AI No.* 291 (1960). *MR* **22** ≠ ≠ 11286.

H. Nikaido, 'Competitive Equilibrium and Fixed Point Theorems, I. A Techanical Note on the Existence Proof for Competitive Equilibrium', *Economic Studies Quart.* **8** Sept. 1962, 54–58.

L. Nirenberg, *Topics in Nonlinear Functional Analysis*, Lectures Notes, New York Univ. 1974.

L. Nirenberg, 'An Abstract Form of the Cauchy–Kovalewski Theorem', *J. Diff. Geom.* **6** (1972) 561–576.

L. Nirenberg, 'An Application of Generalized Degree to a Class of Nonlinear Problems', *Troisième Colloque CBRM d'analyse fonctionnelle,* Vander, Louvain, 1971, 57–74.

J. M. Nocedal, 'On the Method of the Conjugate Gradients for Function Minimization' Ph.D. Thesis, Rice Univ. (Texas) 1978.

R. Nussbaum, 'The Fixed Point Index for Local Condensing Mappings', *Annali di Mat. Pura ed Appl.* **89** (1971a) 217–258. *MR* **47** ≠ ≠ 903.

R. Nussbaum, 'Asymptotic Fixed Point Theorems for Local Condensing Maps', *Math. Ann.* **191** (1971b) 181–195.

R. Nussbaum, 'Estimates for the Number of Solutions of Operator Equations, *Applicable Analysis* **1** (1971c) pp. 183–200.

R. Nussbaum, 'Degree Theory for Local Condensing Maps', *J. Math. Anal. Appl.* **37** (1972a) 741–766.

R. Nussbaum, 'Some Asymptotic Fixed Point Theorems', *Trans. Amer. Math. Soc.* **171** (1972b) 349–374.

R. Nussbaum, 'Generalizing the Fixed Point Index', (to appear).

R. Nussbaum, 'Some Generalizations of the Borsuk–Ulam Theorem', (to appear).

R. Nussbaum, 'The Fixed Point Index and Asymptotic Fixed Point Theorems for *k*-Set Contractions', *Bull. Amer. Math. Soc.* **75** (1969) 490–495. *MR* **39** ≠ ≠ 7589.

R. Nussbaum, 'Fixed Points and Mapping Theorems for Non-linear *k*-Set Contraction Mappings', Ph.D. Thesis, Univ. of Chicago, Chicago, Ill. 1969.

R. B. O'Brien 'Controllability, Stabilization and Contraction Semigroups', Ph.D. Thesis, The George Washington University, Washington D.C. 1977.

C. Olech and A. Plis, 'Monotonicity Assumptions in Uniqueness Criteria for Differential Equations', *Colloq. Math.* **18** (1967) 43–48. *MR* **37** ≠ ≠ 485.

R. Oliver, 'Periodic Fixed Points for Free Diffeomorphisms of $\mathbf{R}^n$', *Notices Amer. Math. Soc.* **26** (1976) A.251.

H. Okamura, 'Sur l'unicité de la solution de dy/d$y = f(x,y)$', *Mem. Coll. Sci. Kyoto,* **17** (1934).

H. Okano, 'On a Class of Complete Spaces and Fixed Point Theorems', *Math. Jap.* **21** (1976) 179–185.

H. Omori, 'On the Group of Diffeomorphisms on a Compact Manifold', *Proc. Sympos. Pure Math.* **15** (1970) Amer. Math. Soc. R. I. 167–173. *MR* **42** ≠ ≠ 6864.

Barret O'Neill, 'Essential Sets and Fixed Points', *Amer. J. Math.* **75** (1953) 497–509.

Z. Opial, 'Nonexpansive and Monotone Mappings in Banach Spaces', *Lectures Notes.* Division of Applied Mathematics, Brown University, Providence, R. I. 1967.

Z. Opial, 'Weak Convergence of the Sequence of Successive Approximations for Nonexpansive Mappings', *Bull. Amer. Math. Soc.* **73** (1967) 591–597. *MR* **35** ≠ ≠ 2183.

V. I. Opoitzev, 'A Converse to the Principle of Contracting Maps', *Uspechi Math. Nauk* **31** (1976) 169–198; *Russian Surveys* **31** (1976) 175–204.

S. M. Ortega and W. C. Rheinboldt, *Iterative Solution of Nonlinear Equations in Several Variables*, Academic Press. New York. N. Y. 1970.

C. L. Outlaw and C. W. Groetsch, 'Averaging Iteration in Banach Spaces', *Bull. Amer. Math. Soc.* **75** (1969) 430–432.

W. J. Padgett, Approximate Solutions of Random Nonlinear Operators by Random Contractors', *Notices Amer. Math. Soc.* **25** (1978) A–150.

T. K. Pal and M. Arti, 'Extension of a Fixed Point Theorem of Rhoades and Čirič', *Proc. Amer. Math. Soc.* **64** (1977) 283–286.

R. Palais, 'Seminar on the Atiyah–Singer Index Theorem', *Ann. Math. Study No.* 57. Princeton Univ. Press, Princeton, 1965.

R. Palais, 'Liusternik–Schnirelman Theory on Banach Manifolds', *Topology* **5** (1966) pp. 115–131.

R. Palais, 'Homotopy Theory of Infinite Dimensional Manifolds', *Topology* **5** (1966) 1–16. *MR* **32** ≠ ≠ 6455.

R. Palais, *Foundations of Global Nonlinear Analysis* , Benjamin, New York, 1968. *MR* **40** ≠ ≠ 2130.

R. Palais, 'Critical Point Theory and the Minimax Principle', *Proc. Sympos. Pure Math.* **15** (1970). Amer. Math. Soc. R. I. 185–212. *MR* **41** ≠ ≠ 9303.

S. Park, 'A Generalization of a Theorem of Janos and Edelstein', *Proc. Amer. Math. Soc.* **66** (1978) pp. 344–346.

G. P. Parker, 'Construction and Differentiable Approximation of Semigroups of Nonlinear Transformations', Ph.D. Thesis, Emory University (Georgia) 1977.

G. Peano, 'Sull' integrabilita delle equazioni differenziali del primo ordine', *Atti R. Accad. Sci. Torino* **21** (1885–1886) 677–685.

G. Peano, 'Demonstration de l'integrabilité des équations différentilles ordinaires', *Math. Ann.* **37** (1890) 182–228.

C. Pearcy and A. L. Shields, 'A Survey of the Lomonosov Technique in the Theory of Invariant Subspaces', *Topics in Operator Theory*, Amer. Math. Soc. Surveys No. 13. Providence R. I. (1974) pp. 219–229.

W. V. Petryshin, 'Constructional Proof of the Lax–Milgram Lemma and its Application to Non-k-p.d. Abstract and Differential Equations', *SIAM J. Numerical Analysis* **2** (1965) 404–420.

W. V. Petryshin, 'Direct and Iterative Methods for the Solution of Linear Operator Equations in Hilbert Spaces', *Trans. Amer. Math. Soc.* **105** (1962) 136–175.

W. V.Petryshin, 'Construction of Fixed Points of Demicompacts Mappings in Hilbert Spaces', *J. Math. Anal. Appl.* **14** (1966) 276–284. *MR* 33 ≠ ≠3147.

W. V. Petryshin, 'On the Extension and the Solution of Nonlinear Operator Equations', *Ill. J. Math.* **10** (1966) 255–275. *MR* 34 ≠ ≠8284.

W. V. Petryshin, 'Remarks on Fixed Points and Their Extensions', *Trans. Amer. Math. Soc.* **126** (1967) 43–53.

W. V. Petryshin, 'On the Approximate-Solvability of Nonlinear Functional Equations in Normed Linear Spaces', *CIME Lectures Notes*, Ipsa, Italy, 1967.

W. V. Petryshin, 'Iterative Construction of Fixed Points of Contractive Type in Banach Spaces', *CIME Lecture Notes*, Ispra, Italy, 1967.

W. V. Petryshin, 'Projection Methods in Nonlinear Numerical Functional Analysis', *J. Math. Anal. Appl.* **17** (1967) 353–372.

W. V. Petryshin, 'On Nonlinear *P*-Compact Operators in Banach Spaces with Application to Constructive Fixed Point Theorems', *J. Math. Anal. Appl.* **15** (1968) 228–242. *MR* 34 ≠ ≠1890.

W. V. Petryshin, 'Further Remarks on Nonlinear *P*-Compact Operators in Banach Spaces', *J. Math. Anal. Appl.* **16** (1968) 243–253.

W. V. Petryshin, 'Nonlinear Equations Involving Noncompact Operators', *Proc. Sympos. Nonlinear Analysis*, Chicago, 1968, Amer. Math. Soc. Vol. 18 Part 1.

W. V. Petryshin and T. S. Tucker, 'On the Functional Equations Involving Nonlinear Generalized *P*-Compact Operators', *Trans. Amer. Math. Soc.* **135** (1969) pp. 343–373. *MR* 40 ≠ ≠804.

W. V. Petryshin, 'Structure of the Fixed Point Set of *k*-Set Contractions', *Arch. Rat. Mech. Ann.* **40** (1971) 312–328. *MR* **42** ≠ ≠8358.

W. V. Petryshin, 'A New Fixed Point Theorem and its Applications', *Bull. Amer. Math. Soc.* **78** (1972) 225–230.

W. V. Petryshin, 'Remarks on Condensing and *k*-Set Contractive Mappings', *J. Math. Anal. Appl.* **39** (1972) 717–741.

W. V. Petryshin, and T. E. Williamson, 'A Necessary and Sufficient Condition for the Convergence of Iterates for Quasi-Nonexpansive Mappings' *Bull. Amer. Math. Soc.* **78** (1972) 1027–1031.

W. V. Petryshin and T. E. Williamson, 'Strong and Weak Convergence of the Sequence of Successive Approximations for Quasi-Nonexpansive Mappings', *J. Math. Anal. Appl.* **43** (1973) 459–497.

W. V. Petryshin and T. E. Williamson, 'A Degree Theory, Fixed Point Theorems and Mapping Theorems for Multivalued Noncompact Mappings', *Trans. Amer. Math. Soc.* **194** (1974) 1–26.

W. V. Petryshin and T. E. Williamson, 'Fixed Point Theorems for Multi-Valued Noncompact Inward Maps', *J. Math. Anal. Appl.* **46** (1974) pp. 756–767.

W. V. Petryshin, 'A Characterization of Strict Convexity of Banach Spaces and Other Uses of Duality Mappings', *J. Funct. Anal.* **6** (1970) 282–291. *MR* **44** ≠ ≠4496.

W. V. Petryshin, 'Fixed Point Theorems for Various Classes of 1-Set Contractive and 1-Ball Contractive Mappings in Banach Spaces', *Trans. Amer. Math. Soc.* **182** (1973) 323–352.

W. F. Pfeffer, 'More on Involutions of a Circle', *Amer. Math. Monthly* **81** (1974) 613–616.

E. Picard, 'Mémoire sur la théorie des équations aux derivées partielles et la méthode des approximations successives', *Journ. de Math.* **6** (1890) 145–210.

D. Pompeiu, 'Sur la continuité des fonctions de variables complexes', *Ann. Fac. de Toulouse* (2) Vol. 7 (1905) 265–315.

L. S. Pontryagin, 'A classification of Continuous Transformations of a Complex into a Sphere', *DAN SSSR* **19** (1938) 147–149.

L. S. Pontryagin, 'Smooth Manifolds and Their Applications in Homotopy Theory', *Trudy Inst. Stecklov* **45** (1955); *Amer. Math. Transl.* **2**, II, (1959) 1–114.

S. I. Pohojaev, 'The Solvability of Nonlinear Equations with Odd Operators, *Functional. Analiz. i ego Priloj.* **1** (1967) 66–73. *MR* 36 $\neq$ $\neq$ 4396.

N. I. Polsky, 'Projection Methods in Applied Mathematics', *DAN SSSR* **143** (1962) 787–790. = *Sov. Math. Doklady* **3** (1963) 488.

H. Poincaré, 'Sur les courbes definies par les équations différentielles', *Journ. de Math.* **2** (1886).

H. Poincaré, 'Sur un théorème de Géometrie', *Rend. Circ. Mat. di Palermo*, **33** (1912) 375–407.

A. J. B. Potter, 'An Elementary Version of the Leray–Schauder Theorem', *J. London Math. Soc.* **5** (1972) 414–416.

A. J. B. Potter, 'A Fixed Point Theorem for Positive k-Set Contractions', *Proc. Edinburgh Math. Soc.* **19** (1974) 93–102.

P. M. Prenter, 'A Weierstrass Theorem for Normed Linear Spaces', *Bull. Amer. Math. Soc.* **74** (1969) 860–862. (See also *MRC Summary Report* $\neq$ $\neq$ 957, Jan. 1969).

P. M. Prenter, A Weierstrass Theorem for Real Separable Hilbert Spaces', *J. Approx. Theory* **3** (1970) 341–351.

P. H. Rabinowitz, 'Variational Methods for Nonlinear eigenvalue Problems', *CIME* ?Varena?ITALY, 1974.

T. Rádo, A Lemma on the Topological Index', *Fund. Math.* **27** (1936) 212–225.

L. Rall, '*Computational Solutions of Nonlinear Equations*, John Wiley, 1969.

N. Rallis, 'Periodic Points and a Fixed Point Index for Symmetric Product Mappings', *Notices Amer. Math. Soc.* **26** (1979) A–282.

E. Rakotch, 'A note on Contraction Mappings', *Proc. Amer. Math. Soc.* **13** (1962) 459–462.

E. Rakotch, 'A Note on α-Locally Contractive Mappings', *Bull. Res. Council Israel*, **40** (1962) 188–191.

B. K. Ray, 'Some Results on Fixed Points and Their Continuity', *Colloq. Math.* **27** (1963) 41–48.

B. K. Ray, 'On Nonexpansive Mappings in a Metric Space', *Nanta Math. J.* **7** (1974) 86–92.

W. O. Ray, 'The Fixed Point Property and Unbounded Sets in Hilbert Spaces', *Notices Amer. Math. Soc.* **26** (1979) A–334.

W. O. Ray, 'The Fixed Point Property and Unbounded Sets in Hilbert Spaces', (to appear).

W. O. Ray, 'Nonexpansive Mappings on Unbounded Convex Domains', *Bull. de L'Academie Polon. Sci, Ser. Math. Astr. Phys.* **26** (1978) 241–245.

W. O. Ray, 'Zeros of Accreative Operators Defined on Unbounded Sets', *Houston Journ. of Math.* **5** (1979) 133–139.

W. O. Ray and W. A. Kirk, see W. A. Kirk and W. O. Ray.

S. Reich, 'Some Remarks, Concerning Contraction Mappings', *Canad. Math. Bull.* **14** (1971) 121–124.

S. Reich, 'Kannan's Fixed Point Theorems', *Boll. Union. Mat. Ital.* **4** (1971) 121–124.

S. Reich, 'Remarks on Fixed Points', *Rend. Accad. Naz. dei Lincei,* **52** (1972) 689–697.

S. Reich, 'Fixed Points of Contractive Functions', *Boll. Union. Mat. Ital.* **5** (1972) 26–42.

S. Reich, 'Fixed Points in Locally Convex Spaces', *Math. Zeitschr.* **125** (1972) 17–31.

S. Reich, 'Fixed Points of Condensing Functions', *J. Math. Anal. Appl.* **53** (1972) 460–467.

S. Reich, 'Remarks on Fixed Points', *Rend. Acad. Naz. dei Lincei,* **53** (1972) 250–254.

S. Reich, 'Fixed Points of Nonexpansive Mappings', *J. London Math. Soc.* **7** (1973) 5–10.

S. Reich, 'On Fixed Point Theorems Obtained From Existence Theorems for Differential Equations', *J. Math. Anal. Appl.* **54** (1976) 26–36.

J. Reinwater, 'Local Uniformity Convexity of Day's Norm on c_0 (6)', *Proc. Amer. Math. Soc.* **22** (1969) 335–339.

I. L. Reilly, 'A Generalized Contraction Principle', *Bull. Austral. Math. Soc.* **10** (1974) 359–363.

C. L. Rennolet, 'Abstract Nonlinear Volterra Integrodifferential Equations of Nonconvolution Type', Ph.D. Thesis, The University of Wisconsin, Madison, Wisconsin 1977.

G. de Rham, *Variètées differentiable*, Herman, Paris, 1955.

B. E. Rhoades, 'Fixed Point Iterations Using Infinite Matrices', *Trans. Amer. Math. Soc.* **196** (1974) 161–176.

B. E. Rhoades, 'Fixed Point Iterations Using Infinite Matrices, II'. *Constructive and Computational Methods for Differential and Integral Equations*, Lectures Notes in Math. No. 430. (1974) pp. 390–395.

B. E. Rhoades, 'Fixed Point Iterations Using Infinite Matrices, III'. *Proc. Conf. Computing Fixed Points with Applications*, Academic Press 1976.

B. E. Rhoades, 'Some Fixed Point Theorems in a Banach Space', *Comment. Math. Univ. St. Pauli* **24** (1976) 13–16.

B. E. Rhoades, 'A Comparison of Various Definitions of Contractive Mappings', *Trans. Amer. Math. Soc.* **226** (1977) 256–290.

B. E. Rhoades, 'Comments on Two Fixed Point Iteration Methods' (to appear).

B. E. Rhoades, 'A Fixed Point Theorem in Metric Spaces' (to appear).

B. E. Rhoades, 'Generalized Contractions' (to appear).

B. E. Rhoades and B. K. Ray, 'Fixed Point Theorems for Mappings with a Contractive Iterate, (to appear).

C. E. Rickart, *General Theory of Banach Algebras*, D. Van Nostrand, 1960.

N. W. Rickert, 'Amenable Groups and Groups with the Fixed Point Property', *Trans. Amer. Math. Soc.* **127** (1967) 221–232.

H. Robbins, 'Some Complements to Brouwer's Fixed Point Theorem', *Israel J. Math.* **5** (1967) 225–226.

B. Robert and P. Schulze, *Modern Mathematics and Economic Analysis*, W. W. Norton Co. Inc. New York, 1973.

R. T. Rockafellar, 'Minimax Theorems and Conjugate Saddle-Functions', *Math. Scand.* **14** (1964) pp. 151–173.

R. T. Rockafellar, 'Leval Sets and Continuity of Conjugate Convex Functions', *Trans. Amer. Math. Soc.* **123** (1966) 46–63.

R. T. Rockafellar, 'Characterization of the Subdifferentials of Convex Functions', *Pacif. J. Math.* **17** (1966) 497–510. *MR* **33** ≠ ≠1769.

R. T. Rockafellar, 'A General Correspondence Between Dual Minimax Problems and Convex Programs', *Pacif. J. Math.* **25** (1968) 597–611.

R. T. Rockafellar, 'Convex Functions, Monotone Operators and Variational Inequalities', *Theory and Applications of Monotone Operators*, Proc. Sympos. NATO, Venice, 1968. Edizioni Oderisi, Gubio (1969) 35–65. *MR* **41** ≠ ≠6028.

R. T. Rockafellar, 'Local Boundedness of Nonlinear Maximal Monotone Operators', *Mich. Math. J.* **16** (1969) 397–407. *MR* **40** ≠ ≠6229.

R. T. Rockafellar, 'On the Maximality of Sums of Nonlinear Monotone Operators', *Trans. Amer. Math. Soc.* **149** (1970) 75–88. *MR* **43** ≠ ≠7984.

R. T. Rockafellar, 'On the Maximal Monotonicity of Subdifferential Operators', *Pacif. J. Math.* **33** (1970) 209–216. *MR* **41** ≠ ≠7432.

R. T. Rockafellar, 'Monotone Operators Associated with Saddle Functions and Minimax Problems', *Proc. Sympos. Pure Math.* Vol. 18. Part 1. Amer. Math. Soc. R. I. (1970) 241–250. *MR* **44** ≠ ≠3159.

R. T. Rockafellar, *Convex Analysis*, Princeton Univ. Press. Princeton 1970.

I. Rosenholz, 'Evidence of a Conspiracy among Fixed Point Theorems', *Proc. Amer. Math. Soc.* **53** (1976) 213–216.

I. Rosenholz, 'On a Fixed Point Problem of D. S. Smart', *Proc. Amer. Math. Soc.* **55** (1976) 252.

S. F. Roehring, 'Fixed points in l_1', *Notices Amer. Math. Soc.* **26** (1979) A–306.

E. Rothe, 'Zur theorie der topologischen Ordnung und der Vektorfelder in Banachschen Räumen', *Compositio Math.* **5** (1937) 177–197.

E. Rothe, 'Mapping Degree in Banach Spaces and Spectral Theory', *Math. Zeitschr.* **63** (1955) 115–218.

E. Rothe, 'Gradient Mapping in Hilbert Spaces', *Ann. Math.* **47** (1946) 580–592.

E. Rothe, 'Gradient Mappings and Extrema in Banach Spaces', *Duke Math. J.* **15** (1948) 421–431.

E. Rothe, 'Remarks on the application of Gradient Mappings to the Calculus of Variations and the Connected Boundary Value Problems in Partial Differential Equations', *Comm. Pure Appl. Math.* **9** (1956) 551–568. *MR* **18** ≠ ≠808.

E. Rothe, 'Some Remarks on Vector Fields in Hilbert Space', *Proc. Sympos. Pure Math.* Vol. 18. Part 1. Amer. Math. Soc. R. I. (1970), pp. 251–269.

V. N. Sadovski, 'A Fixed Point Principle'. *Function Analiza i ego Priloz*, **1** (1967) 74–76; *Functional Analysis and its Appl.* **1** (1967) 151–153 *MR* **35** ≠ ≠2184.

V. N. Sadovski, Local Existence Theorems for Ordinary Differential Equations in Banach Spaces', *Probl. Mat. Analiza Slozn. System.* No. 1, Voronezh (1967) 70–74.

V. N. Sadovski, 'On Measure of Noncompactness and Condensing Operators', *Probl. Mat. Analiza Syst. Slozn. System.* No. 2, Voronezh (1968) 89–119.

V. N. Sadovski, 'On the Local Solubility of Ordinary Differential Equations in Banach Spaces', *Probl. Mat. Analiza Slozn. System.* No. 3, Voronezh (1968) 232–243.

V. N. Sadovski, 'Some Remarks on Condensing Operators and Measures of Noncompactness', *Trudy Mat. Fac. Gos. Univ. No.* 1 (1970) 112–124.

V. N. Sadovski, 'On Differential Equations with Uniformly Continuous Right-Hand Side', *Trudy Mat. Fac. Gos. Univ. No.* 1 (1970) 128–136.

V. N. Sadovski, 'Application of Topological Methods in the Theory of Periodic Solutions of

Nonlinear Differential-Operator Equations of Neutral Type', *Dokl. Akad. Nauk SSSR* **200** (1971) 1037–1040. = *Soviet Phys. Dokl.* **12** (1971).

V. N. Sadovski, 'Limit-Compact and Condensing Operators,' *Russian Math. Surveys* **27** (1972) 85–155; *Uspechi Mat. Nauk* **27** (1972) 81–146.

G. Sansone and R. Conti, *Equazioni differenziali non lineari*, Ed. Cremonesi, Roma, 1956.

A. Sard, 'The Measure of Critical Values of Differentiable Maps', *Bull. Amer. Math. Soc.* **48** (1942) 883–890.

H. Scarf, 'The Approximation of Fixed Points of Continuous Mappings', *SIAM Journal of Appl. Math.* **15** (1967) 1328–1343.

H. Scarf and T. Hansen, *The Computation of Economic Equilibria*, New Haven, Connecticut, 1973.

J. Schauder, 'Zur Theorie stetiger Abblidungen in Funktionalräumen', *Math. Z.* **26** (1927) 47–65, 417–431.

J. Schauder, 'Der Fixpunktsatz in Funktionalräumen', *Studia Math.* **2** (1930) 171–180.

J. Schauder, 'Invarianz des Gebietes in Funktionalräumen', *Studia Math.* **1** (1929) 123–139.

J. Schauder, Über den Zusammenhang zwischen der Eindeutigkeit und Losbarkeit partiellen Differentialgleichungen zweiter ordnung vom elliptischen Typus', *Math. Ann.* **106** (1932) 661–721.

J. Schauder and J. Leray, *See* J. Leray and J. Schauder.

J. Schwartz, *Nonlinear Functional Analysis*, Lectures Notes, Courant Inst. of Math. 1963–1964.

J. Schwartz, 'On Nash's Implicit Function Theorem', *Comm. Pure Appl. Math.* **13** (1960) 509–530. *MR* 22 ≠ ≠ 4971.

J. Schwartz, 'Compact Analytic Mappings of *B*-Spaces and a Theorem of J. Cronin', *Comm. Pure Appl. Math.* **16** (1963) 253–260. *MR* 29 ≠ ≠ 481.

J. Schwartz, 'Generalizing the Liusternik–Schnirelman Theory of Critical Points', *Comm. Pure Appl. Math.* **17** (1964) 307–315. *MR* 29 ≠ ≠ 4069.

J. Schwartz, 'Intersection-Theoretic Principles for the Existence of Critical Points and Fixed Points',', *Lectures Series on Differential Equations* Vol. 1, Van Nostrand, New York, pp. 123–146. *MR* 46 ≠ ≠ 906.

J. Schwartz, 'The Formula for Change of Variable in a Multiple Integral', *Amer. Math. Monthly*, **61** (1954) 81–85.

J. Schwartz, *Lectures on the Mathematical Methods in Analytical Economics*, Gordon and Breach Publishers, New York, 1961.

B. Schweizer and A. Sklar, 'Statistical Metric Spaces', *pacif. J. Math.* **10** (1960) 313–334.

I. J. Schoenberg, 'On a Theorem of Kirszbraun and Valentine', *Amer. Math. Monthly* **60** (1953) 620–622.

S. O. Schönbeck, 'Extension of Nonlinear Contractions', *Bull. Amer. Math. Soc.* **72** (1966) 99–101.

S. O. Schönbeck, 'On the Extension of Lipschitzian Maps', *Ark. Mat.* **7** (1967) 201–207.

V. M. Sehgal, 'A Fixed Point Theorem for Mappings with a Contractive Iterate', *Proc. Amer. Math. Soc.* **23** (1969) 631–644.

V. M. Sehgal, 'On Fixed and Periodic Points for a Class of Mappings', *J. London Math. Soc.* **5** (1972) 571–576.

V. M. Sehgal and B. Reid, 'Fixed Points of Contraction Mappings on Probabilistic Metric Spaces', *Math. Syst. Theory* **6** (1972) 97–100.

V. M. Sehgal, 'A Simple Proof of a Theorem of Ky Fan', *Proc. Amer. Math. Soc.* **63** (1977) 368–369.

T. Seki, 'An Elementary Proof of Brouwer's Fixed Point Theorem', *Tôhoku Math. J.* **9** (1957) 105–109.

A. N. Serstnev, 'On Probabilistic Generalization of Metric Spaces', *Ucen. Zap. Kazan. Gos. Univ.* **124** (1964) 3–11.

H. Sherwood, 'Complete Probabilistic Metric Spaces', *Z. Wahr. Verw. Geb.* **20** (1971) 117–128.

M. Sion, 'On General Minimax Theorems', *Pacif. J. Math.* **8** (1958) 171–176.

R. C. Sine, 'Fixed Points for Certain Nonlinear Contraction Semigroups', *Notices Amer. Math. Soc.* **25** (1978) A–108.

S. P. Singh, 'Some Results on Fixed Point Theorems', *Yokohama Math. J.* **17** (1969) 61–64.

M. Shinbrot, 'A Fixed Point Theorem and Some Applications', *Arch. Rat. Mech. Anal.* **17** (1964) 255–271.

J. Siegel, 'A New Proof of Caristi's Fixed Point Theorem', *Proc. Amer. Math. Soc.* **66** (1977) 54–56.

P. Soardi, 'Su un problema di punto unito di S. Reich', *Boll. Union. Mat. Ital.* **4** (1971) 841–845.

P. Soardi, 'Existence of Fixed Points of Nonexpansive Mappings in Certain Banach Lattices', *Proc. Amer. Math. Soc.* **73** (1979) 25–29.

G. E. Skhilov, 'On Rings with Uniform Convergence', *Ukrain. Math. J.* **3** (1951).

S. Smale, 'Morse Theory and Nonlinear Generalization of the Dirichlet Problem', *Ann. Math.* **80** (1964) 382–396. *MR* **29** $\neq$ $\neq$ 2820.

S. Smale, 'An Infinite Dimensional Version of Sard's Theorem', *Amer. J. Math.* **87** (1965) 861–866.

D. R. Smart, *Fixed Point Theorems*, Cambridge Univ. Press, 1974.

R. E. Smithson, 'Multifunctions', *Nieuw Archief voor Wiskunde* **20** (3) (1972) pp. 31–53.

R. E. Smithson, 'Fixed Points in Partially Ordered Sets', *Pacif. J. Math.* **45** (1973) 363–367.

R. L. Smith, 'Periodic Limits of Solutions of Volterra Integral Equations', Ph.D. thesis Carnegie Mellon Univ. (Pensylvania) 1977.

F. Stenger, 'Computing the Topological Degree of a Mapping in $\mathbf{R}^n$', *Num. Math.* **25** (1975) 23–38.

W. A. Straus, 'Further Applications of monotone Operators Methods to Partial Differential Equations', *Proc. Sympos. Pure Math.* Vol. 18 Part. 1. R. I. 1970. pp. 282–288. *MR* **42** $\neq$ $\neq$ 8113.

D. Strawther and S. Gudder, 'A Characterization of Strictly Convex Spaces', *Proc. Amer. Math. Soc.* **47** (1975) 268.

M. Stynes, Ph.D. Dissertation, Univ. of Oregon, Corvallis 1977.

M. Stynes, 'An n-Dimensional Bisection Method for Solving Systems of n Equations in n Unknowns' (to appear).

M. Stynes, 'An Algorithm for Numerical Calculation of Topological Degree', *Appl. Anal.* **9** (1979) 63–77.

M. Stynes, 'Brouwer's Fixed Point Theorem: A Proof with a Picture'.

D. A. Sybley, 'A Metric for the Weak Convergence of Distributions', *Rocky Mountain J. Math.* **1** (1971) 427–430.

S. Szoufla, 'Measure of Noncompactness and Ordinary Differential Equations in Banach Spaces', *Bull. Acad. Polon. Sci.* **19** (1971) 831–835.

M. Szurek, 'Some Topological Theorems on the Fixed Point Property', *Bull. Acad. Polon. Sci.* **25** (1977) 549–553.

S. Swaminathan (Editor) *Fixed Point Theory and its Applications*. Academic Press 1976.

W. Takahashi, 'Fixed Points Theorems for Amenable Semigroups of Nonexpansive Mappings', *Kodai Math. Sem. Report* **21** (1969) 383–386.

W. Takahashi, 'A Convexity in Metric Space and Nonexpansive Mappings', *Kodai Math. Sem. Rep.* **22** (1970) 142–149.

W. Takahashi, 'Nonlinear Variational Inequalities and Fixed Point Theorems', *J. Math. Soc. Japan* **28** (1976) 168–181.

W. Takahashi, 'Nonlinear Complementarity Problem and Systems of Convex Inequalities', *J. Opt. The. Appl.* **24** (1978) 499–506.

W. Takaheshi, *Recent Results in Fixed Point Theory*, Tokyo Institute of Technology, Dept. of Information Sciences, Series A, Jan. 1979, No. A–61.

D. M. Taliaero, 'Functional Equations in Linear Spaces', Ph.D. Thesis, Emory University (Georgia), 1977.

L. A. Talman, 'A Note on Kellogg's Uniqueness Theorem for Fixed Points', *Proc. Amer. Math. Soc.* **69** (1978) 248–250.

A. Tarski, 'A Lattice Theoretic Fixed Point Theorem and its Applications', *Pacif. J. Math.* **5** (1955) 285–309.

L. Tartar, 'Inequations Quasi-Variationnelles Abstraites', *C.R. Acad. Paris Ser. A.* **278** (1974) pp. 1193–1196.

L. Tartar, 'Equations with Order Preserving Properties', *MRC Techn. Report* $\neq$ $\neq$ 1580. Univ. of Wisconsin, 1976.

A. E. Taylor, *Introduction to Functional Analysis*, John Wiley, 1958.

J. G. Taylor, 'Topological Degree of Noncompact Mappings', *Proc. Camb. Math. Soc.* **63** (1967) 335–347. *MR* **36** $\neq$ $\neq$ 863.

F. Terkelsen, 'Some Minimax Theorems', *Math. Scand.* **31** (1972) 405–413.

J. W. Thomas, 'A Bifurcation Theorem for k-Set Contractions', *Pacif. J. Math.* **44** (1973) 749–756.

J. W. Thomas and T. O'Neil, 'The Calculation of the Topological Degree by Quardrature', *SIAM J. Numer. Anal.* **12** (1975) 673–680.

R. B. Thompson, 'A Unified Approach to Local and Global Fixed Point Indices', *Advances in Math.* **3** (1969) 1–72. *MR* **40** $\neq$ $\neq$ 891.

R. B. Thompson, 'On the Semi-Complexes of F. Browder', *Bull. Amer. Math. Soc.* **773** (1967) 531–536. *MR* **35** $\neq$ $\neq$ 4902.

R. B. Thompson, 'A Metatheorem for Fixed Point Theories', *Comment. Mat. Univ. Carolin.* **11** (1970).

R. B. Thompson, 'Retracts of Semi-Complexes', *Ill. J. Math.* **15** (1971) 258–272. **MR** **43** $\neq$ $\neq$ 4028.

C. Tompkins, 'Sperner's Lemma and Extensions', *Appl. Combinatorial Math.* Edited by E. Beckenbach, J. Wiley (1966) 415–455.

E. Trafaldar, 'An Approach to Fixed Points Theorems in Uniform Spaces', *Trans. Amer. Math. Soc.* **191** (1974) 209–225.

F. Tricomi, 'Una teorema sulla convergenza delle successioni formate delle successive iterate di una funzione di una variabile reale', *Giorn. Mat. Bataglini* **54** (1916) 1–9.

A. N. Tychonoff, 'Ein Fixpunktsatz', *Math. Ann.* **111** (1935) 767–776.

A. N. Tychonoff, 'On the Stability of the Functional Optimization Problem', *USSR Computational Math. Phys.* **6** (1966).

J. F. Toland, 'Asymptotic Linearity and Nonlinear Eigenvalues Problems', *Quart. J. Math.* **24** (1973) 241–250.

A. J. Tromba, 'Degree Theory on Banach Manifolds', Ph.D. Thesis, Princeton Univ. 1968.

A. J. Tromba, 'Euler–Poincaré Index Theory on Banach Manifolds', *Annali della Scuola Norm. Sup. Pisa* **IV** (1975) pp. 89–106.

A. J. Tromba, 'Degree Theory and Nonlinear Partial Differential Equations', *Global Analysis and Its Applications.* IEA' SMR-11/62, pp. 209–215.

T. S. Tucer, 'Leray–Schauder Theorem for P-Compact Operators and its Consequences', *J. Math. Anal. Appl.* **23** (1972) 355–364.

A. Tucker, 'Some Properties of the Disc and Sphere', *Proc. First Canad. Math. Congress*, 1945, pp. 285–309.

U. Urabe, 'Convergence of numerical iterations of equations', *J. Sci. Hiroshima Univ.* A **19** (1956), 479–489.

W. R. Utz, 'The Equation $f'(x) = af(g(x))$', *Bull. Amer. Math. Soc.* **71** (1965) 138.

H. Uzawa, 'Competitive Equilibrium and Fixed Point Theorems, II. Walras Existence Theorem and Brouwer's Fixed Point Theorem', *Economic Studies Quart.* **8** (Sept. 1962) 59–62.

M. M. Vainberg and I. V. Sragin, 'Nonlinear Operators and Hammerstein's Equations in Orlicz's Spaces', *Doklady Akad. Nauk SSSR* **128** (1959) 9–12. *MR* **21** $\neq$ $\neq$ 7414.

M. M. Vainberg and R. I. Kachurovskii, 'On the Variational Theory of Nonlinear Operator Equations', *Doklady Akad. Nauk SSSR* **129** (1959) 1199–1202. *MR* **22** $\neq$ $\neq$ 4930.

M. M. Vainberg, 'On the Convergence of the Method of Steepest Descent for Nonlinear Equations' *Doklady Akad. Nauk SSSR* **130** (1960) 9–12. = *Soviet Math. Dokl.* **1** (1960) 1–4. *MR* **25** $\neq$ $\neq$ 751.

M. M. Vainberg, 'On the Convergence of the Process of Steepest Descent for Nonlinear Equations', *Mat. Z.* **2** (1961) 201–220. *MR* **23** $\neq$ $\neq$ A 4026.

M. M. Vainberg and V. A. Tregonin, 'The Method of Lyapunov–Schmidt in the Theory of Non-Linear Equations and Their Further Development', *Russian Surveys* **17** (1962) 1–60.

M. M. Vainberg, *Variational Methods for the Study of Nonlinear Operators*, Holden Day, San Francisco, Calif. 1964. *MR* **19** $\neq$ $\neq$ 567 (of Russian edition).

M. M. Vainberg and P. G. Aizengendler, 'Methods of Developement in the Bifurcation Theory of Solutions', *Progress in Math.* Vol. 2. Plenum Press, New York, 1968.

F. A. Valentine, 'A Lipschitz Condition Preserving Extension for a Vector Function', *Amer. J. Math.* **67** (1945) 83–93.

F. A. Valentine, *Convex Sets*, McGraw Hill Comp. 1960.

Florin Fasilescu, 'Essai sur les Fonctions multiformes de variables réelles', Thesis, Gauthier Villars, Paris, 1925.

Florin Fasilescu, 'Functiunile multiforme de variabile reale si analiza clasica', *Bull. Math. Soc. Roum. des Sci.* **30** (1927) 110–112.

Florin Fasilescu, 'Sur les valeurs limites des fonctions harmoniques', *C.R. Acad. Paris* **184** (1927) 434–436.

G. M. Vainikko and B. N. Sadovskii, 'On the Degree of (Ball)-Condensing Vector Fields', *Probl. Mat. Analiz. Slozn. Systems* **2** (1968) 84–88.

G. Vidossich, 'On Peano Phenomenon', *Boll. Union. Mat. Ital.* **3** (1970) 33–42.

G. Vidossich, 'Applications of Topology to Analysis; On the Topological Properties of the Set of Fixed Points of Nonlinear Operators', *Conf. Sem. Univ. di Bari* No. 126 (1971) 1–62.

M. I. Visik, 'Solutions of Boundary Value Problems for Quasi-Linear Parabolic Equations of Arbitrary Order', *Mat. Sbornik*, **51** (1962) 289–325. English Transl. in *Amer. Math. Transl.* **65** (1967) 1–40. *MR* **28** ≠ ≠ 361.

L. P. Vlasov, Chebyshev Sets in Banach Spaces', *DAN* **2** (1961) 1373–1374 = *Soviet Math. Doklady* **141** (1961) 19–20.

M. Volpato, 'Sugli elementi uniti di transformazioni funzionali: un problem ai limit per una classe di equazioni alle derivate parziali di tipo iperbolico', *Ann. Univ. di Ferrara II*, **8** (1953) 93–109.

C. T. C. Wall, *A Geometric Introduction to Topology*, Addison Wesley Corp., 1972.

L. Walras, *Elements of Pure Economics*, Allen and Unwin Ltd. London, 1964.

T. van der Walt, *Fixed and Almost Fixed Points*, Mathematical Centre Tracts No. 1, 1967.

L. E. Ward, 'A Fixed Point Theorem for Chained Spaces', *Pacif. J. Math.* **9** (1959) 1273–1278.

D. Westreich, 'Banach Space Bifurcation Theory', *Trans. Amer. Math. Soc.* **171** (1972) 135–156.

P. R. Weiss, 'Numerical Solutions of Volterra Integral Equations', Ph.D. Thesis, Carnegie Mellon University (Pennsylvania). 1978.

E. F. Whittlesey, 'Fixed Points and Antipodal Points', *Amer. Math. Monthly* **70** (1963) 807–821.

S. P. Williams, 'A Connection Between the Cesari and Leray–Schauder Methods', *Mich. Math. J.* **15** (1968) 441–448.

J. P. Williams, 'Spectra of Products and Numerical Ranges', *J. Math. Anal. Appl.* **17** (1967) 214–220. *MR* **34** ≠ ≠ 3341.

R. Wintner, 'A Numerical Galerkin Method for a Parabolic Control Problem', Ph.D. Thesis Cornell Univ. New York, 1977.

Chi Son Wong, 'Fixed Point Theorems for Nonexpansive Mappings', *J. Math. Anal. Appl.* **37** (1972) 142–150.

Chi Son Wong, 'A Fixed Point Theorem for a Class of Mappings', *Math. Ann.* **204** (1973) 97–103.

Chi Son Wong, 'Common Fixed Point of Two Mappings', *Pacif. J. Math.* **48** (1973) 299–312.

Chi Son Wong, 'Generalized Contractions and Fixed Point Theorems', *Proc. Amer. Math. Soc.* **42** (1974) 409–417.

Chi Son Wong, 'Fixed Point Theorems for Generalized Nonexpansive Mappings', *J. Austral. Math. Soc.* **18** (1974) 265–276.

Chi Son Wong, 'On a Fixed Point Theorem of Contractive Type', *Proc. Amer. Math. Soc.* **57** (1976) 283–284.

Chi Son Wong, 'Maps of Contractive Type', *Fixed Point Theory and Its Applications*, Pergamon Press, 1976.

J. S. Wong, 'Mappings of Contractive Type on Abstract Spaces', *J. Math. Anal. Appl.* **37** (1972) 331–340.

J. S. Wong, 'A Note on Subadditive Functions', *Proc. Amer. Math. Soc.* **44** (1974) 106.

J. S. Wong, 'Two Extensions of the Banach Contraction Principle', *J. Math. Anal. Appl.* **22** (1968) 438–443. *MR* **37** ≠ ≠ 4682.

A. Wouk, 'Direct Iteration, Existence and Uniqueness in Non-Linear Integral Equations', *Proc. Sympos. Univ. of Wisconsin*, Wisconsin, 1964.

S. Ymamura, 'Some Fixed Point Theorems in Locally Convex Spaces', *Yokohama Math. J.* **11** (1953) 5–12. *MR* **29** ≠ ≠ 5095.

C. L. Yen, 'Remark on Common Fixed Points', *Tamkang J. Math.* **3** (1972) 95–96.

C. L. Yen and K. J. Chung, 'A Theorem on Fixed Points and Periodic Points', *Tamkang J. Math.* **5** (1974) 235–239.

T. Yoshizawa, 'Stability Theory by Liapunov's Second Method', *Publ. Math. Soc. Japan* No. 9. Tokyo (1966). *MR* **34** ≠ ≠ 7896.

K. Yosida, *Functional Analysis*, Springer Verlag, New York, 1965.

G. S. Young, 'Fixed Point Theorems for Arcwise Connected Continua', *Proc. Amer. Math. Soc.* **11** (1960) 880–884.

P. Zabreiko, 'A Theorem for Semiadditive Functionals', *Functional Analysis and its Applications* **3** (1) 86–88; English Transl. **3** (1969) 70–72.

P. Zabreiko, R. I. Kachurowskii and M. A, Krasnoselskii, 'On a Fixed Point Principle for Operators in a Hilbert Space', *Functional Analysis and its Appl.* **1** (1967) 93–94. *MR* **35** ≠ ≠ 3505.

P. Zabreiko and M. A. Krasnoselskii, 'On a Method for Obtaining New Fixed Point Principles', *Doklady Akad. Nauk SSSR* **176** (1967) 1233–1235. = *Soviet Math. Doklady* **8** (1967) 1297. *MR* **36** ≠ ≠ 3183.

P. Zabreiko and I. B. Ledovskaia, 'Existence Theorems for Equations in a Banach Space and the Principle of Averaging', *Probl. Matem. Anal. Slojzn. System.* No. 3 ?Voronezh (1968) 122–136.

P. Zabreiko and Yu. I. Fetisov, 'On a Method of Small Parameter for Hyperbolic Equations', *Differentialnye Uravnenya* (1972) = *Differential Eq.* (1972).

E. Zarantonello, 'Solving Functional Equations by Contractive Mappings', *Math. Res. Center Re. No.* 160. Madison, Wisconsin, 1960.

E. Zarantonello, 'The Closure of the Numerical Range Contains the Spectrum', *Bull. Amer. Math. Soc.* **70** (1964) 781–787. *MR* **30** ≠ ≠ 3389.

E. Zarantonello, 'The Closure of the Numerical Range Contains the Spectrum', *Pacif. J. Math.* **22** (1967) 575–595. *MR* **37** ≠ ≠ 4657.

V. Zizler, 'Banach Spaces with Differentiable Norms', *Comment. Math. Univ. Carolin.* **9** (1968) 415–440.

V. Zizler, 'Some Notes on Various Roundity and Smoothness Properties of Separable Banach Spaces', *Comment. Math. Univ. Carolin.* **10** (1969) 195–206.

V. Zizler, 'On Some Rotundity and Smoothness Properties of Banach Spaces', *Rozprawy Math. Dissert. Mat. No.* **87** (1971).

Index

Mathematics and Its Applications

Managing Editor:

M. HAZEWINKEL
Department of Mathematics, Erasmus University, Rotterdam, The Netherlands

Editorial Board: